CHILTON'S™

ELECTRIC COOLING FAN, ACCESSORY DRIVE BELT & WATER PUMP SERVICE MANUAL

COVERS ALL 1995-99 U. S. AND CANADIAN VEHICLES DOMESTIC AND IMPORTED

President
Dean F. Morgantini, S.A.E.

Vice President–Finance
Barry L. Beck

Vice President–Sales
Glenn D. Potere

Executive Editor
Kevin M. G. Maher, A.S.E.

Production Manager
Ben Greisler, S.A.E.

Production Assistant
Melinda Possinger

Project Managers
George B. Heinrich III, A.S.E., S.A.E., Will Kessler, A.S.E., S.A.E., James R. Marotta, A.S.E., S.T.S., Richard Schwartz, A.S.E., Todd W. Stidham, A.S.E., Ron Webb

Schematics Editor
Christopher G. Ritchie

Editors
Christopher Bishop, Leonard Davis, A.S.E., S.T.S., Matthew E. Frederick, A.S.E., S.A.E., Frank Keytanjian, A.S.E., S.A.E., Thomas A. Mellon, A.S.E., S.A.E., Eric Michael Mihalyi, A.S.E., S.A.E., S.T.S., Christine Nuckowski, S.A.E.

CHILTON™ *Automotive Books*
PUBLISHED BY **W. G. NICHOLS, INC.**

Manufactured in USA, © 1999 W. G. Nichols, 1020 Andrew Drive, West Chester, PA 19380
ISBN 0-8019-9126-9
Library of Congress Catalog Card No. 98-74942
1234567890 8765432109

HOW TO USE THIS MANUAL

Locating Information

The Table of Contents, located at the front of this manual, lists all contents in the order in which they appear in the manual.

To find where a particular model specific section is located in the book, you need only look in the Table of Contents. Once you have found the proper section, you may wish to find where specific procedures are located in that section. Either use those listed in the Table of Contents or turn to the Index at the front of each of the three sections. Each section index is an alphabetical listing of all the procedures within the particular section and their page numbers.

Safety Notice

Proper service and repair procedures are vital to the safe, reliable operation of all motor vehicles, as well as the personal safety of those performing the repairs. This manual outlines procedures for servicing and repairing vehicles using safe effective methods. The procedures contain many NOTES, WARNINGS and CAUTIONS which should be followed along with standard safety procedures to eliminate the possibility of personal injury or improper service which could damage the vehicle or compromise its safety.

It is important to note that repair procedures and techniques, tools and parts for servicing vehicles, as well as the skill and experience of the individual performing the work vary widely. It is not possible to anticipate all of the conceivable ways or conditions under which vehicles may be serviced, or to provide cautions as to all of the possible hazards that may result. Standard and accepted safety precautions and equipment should be used when handling toxic or flammable fluids, and safety goggles or other protection should be used during cutting, grinding, chiseling, prying, or any other process that can cause material removal or projectiles.

Some procedures require the use of tools specially designed for a specific purpose. Before substituting another tool or procedure, you must be completely satisfied that neither your personal safety, nor the performance of the vehicle will be endangered.

Although information in this manual is based on industry sources and is as complete as possible at the time of publication, the possibility exists that some vehicle manufacturers made later changes which could not be included here. Information on very late models may not be available in some circumstances. While striving for total accuracy, NP/Chilton cannot assume responsibility for any errors, changes, or omissions that may occur in the compilation of this data.

Part Numbers

Part numbers listed in this book are not recommendations by NP/Chilton for any product by brand name. They are references that can be used with interchange manuals and aftermarket supplier catalogs to locate each brand supplier's discrete part number.

Special Tools

Special tools are recommended by the vehicle manufacturer to perform their specific job. Use has been kept to a minimum, but where absolutely necessary, they are referred to in the text by the part number of the tool manufacturer. These tools may be purchased, under the appropriate part number, from your local dealer or regional distributor, or an equivalent tool can be purchased locally from a tool supplier or parts outlet. Before substituting any tool for the one recommended, read the previous Safety Notice.

Acknowledgments

NP/Chilton wishes to express its appreciation to all manufacturers involved in the production of this book. A special thanks goes to Lincoln Automotive Products, 1 Lincoln Way, St. Louis, MO 63120, for providing the professional shop equipment used in our tear-down facility, including jacks, engine stands, fluid and lubrication tools, and shop presses.

Copyright Notice

Table of Contents

Table of Contents

Section 3—Water Pumps

Table of Contents

Table of Contents

Section 3—Water Pumps (cont.)

Table of Contents

Get ready for ASE testing with Motor Age Self Study Guides.

Each training unit contains a complete description of the ASE Task Analysis and Test Specifications, and covers the subject areas of the corresponding ASE test question group. Also included are sample ASE test questions. In addition, each book includes a special glossary, sample questions and an expanded answer analysis to increase your knowledge of the subject.

AA Car & Light Truck
A1 Engine Repair
A2 Automatic Transmission/Transaxle
A3 Manual Drive Train & Axles
A4 Suspension & Steering
A5 Brakes
A6 Electrical/Electronic Systems
A7 Heating & A/C
A8 Engine Performance

Parts Specialist
P1 Medium/Heavy Parts Specialist
P2 Automobile Parts Specialist

Advanced Level
L1 Advanced Engine Performance Specialist
L2 Med/Hvy Vehicle Electronic Diesel Engine Diagnosis Specialist
F1 Light Vehicle Compressed Natural Gas

ALSO AVAILABLE:
TT Medium/Heavy Truck Service: T1 Gasoline Engines, T2 Diesel Engines, T3 Drive Train, T4 Brakes, T5 Suspension & Steering, T6 Electrical/Electronic Systems, T7 Heating, Ventilation & A/C, T8 **Preventive Maintenance Inspection (PMI), & MM Engine Machinist (M1, M2, M3),**
BB Collision Repair/Paint & Refinish: B2 Paint & Refinishing, B3 Non-Structural Analysis & Damage Repair, B4 Structural Analysis & Damage Repair, B5 Mechanical & Electrical Components, B6 Damage Analysis & Estimating

Name: _____
          first   middle   last

Company _____

Address: _____ Apt. # _____

City: _____ State: _____ Zip: _____

Phone (DAYTIME): _____ Fax _____

For pricing and shipping information, fax this form to Trudy Kolb, 610-964-4251

ELECTRIC COOLING FANS

1

GENERAL INFORMATION

A basic vehicle cooling system consists of a radiator, water pump, thermostat, electric or engine-driven cooling fan, and hoses. Electric cooling fans are common on today's vehicles due to engine compartment space limitations or engine layout. Electric cooling fans operate in either a pusher or a puller capacity. A pusher type fan is typically mounted on the front of the radiator assembly and forces air through the radiator, whereas a puller type fan is mounted on the engine side of the radiator and draws air through the grill and radiator assembly. Vehicles that utilize a transversely-mounted engine will always be equipped with at least one electric cooling fan (most having two), because none of the engine pulleys are inline with the radiator air-flow.

There are generally two types of electric cooling fans: primary cooling fans and secondary cooling fans. Primary cooling fans are typically of the puller style. Vehicles that do not incorporate an engine-driven mechanical cooling fan will utilize a primary cooling fan. The secondary cooling fan, also known as a A/C condenser fan

or auxiliary cooling fan by certain manufacturers, could be of either a pusher or a puller style. Vehicles equipped with A/C will either utilize the radiator cooling fan or a separate fan as the A/C condenser cooling fan (which performs the same function as an auxiliary cooling fan on vehicles with a primary mechanical fan). The engine control computer that receives inputs from various sensors in the engine compartment commonly controls electric cooling fans. The engine control computer receives inputs from the engine coolant temperature sensors and A/C system pressure switches, then actuates the necessary cooling fan relays to engage the applicable cooling fan for the condition. On models equipped with only one electric primary cooling fan, the fan can operate at two speeds: low speed and high speed. The low speed condition is enabled when the engine begins to heat up or when the A/C is engaged. As the engine demands more cooling, the cooling fan will be stepped-up to high speed.

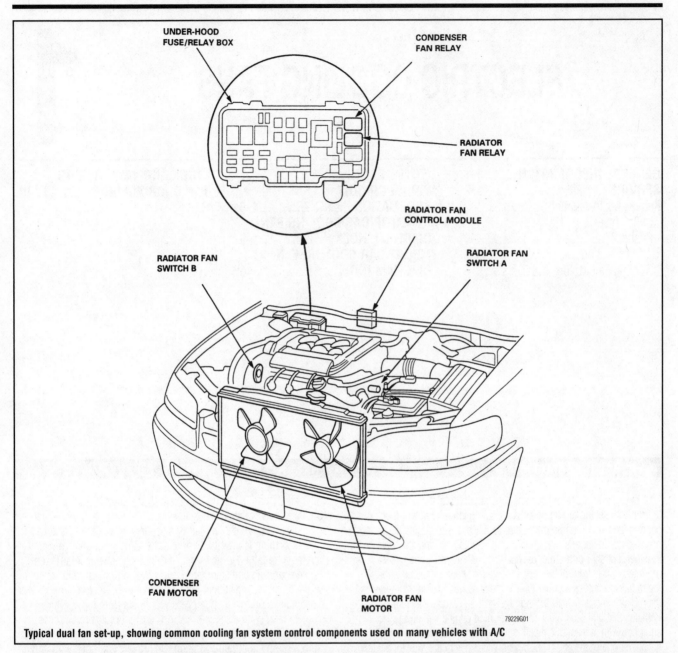

Typical dual fan set-up, showing common cooling fan system control components used on many vehicles with A/C

SERVICE

Due to the wide variety of vehicle manufacturers and suppliers of electric cooling fans it is almost impossible to cover every specific combination of cooling fan and model. The following procedures will cover the most common types of mountings and troubleshooting techniques.

Removal & Installation

PULLER TYPE

➡**It may be simpler to remove the cooling fan(s) with the radiator as an assembly.**

1. Disconnect the negative battery cable.
2. Inspect the cooling fan and take note of any wires, hoses or A/C lines which may hamper fan removal. Also at this time, decide whether it is necessary to remove the fan along with the radiator or not.

3. Position aside all wires, hoses and A/C lines for fan removal. It may not always be possible to create enough clearance for fan removal by simply moving these obstructions aside; often they must be disconnected. If any cooling system lines must be disconnected, drain and recycle the engine coolant. If any of the A/C lines must be disconnected, the A/C system will need to be discharged and evacuated by a MVAC-trained technician using an approved recovery machine.

4. Disengage the cooling fan wiring harness connector.
5. If the fan can be removed without the radiator, perform the following:

   a. Loosen the mounting fasteners. Usually there are two nuts or bolts along the top edge of the cooling fan shroud and either two retaining clips or bolts along the bottom edge.

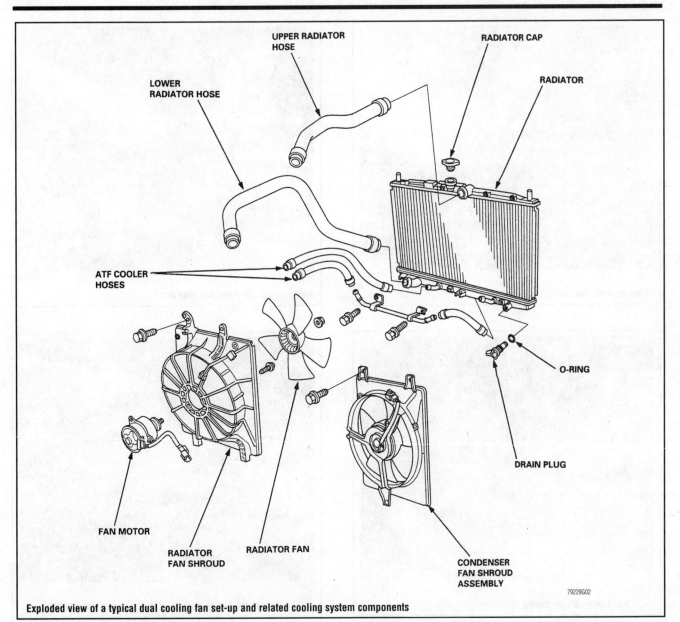

UPPER RADIATOR HOSE

RADIATOR CAP

RADIATOR

LOWER RADIATOR HOSE

ATF COOLER HOSES

O-RING

DRAIN PLUG

FAN MOTOR

RADIATOR FAN SHROUD

RADIATOR FAN

CONDENSER FAN SHROUD ASSEMBLY

Exploded view of a typical dual cooling fan set-up and related cooling system components

To remove a common puller type cooling fan, first detach any braces (1), wires (2) or other obstructions . . .

. . . including cooling system hoses, to allow fan removal

Disengage the fan wiring connector(s) . . .

Separate the fan from the radiator . . .

. . . and loosen all fan mounting fasteners

. . . , then lift the fan up and out of the engine compartment

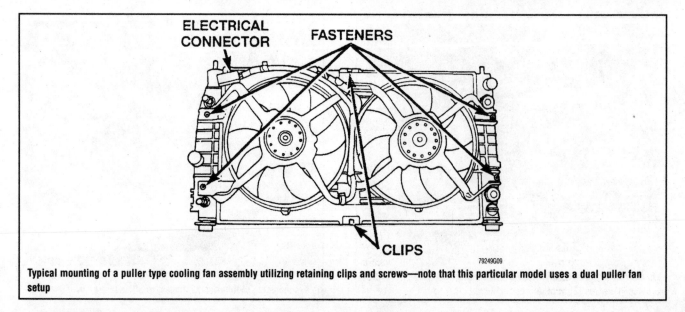

Typical mounting of a puller type cooling fan assembly utilizing retaining clips and screws—note that this particular model uses a dual puller fan setup

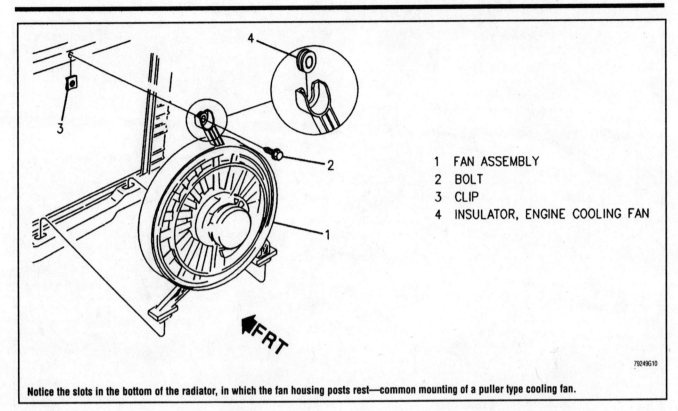

1 FAN ASSEMBLY
2 BOLT
3 CLIP
4 INSULATOR, ENGINE COOLING FAN

Notice the slots in the bottom of the radiator, in which the fan housing posts rest—common mounting of a puller type cooling fan.

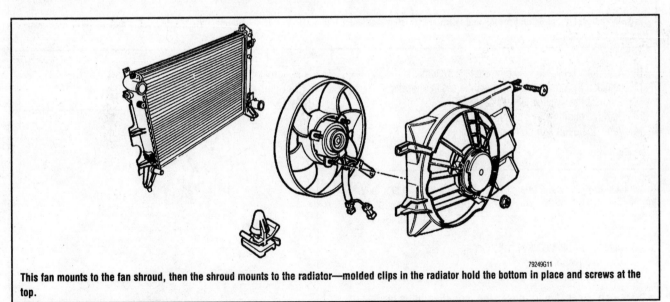

This fan mounts to the fan shroud, then the shroud mounts to the radiator—molded clips in the radiator hold the bottom in place and screws at the top.

b. Carefully lift the fan up and out of the engine compartment, making sure that no wires or hoses get hung up on it.

6. If it is necessary to remove the radiator for fan removal, perform the following:

a. Disconnect all cooling system hoses from it after draining the cooling system.

b. Locate all of the radiator mounting fasteners (usually two or more nuts or bolts along the top, possibly two along the bottom).

➡**Quite a few radiators are secured along the bottom by two posts which fit into rubber grommets. The rubber grommets help isolate the radiator from harsh vibrations in the frame. If no nuts or bolts can be located along the bottom of the**
radiator, chances are that the radiator is secured with the posts and grommets.

c. Lift the radiator and cooling fan up and out of the engine compartment together.

d. Separate the cooling fan from the radiator by removing the attaching fasteners.

To install:

7. If applicable, install the cooling fan on the radiator.

8. Install the cooling fan and shroud assembly (also the radiator if necessary). Tighten the fan shroud mounting bolts.

9. Reattach all wires, hoses and A/C lines as applicable. If the A/C lines were detached, the system must be evacuated and recharged by a MVAC-trained technician.

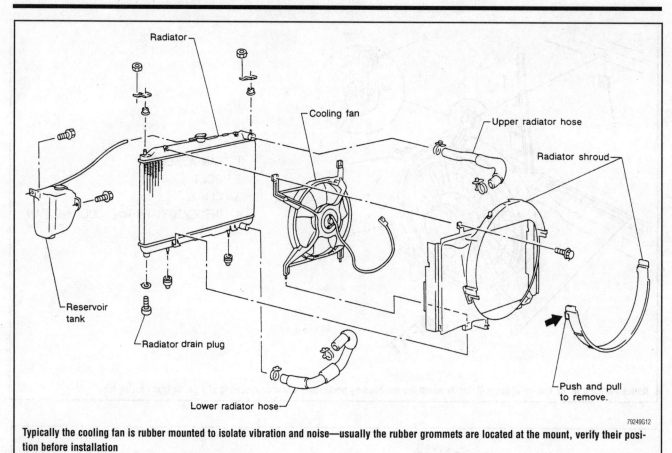

Typically the cooling fan is rubber mounted to isolate vibration and noise—usually the rubber grommets are located at the mount, verify their position before installation

10. If drained, refill and bleed the cooling system.
11. Reattach the cooling fan electrical harness connector.
12. Connect the negative battery cable.
13. Start the engine and check for leaks.
14. Verify the operation of the cooling fan(s).

PUSHER TYPE

Vehicles that utilize the pusher type of electric cooling fan, may require the removal of the grilles and/or upper radiator shroud in order to gain access the fasteners that mount the fan assembly in the vehicle.
1. Disconnect the negative battery cable.
2. Access the cooling fan.
3. Label and disconnect the cooling fan electrical harness.

➡ **It may be necessary to loosen the mounting bolts for the A/C condenser to the body**

4. Remove the fasteners that mount the cooling fan to the A/C condenser or radiator.
5. Lift the cooling fan out of the vehicle.
To install:
6. Insert the cooling fan into the vehicle.
7. Mount the cooling fan to the A/C condenser or radiator
8. Connect the cooling fan electrical harness.

9. If removed, install any shrouding or grills.
10. Connect the negative battery cable.

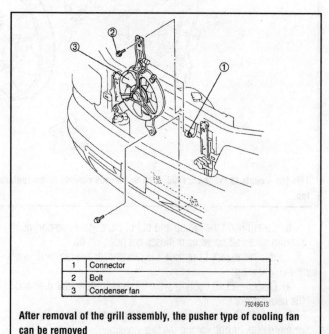

1	Connector
2	Bolt
3	Condenser fan

After removal of the grill assembly, the pusher type of cooling fan can be removed

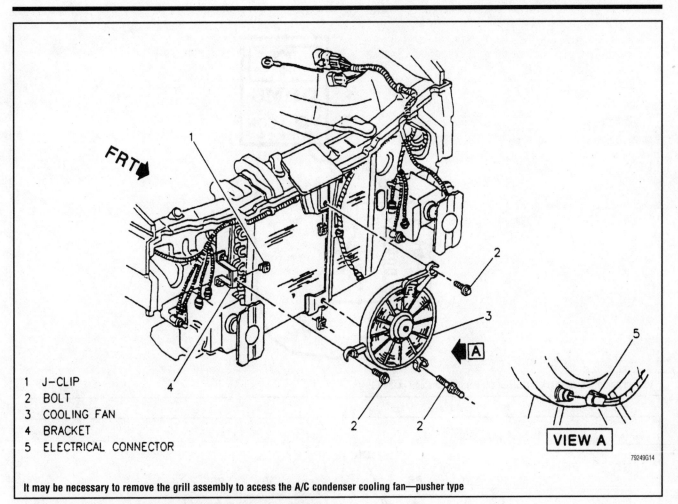

FRT

1 J—CLIP
2 BOLT
3 COOLING FAN
4 BRACKET
5 ELECTRICAL CONNECTOR

VIEW A

79249G14

It may be necessary to remove the grill assembly to access the A/C condenser cooling fan—pusher type

Troubleshooting

When diagnosing an inoperative cooling fan it may be necessary to use a diagnostic scan tool to monitor engine coolant temperature and the engine control computer.

1. Perform a visual inspection of the cooling fan. If the fan does not turn with ease, the fan motor is seized and needs to be replaced.
2. Check all the fuses and fusible links related to the cooling fan circuit.
3. Check the integrity of the electrical connections related to the cooling fan circuit.
4. Check the cooling fan motor.
5. Check the relays associated with the cooling fan circuit.
6. Using a scan tool, determine if the engine control computer is calling for the fan to activate.

COOLING FAN MOTOR

1. Disconnect the negative battery cable.
2. Disengage the cooling fan motor connector.
3. Identify and label the ground and the power terminals of the cooling fan connector using the wiring diagrams provided.
4. Using jumper leads with a fuse in series, apply battery voltage to the appropriate terminals of the cooling fan.

5. The cooling fan should operate. If not, replace the cooling fan.

If the cooling fan functions properly during this test, proceed to the cooling fan relay test.

COOLING FAN RELAY

1. Turn the ignition **OFF**.
2. Remove the relay.
3. Locate the two terminals on the relay, which are connected to the coil windings. Check the relay coil for continuity. Connect the common meter lead to terminal 85 and positive meter lead to terminal 86. There should be continuity. If not, replace the relay.
4. Check the operation of the internal relay contacts.
    a. Connect the meter leads to terminals 30 and 87. Meter polarity does not matter for this step.
    b. Apply positive battery voltage to terminal 86 and ground to terminal 85. The relay should click as the contacts are drawn toward the coil and the meter should indicate continuity. Replace the relay if your results are different.

If the relay functions properly during this test, inspect the coolant temperature sensor and the cooling fan system wiring for defects.

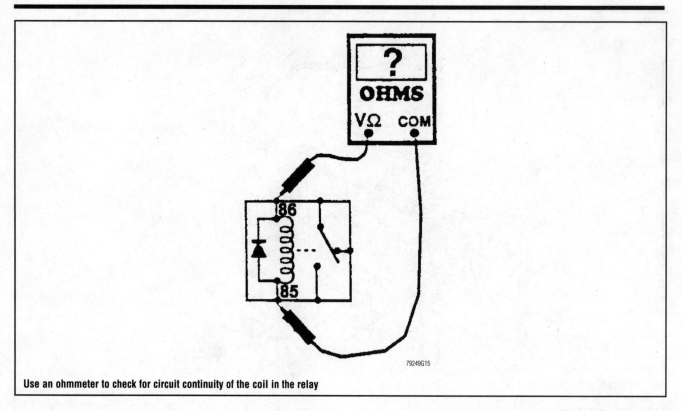

Use an ohmmeter to check for circuit continuity of the coil in the relay

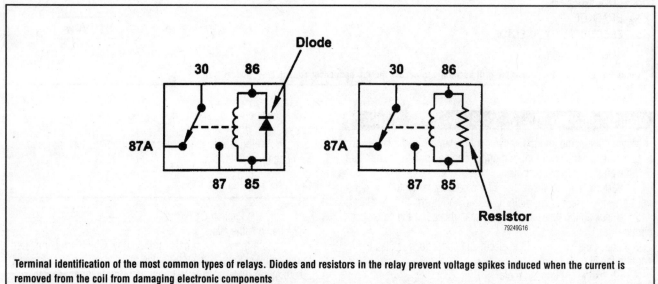

Terminal identification of the most common types of relays. Diodes and resistors in the relay prevent voltage spikes induced when the current is removed from the coil from damaging electronic components

WIRING CIRCUIT SCHEMATICS

➡The wiring schematics are divided into three subsections, (domestic cars, import cars, trucks and vans), each with its own index. Please refer to the appropriate index for your particular application.

DOMESTIC CAR
COOLING FAN DIAGRAM INDEX

MANUFACTURER

Model and Engine	Diagram
Chrysler Corp.	
Avenger, Sebring Coupe 2.0L	1
Avenger, Sebring Coupe 2.5L	2
Breeze, Cirrus, Sebring Convertible, Stratus 2.0L/ 2.4L/ 2.5L	
1995-96 Models	3
1997-99 Models	4
Concord, Intrepid, LHS, New Yorker, Vision 2.7L/ 3.2L/ 3.3L 3.5L	5
Neon 2.0L	
1995-96 Models	6
1997-99 Models	7
Talon 2.0L (Non-turbo)	
1995-97 Models with A/T	8
1995-97 Models with M/T	9
1998 Models	10
Talon 2.0L (Turbo)	
Models with A/T	11
Models with M/T	12
Ford Motor Co.	
Ford Aspire 1.3L	13
Ford Continental 4.6L	
1995-96 Models	14
1997-99 Models	15
Ford Contour, Mystique	
1995-97 2.0L	16
1995-97 2.5L	17
1998-99 2.0L and 2.5L	18
Ford Crown Victoria, Grand Marquis 4.6L	
1995-97 Models	19
1998-99 Models	20
Ford Escort, Tracer, ZX2	
1995-96 Models with A/T	21
1995-96 Models with M/T	22
1997-99 Models	23
Ford Mark VIII 4.6L	
1995-96 Models	24
1997-99 Models	25
Ford Mustang	
1995 Models	26
1996-99 Models with 3.8L	27
1996-99 Models with 4.6L	28
Ford Probe 2.0L/ 2.5L	29
Ford Taurus, Sable	
1995 Models with 3.0L, 3.0L SHO and 3.8L	30
1995 Models with 3.2L SHO	31
1996-97 Models with 3.0L and 3.4L	32

DOMESTIC CAR
COOLING FAN DIAGRAM INDEX

MANUFACTURER

Model and Engine	Diagram
Ford Motor Co. (cont.)	
1998-99 Models with 3.0L and 3.4L	33
Ford Thunderbird, Cougar XR7 3.8L/ 4.6L	34
1995 Models	35
1996-99 Models	
Lincoln Town Car 4.6L	36
1995-97 Models	37
1998-99 Models	
General Motors	
A Body (Century, Cutlass Ciera, Cruiser) 2.2L/ 3.1L	38
B Body (Caprice, Impala SS, Roadmaster) 4.3L/ 5.7L (w/o Mechanical Fan)	39
B Body (Caprice, Impala SS, Roadmaster) 4.3L/ 5.7L (w/ Mechanical Fan)	40
B Body (Fleetwood) 5.7L (w/o Mechanical Fan)	41
B Body (Fleetwood) 5.7L (w/ Mechanical Fan)	42
C & H Bodies (Bonneville, Eighty-Eight, Ninety-Eight, Park Ave., Le Sabre, LSS, Regency) 3.8L	43
E & K Bodies (DeVille, ElDarado, Seville) 4.6L/ 4.9L	
1995-96 Models	44
1997 Models	45
1998-99 Models	46
F Body (Camaro, Firebird) 3.4L (w/ C41)	47
F Body (Camaro, Firebird) 3.4L (w/ C60)	48
F Body (Camaro, Firebird) 3.8L	49
F Body (Camaro, Firebird) 5.7L	
1995 Models with C41 option	47
1995 Models with C60 option	48
1996-97 Models	49
1998-99 Models	50
G Body (Aurora, Riviera) 3.8L	
1995 Models with 3.8L	51
1996 Models with 3.8L	52
1997-99 Models with 3.8L	53
1995 Models with 4.0L	51
1996-99 Models with 4.0L	54
J Body (Cavalier, Sunfire) 2.2L/ 2.3L/ 2.4L	55
L Body (Beretta, Corsica) 2.2L/ 3.1L	56
L/N Bodies (Cutlass, Malibu) 2.4L/ 3.1L	57
N Body (Acheiva, Grand Am, Skylark) 2.3L/ 2.4L/ 3.1L	58
V Body (Catera) 3.0L	59
W Body (Lumina, Monte Carlo, Grand Prix, Cutlass Supreme, Regal, Intrigue) 3.1L/ 3.4L	
1995-96 Models (except Cutlass Supreme)	60
1995-97 Cutlass Supreme	60
1997-99 Models (except Cutlass Supreme)	61

91261C02

DOMESTIC CAR
COOLING FAN DIAGRAM INDEX

MANUFACTURER
Model and Engine   Diagram

General Motors (cont.)

Model and Engine	Diagram
Y Body (Corvette) 5.7L	
1995-96 Models	62
1997-99 Models	63
Geo/Chevrolet	
Metro 1.0L/ 1.3L (w/o A/C)	64
Metro 1.0L/ 1.3L (w/ A/C)	65
Prism 1.6L/ 1.8L (w/ A/C)	66
Prism 1.6L/ 1.8L (w/o A/C)	67
Saturn	
1.9L	68

91261C03

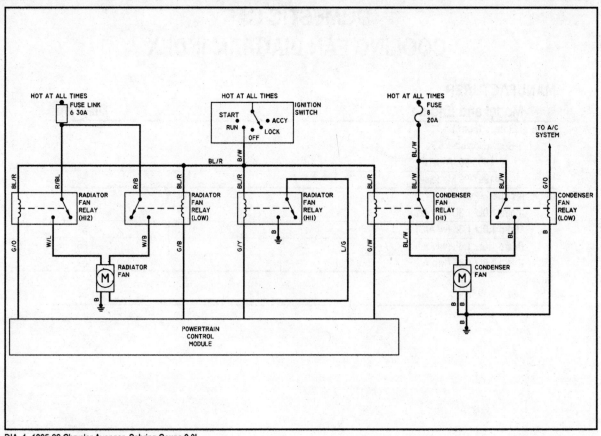

DIA. 1- 1995-99 Chrysler Avenger, Sebring Coupe 2.0L

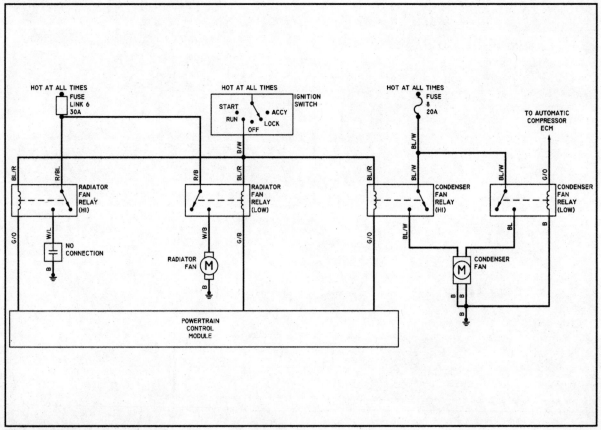

DIA. 2 - 1995-99 Chrysler Avenger, Sebring Coupe 2.5L

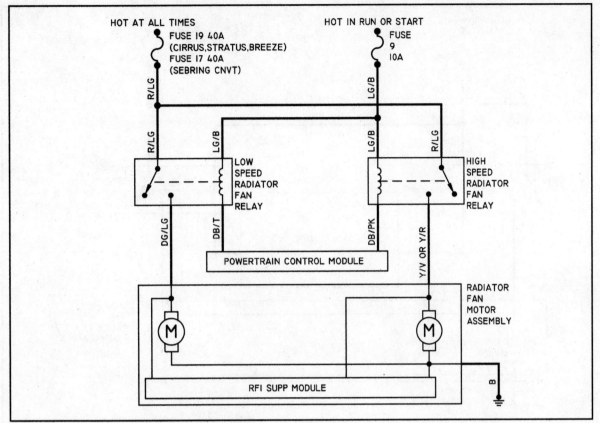

DIA. 3 - 1995-96 Chrysler Breeze, Cirrus, Sebring Convertible, Stratus
2.0L / 2.4L / 2.5L

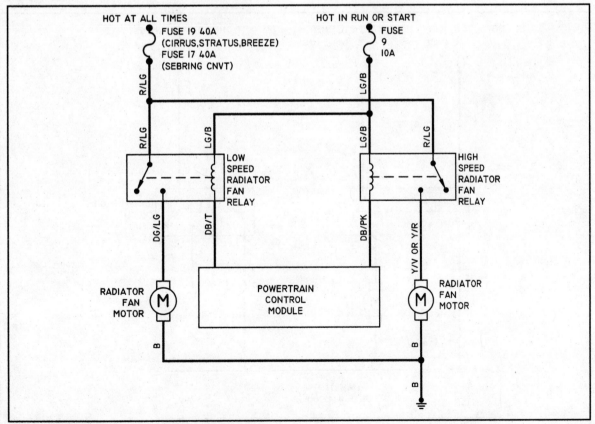

DIA. 4 - 1997-99 Chrysler Breeze, Cirrus, Sebring Convertible, Stratus
2.0L / 2.4L / 2.5L

79229W02

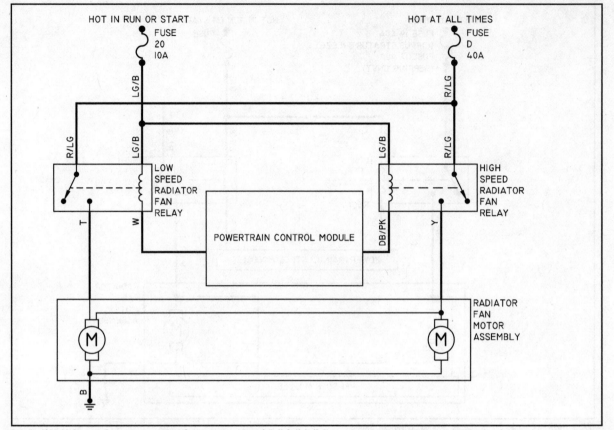

DIA. 5 - 1995-99 Chrysler Concorde, Intrepid, LHS, New Yorker, Vision 2.7L/3.2L/3.3L/3.5L

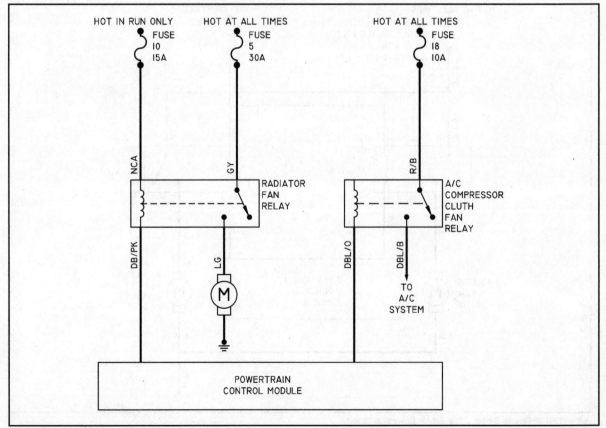

DIA. 6 - 1995-96 Dodge Neon 2.0L

79229W03

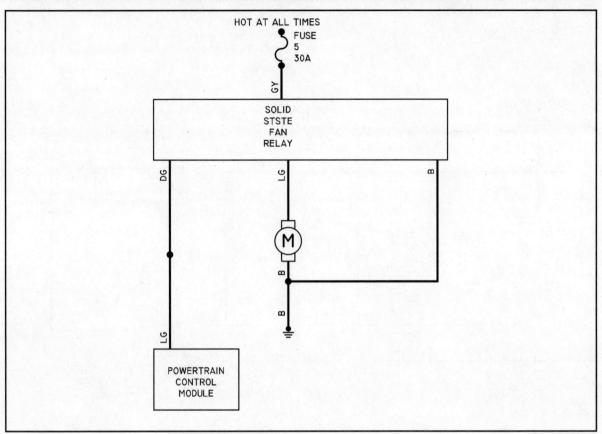

DIA. 7 - 1997-99 Dodge Neon 2.0L

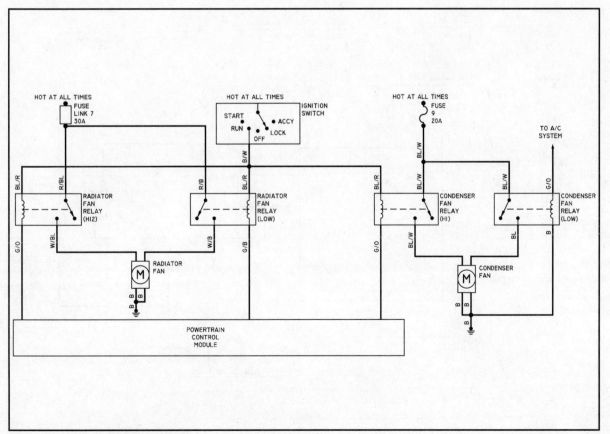

DIA. 8 - 1995-97 Eagle Talon 2.0L (Non-turbo A/T)

79229W04

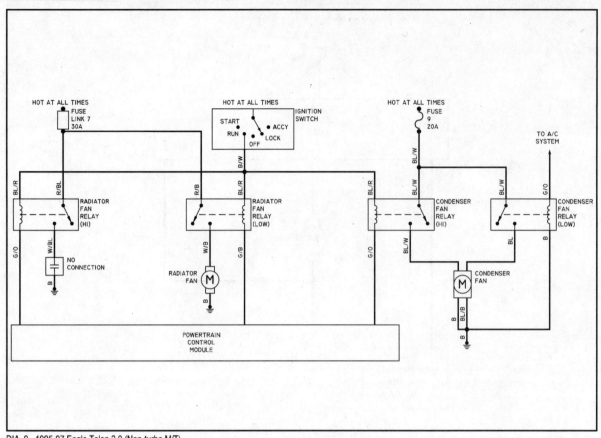

DIA. 9 - 1995-97 Eagle Talon 2.0 (Non-turbo M/T)

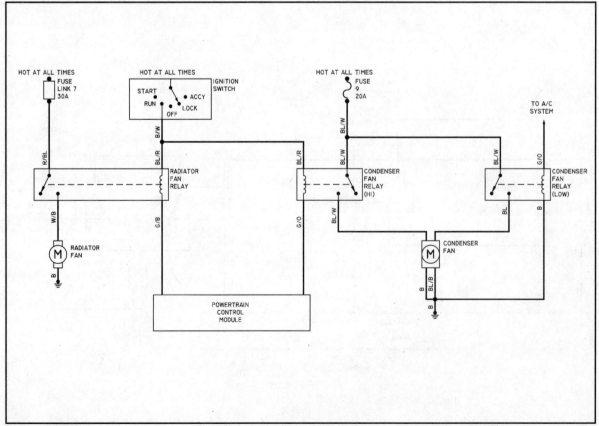

DIA. 10 - 1998 Eagle Talon 2.0L (Non-turbo)

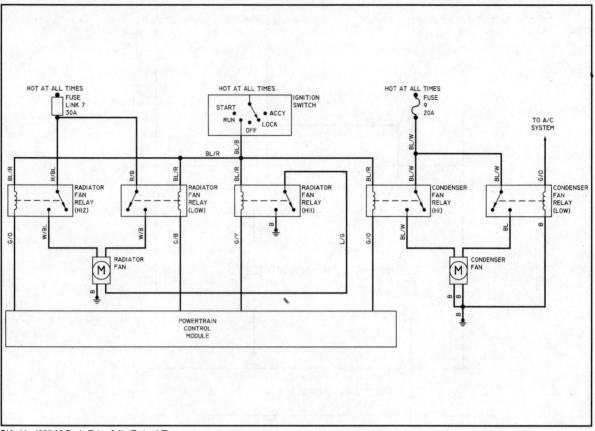

DIA. 11 - 1995-98 Eagle Talon 2.0L (Turbo A/T)

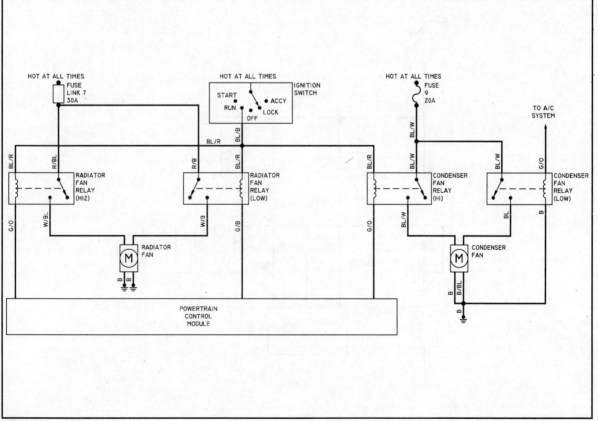

DIA. 12 - 1995-98 Eagle Talon 2.0L (Turbo M/T)

79229W06

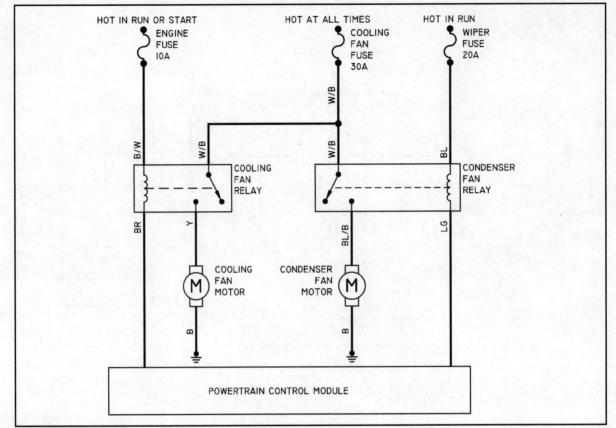

DIA. 13 - 1995-97 Ford Aspire 1.3L

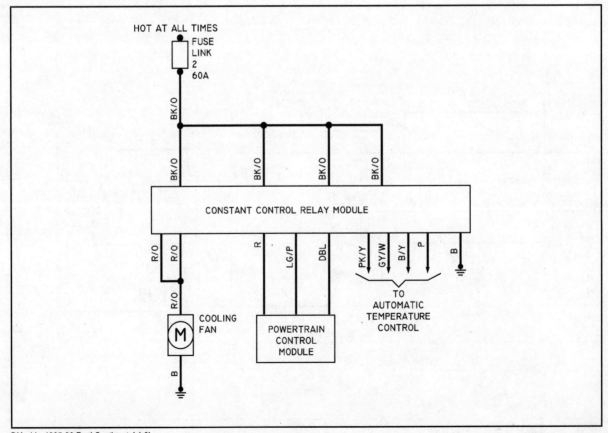

DIA. 14 - 1995-96 Ford Continental 4.6L

79229W07

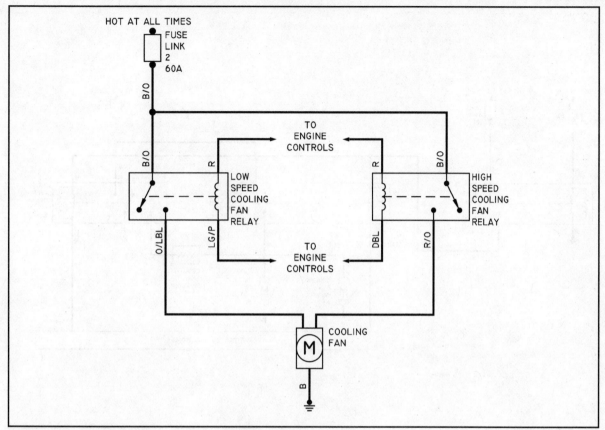

DIA. 15 - 1997-99 Ford Continental 4.6L

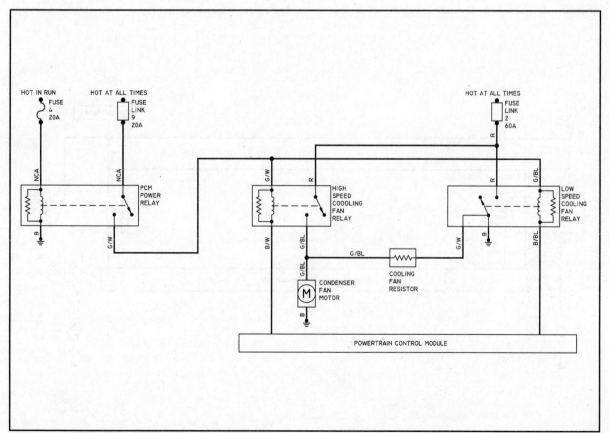

DIA. 16 - 1995-97 Ford Contour, Mystique 2.0L

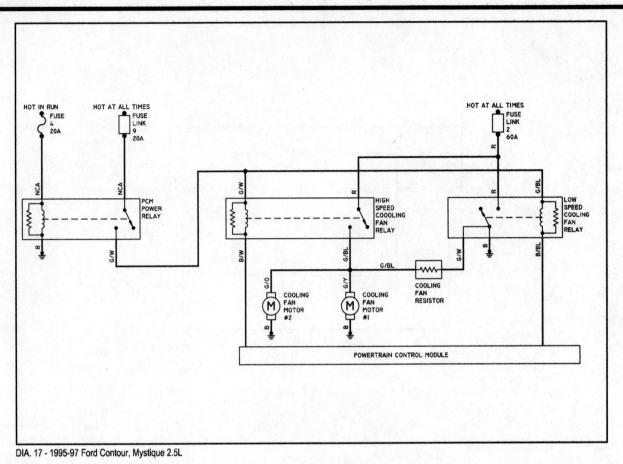

DIA. 17 - 1995-97 Ford Contour, Mystique 2.5L

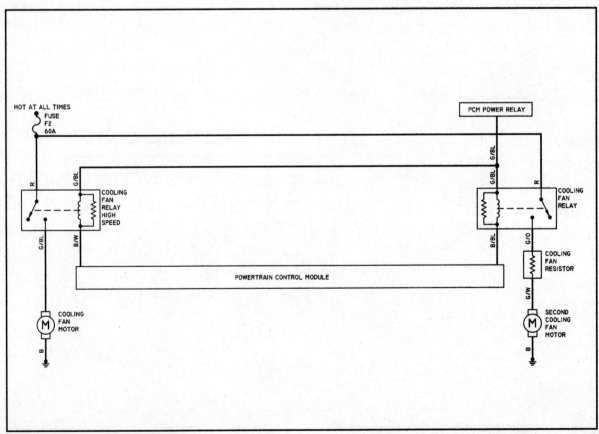

DIA. 18 - 1998-99 Ford Contour, Mystique 2.0L / 2.5L

79229W09

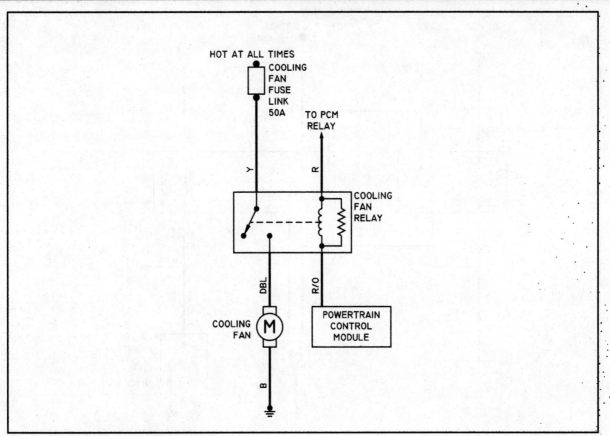

DIA. 19 - 1995-97 Ford Crown Victoria, Grand Marquis 4.6L

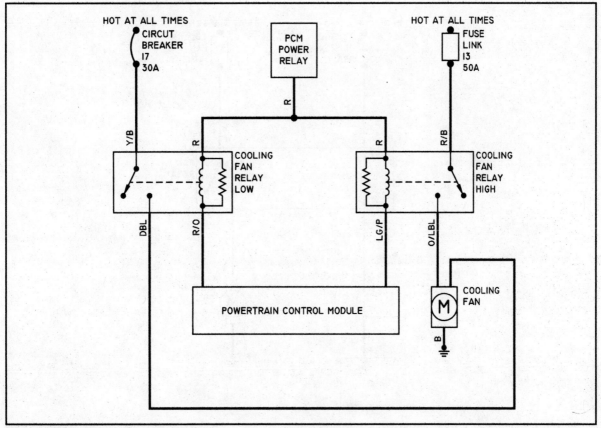

DIA. 20 - 1998-99 Ford Crown Victoria, Grand Marquis 4.6L

79229W10

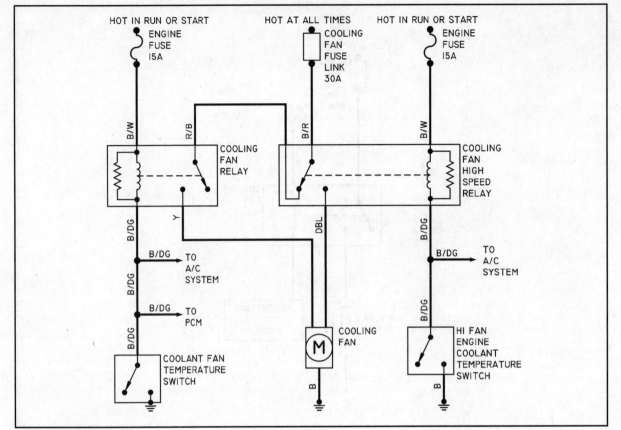

DIA. 21 - 1995-96 Ford Escort, Tracer 1.8L A/T

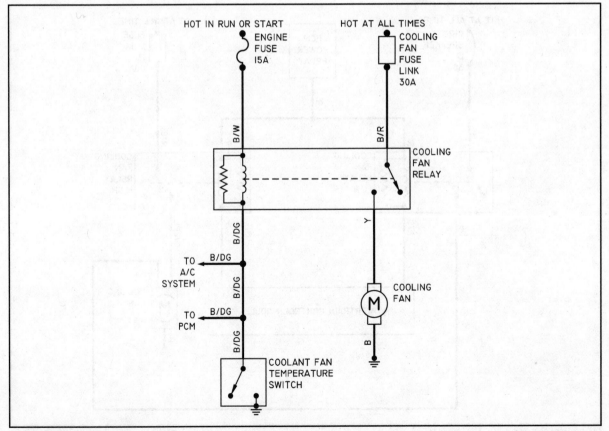

DIA. 22 - 1995-96 Ford Escort, Tracer 1.8L M/T

79229W11

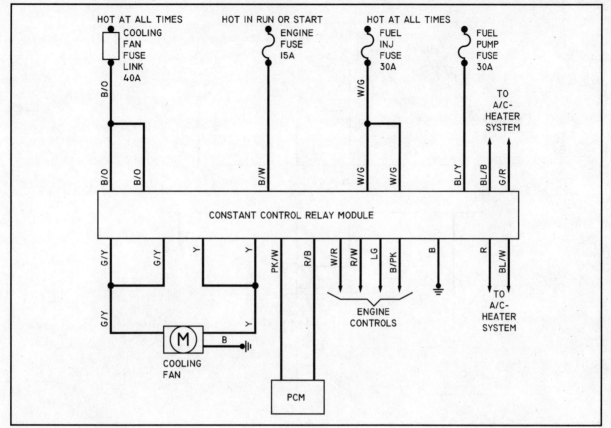

DIA. 23 - 1997-99 Ford Escort, Tracer, ZX2 1.8L / 1.9L

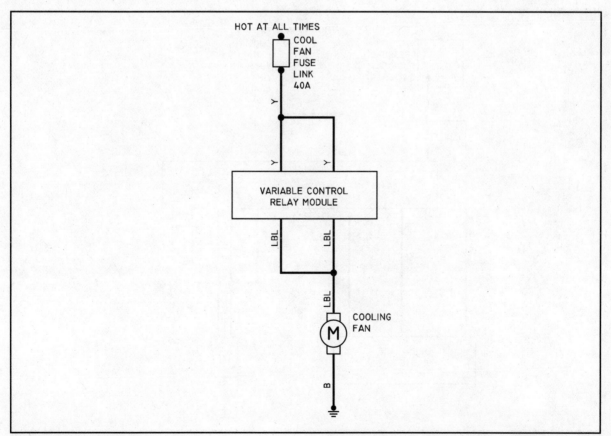

DIA. 24 - 1995-96 Ford Mark VIII 4.6L

79229W12

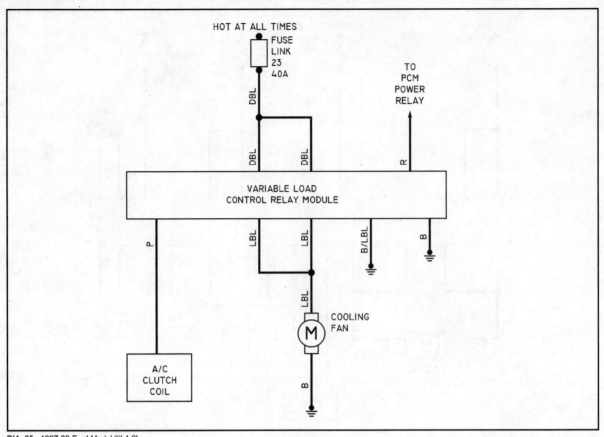

DIA. 25 - 1997-99 Ford Mark VIII 4.6L

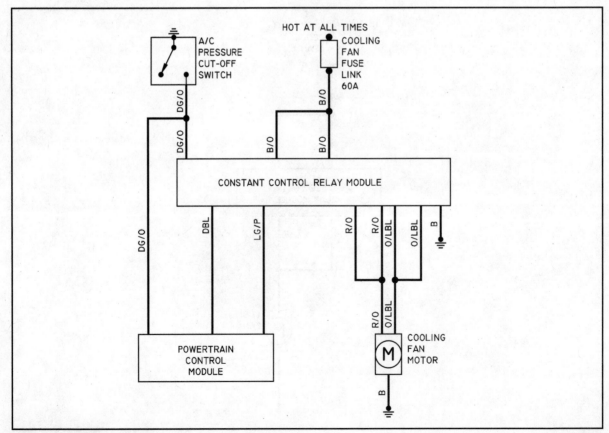

DIA. 26 - 1995 Ford Mustang 3.8L / 5.0L

79229W13

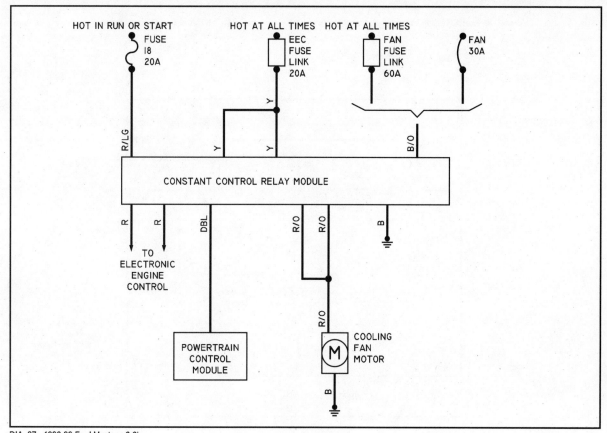

DIA. 27 - 1996-99 Ford Mustang 3.8L

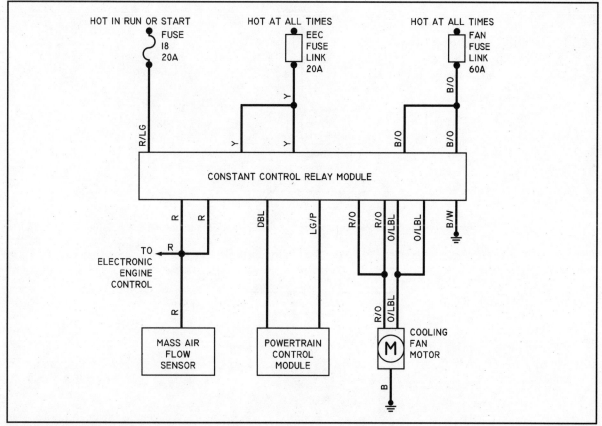

DIA. 28 - 1996-99 Ford Mustang 4.6L

79229W14

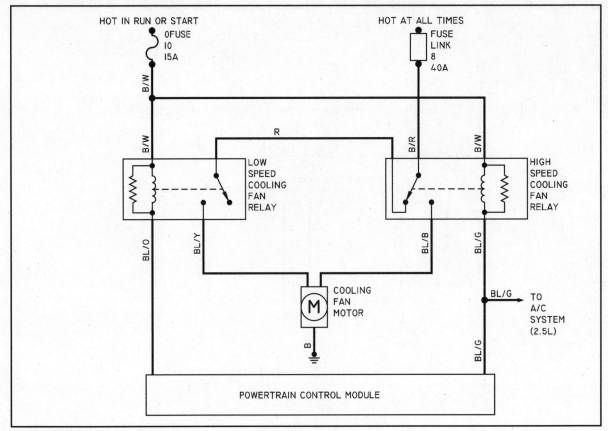

DIA. 29 - 1995-97 Ford Probe 2.0L / 2.5L

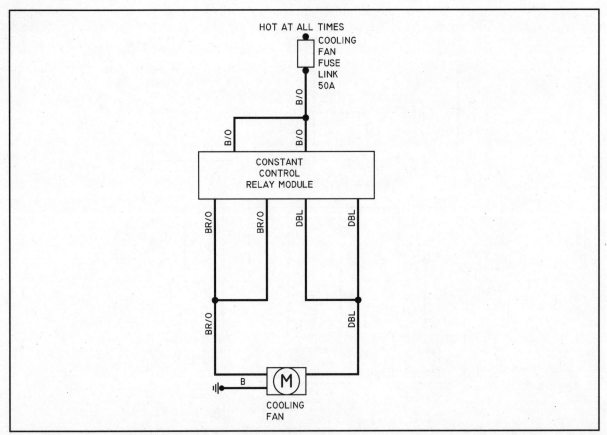

DIA. 30 - 1995 Ford Taurus, Sable 3.0L / 3.0L SHO / 3.8L

79229W15

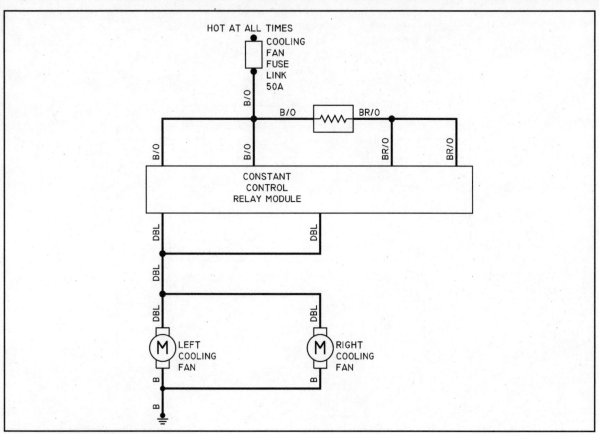

DIA. 31 - 1995 Ford Taurus, Sable 3.2L SHO

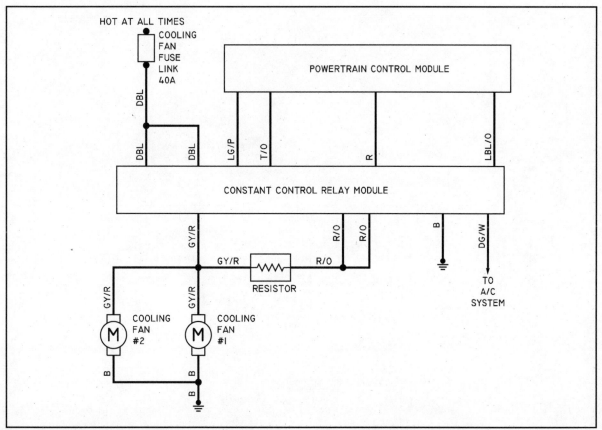

DIA. 32 - 1996-97 Ford Taurus, Sable 3.0L / 3.4L

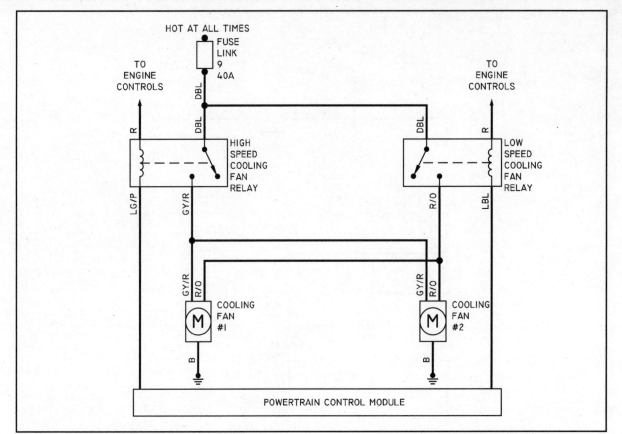

DIA. 33 - 1998-99 Ford Taurus, Sable 3.0L / 3.4L

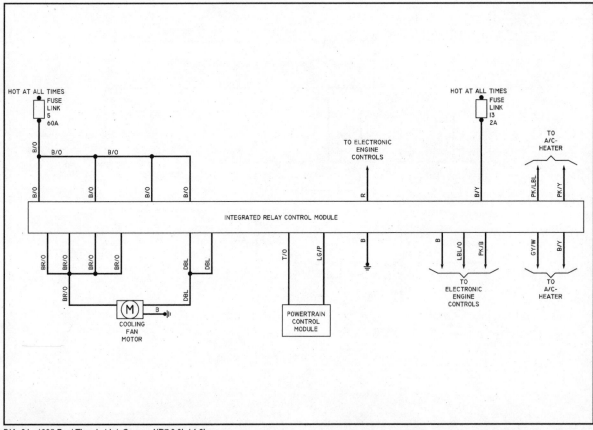

DIA. 34 - 1995 Ford Thunderbird, Cougar, XR7 3.8L / 4.6L

79229W17

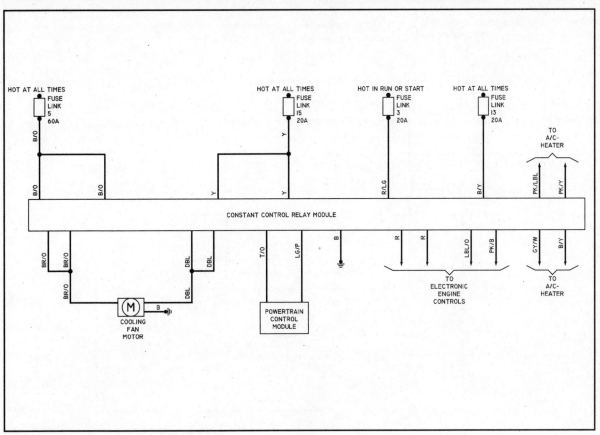

DIA. 35 - 1996-99 Ford Thunderbird, Cougar XR7 3.8L / 4.6L

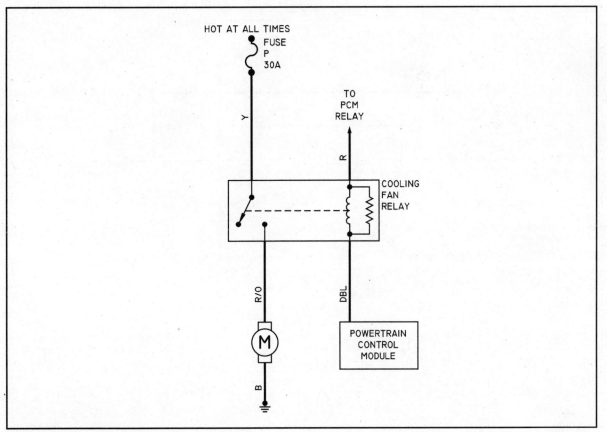

DIA. 36 - 1995-97 Lincoln Town Car 4.6L

79229W18

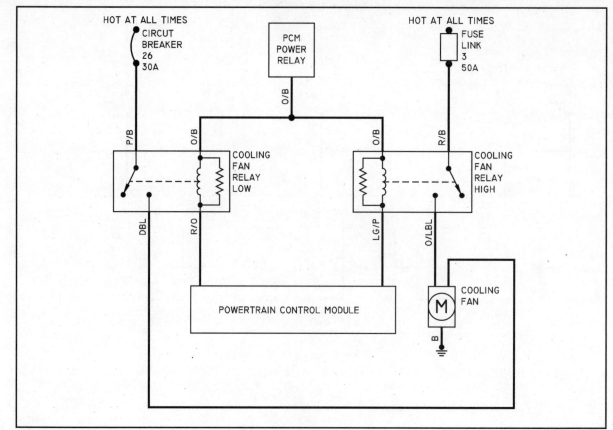

DIA. 37 - 1998-99 Lincoln Town Car 4.6L

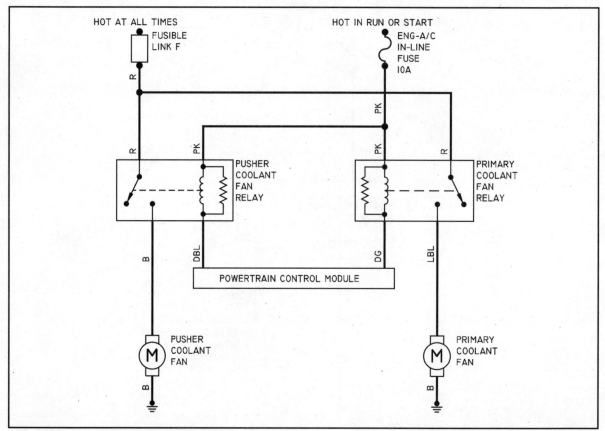

DIA. 38 - 1995-96 GM A Body (Century, Cutlass Ciera, Cutlass Cruiser) 2.2L / 3.1L

79229W19

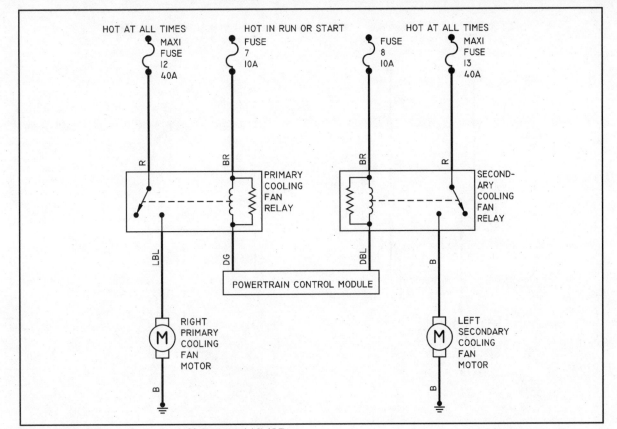

DIA. 39 - 1995-96 GM B Body (Caprice, Impala SS, Roadmaster) 4.3L / 5.7L
(W/O Mech Fan)

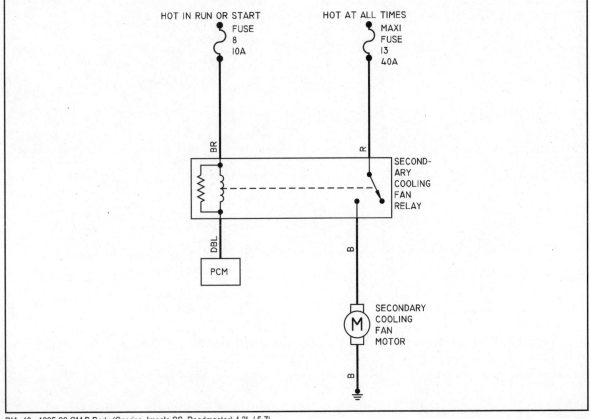

DIA. 40 - 1995-96 GM B Body (Caprice, Impala SS, Roadmaster) 4.3L / 5.7L
(W/ Mech Fan)

79229W20

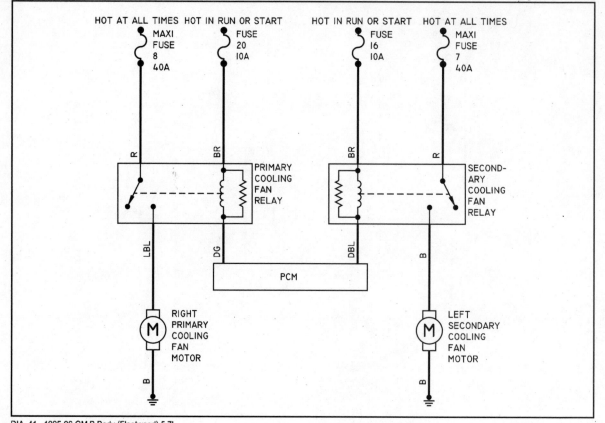

DIA. 41 - 1995-96 GM B Body (Fleetwood) 5.7L
(W/O Mech Fan)

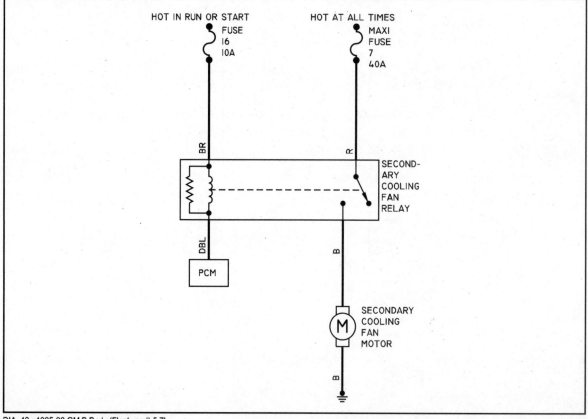

DIA. 42 - 1995-96 GM B Body (Fleetwood) 5.7L
(W/ Mech Fan)

79229W21

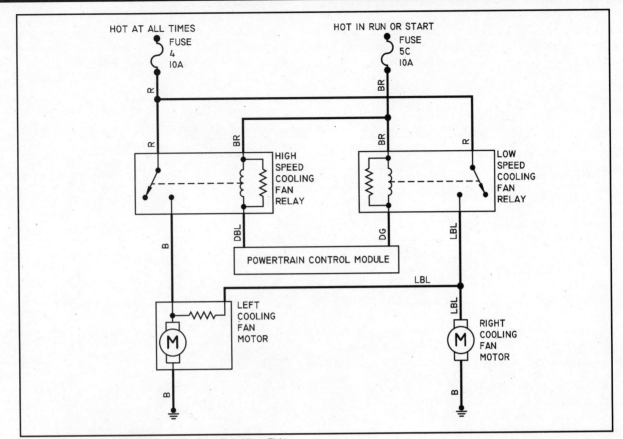

DIA. 43 - 1995-99 GM C & H Bodies (Bonneville, Eighty-Eight, Ninety-Eight, Park Ave, Le Sabre, LSS, Regency) 3.8L

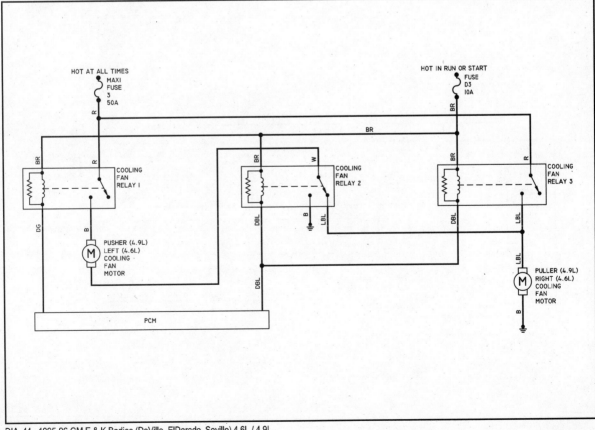

DIA. 44 - 1995-96 GM E & K Bodies (DeVille, ElDarado, Seville) 4.6L / 4.9L

79229W22

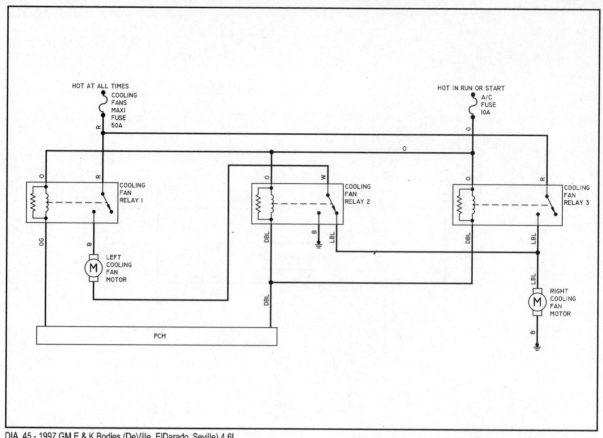

DIA. 45 - 1997 GM E & K Bodies (DeVille, ElDarado, Seville) 4.6L

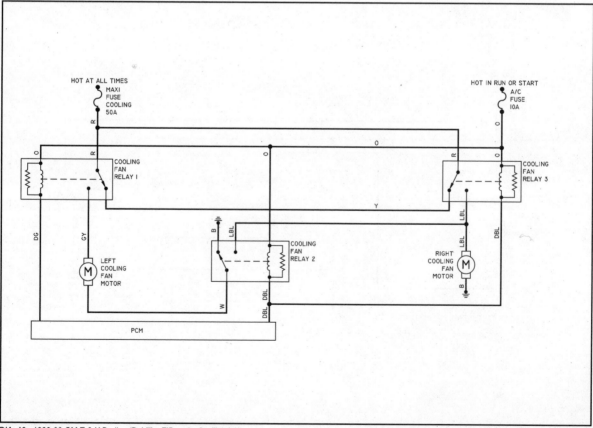

DIA. 46 - 1998-99 GM E & K Bodies (DeVille, ElDarado, Seville) 4.6L

79229W23

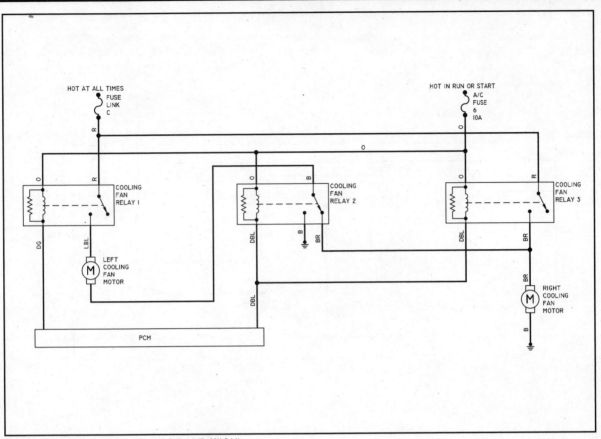

DIA. 47 - 1995 GM F Body (Camaro, Firebird) 3.4L / 5.7L (W/ C41)

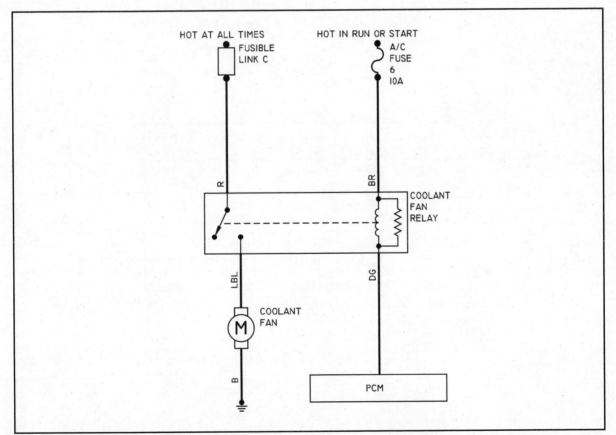

DIA. 48 - 1995 GM F Body (Camaro, Firebird) 5.7L (W/ C60)

79229W24

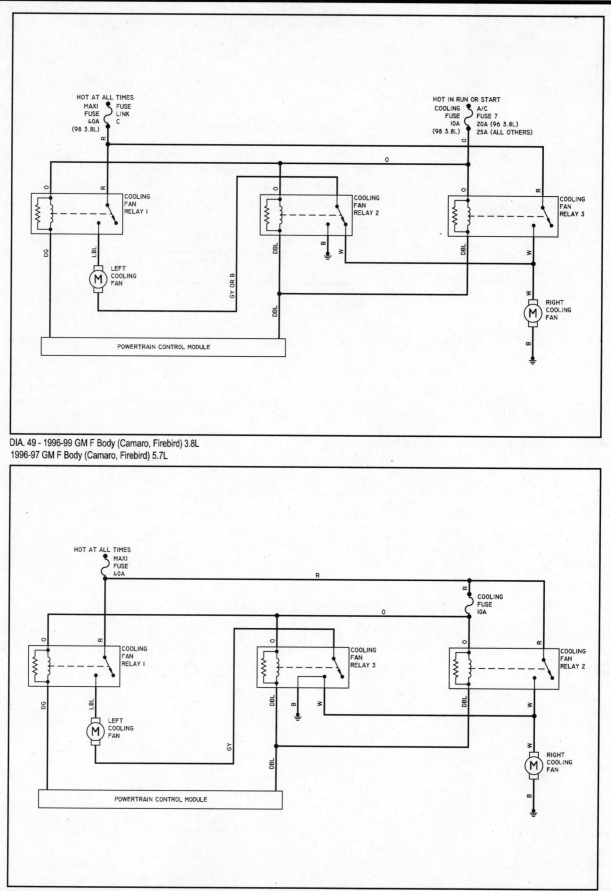

DIA. 49 - 1996-99 GM F Body (Camaro, Firebird) 3.8L
1996-97 GM F Body (Camaro, Firebird) 5.7L

DIA. 50- 1998-99 GM F Body (Camaro, Firebird) 5.7L

79229W25

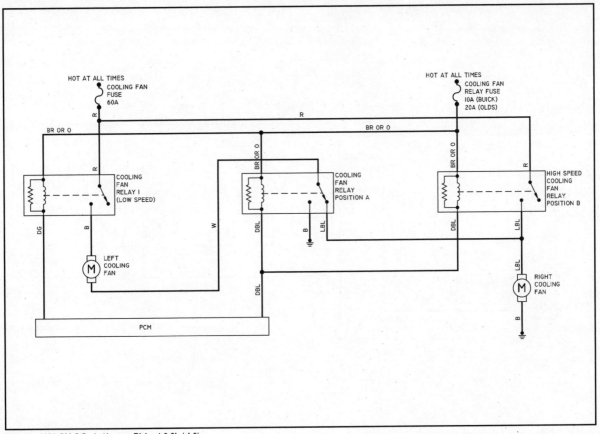

DIA. 51 - 1995 GM G Body (Aurora, Riviera) 3.8L / 4.0L

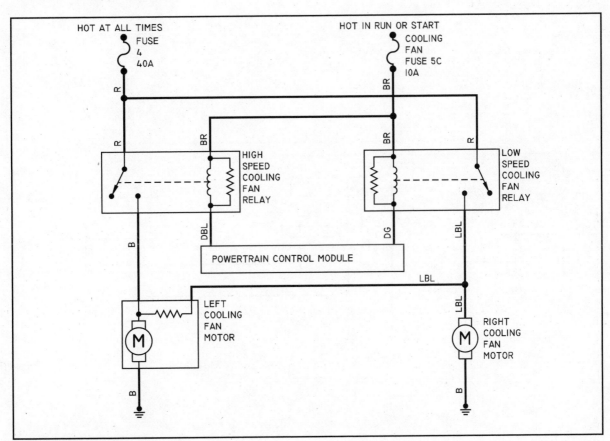

DIA. 52 - 1996 GM G Body (Aurora, Riviera) 3.8L

79229W26

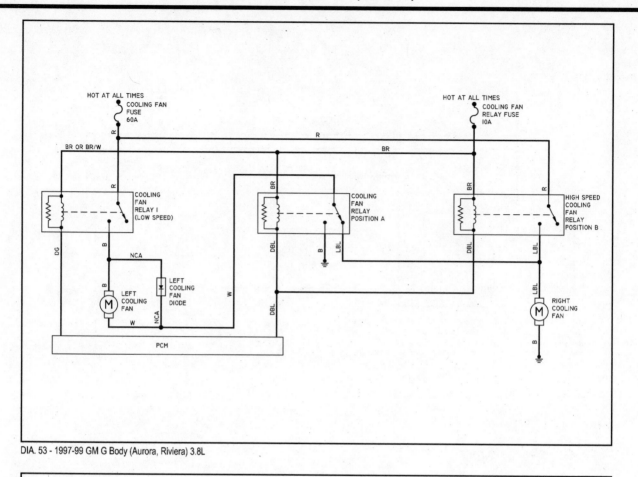

DIA. 53 - 1997-99 GM G Body (Aurora, Riviera) 3.8L

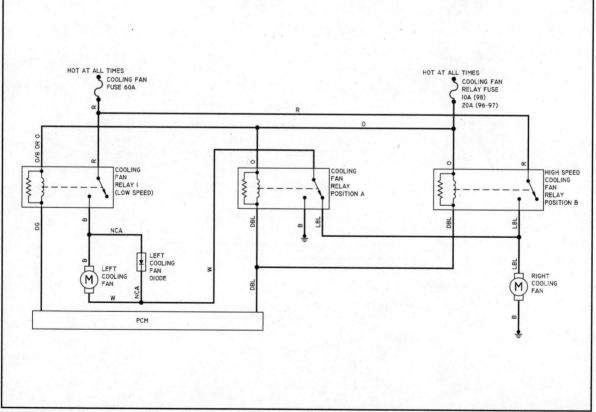

DIA. 54 - 1996-99 GM G Body (Aurora, Riviera) 4.0L

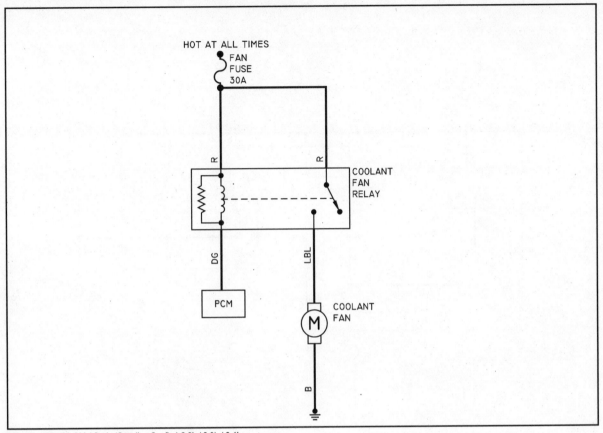

DIA. 55 - 1995-99 GM J Body (Cavalier, Sunfire) 2.2L / 2.3L / 2.4L

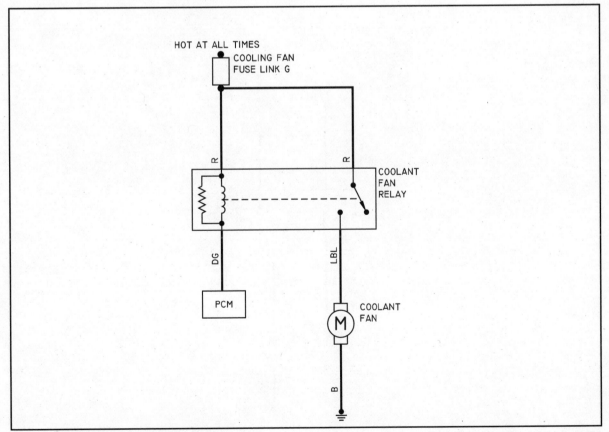

DIA. 56 - 1995-96 GM L Body (Beretta, Corsica) 2.2L / 3.1L

79229W28

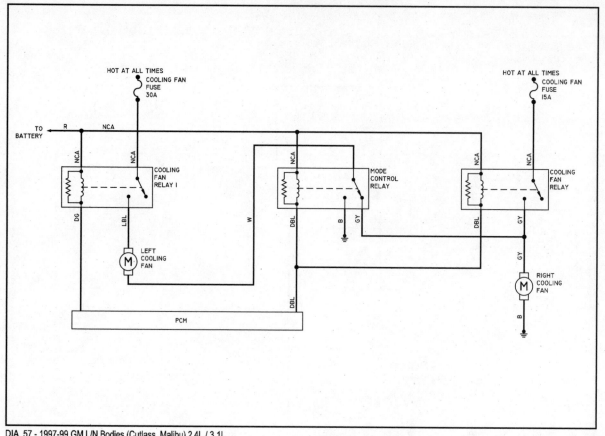

DIA. 57 - 1997-99 GM L/N Bodies (Cutlass, Malibu) 2.4L / 3.1L

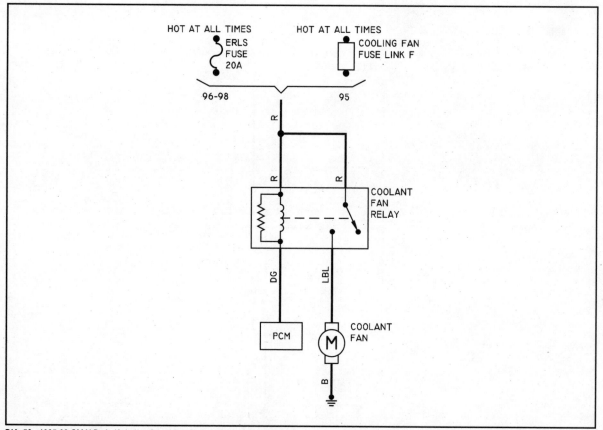

DIA. 58 - 1995-99 GM N Body (Acheiva, Grand Am, Skylark) 2.3L / 2.4L / 3.1L

79229W29

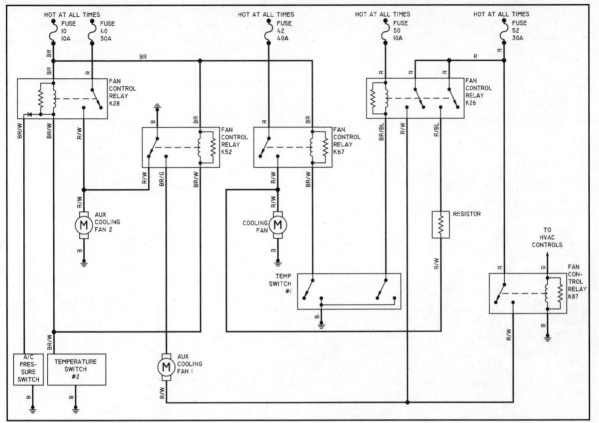

DIA. 59 - 1997-99 GM V Body (Catera) 3.0L

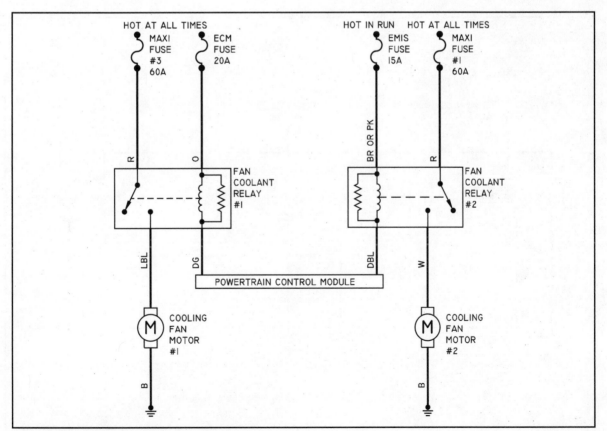

DIA. 60 - 1995-96 GM W Body (Lumina, Monte Carlo, Grand Prix,
Cutlass Supreme, Regal & 1997 Cutlass Supreme) 3.1L / 3.4L

79229W30

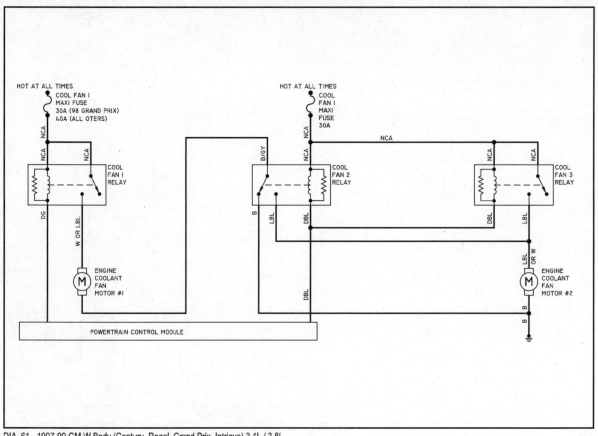

DIA. 61 - 1997-99 GM W Body (Century, Regal, Grand Prix, Intrigue) 3.1L / 3.8L

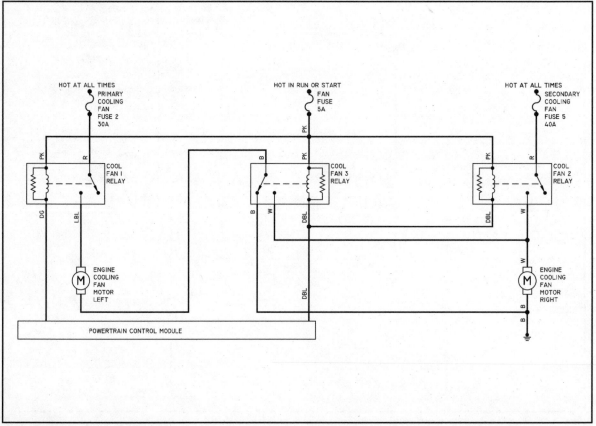

DIA. 62 - 1995-96 GM Y Body (Corvette) 5.7L

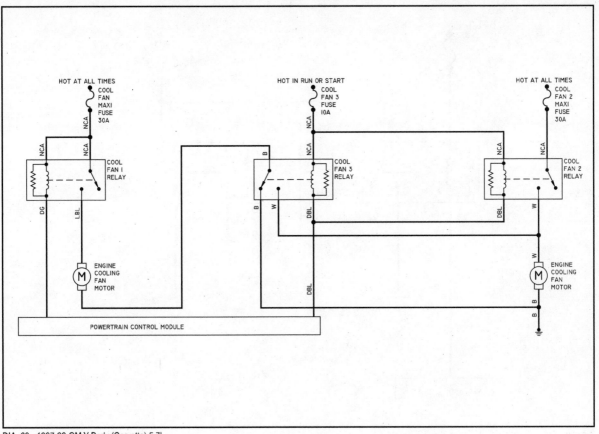

DIA. 63 - 1997-99 GM Y Body (Corvette) 5.7L

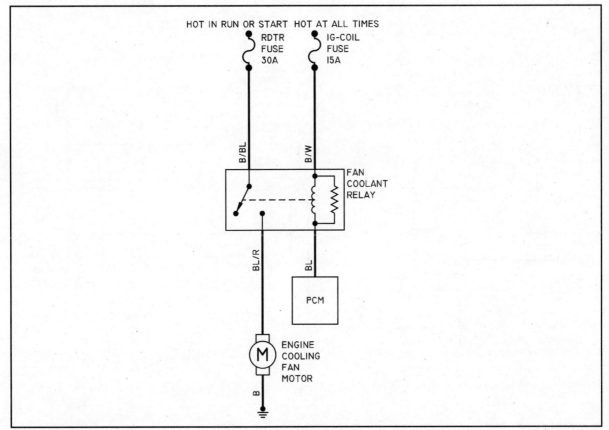

DIA. 64 - 1995-99 Geo/Chevrolet METRO 1.0L/1.3L (W/O A/C)

79229W32

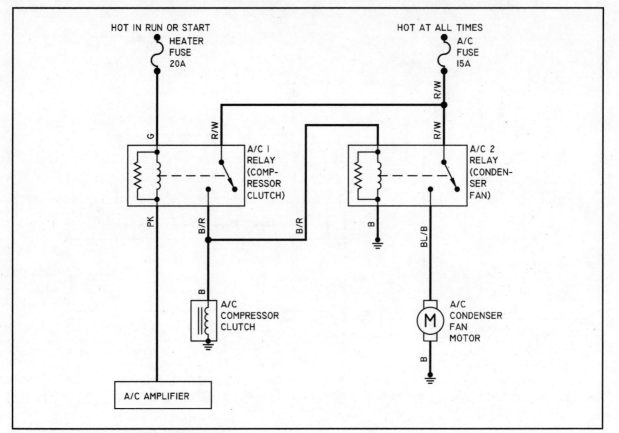

DIA. 65 - 1995-99 Geo/Chevrolet METRO 1.0L/1.3L (W/ A/C)

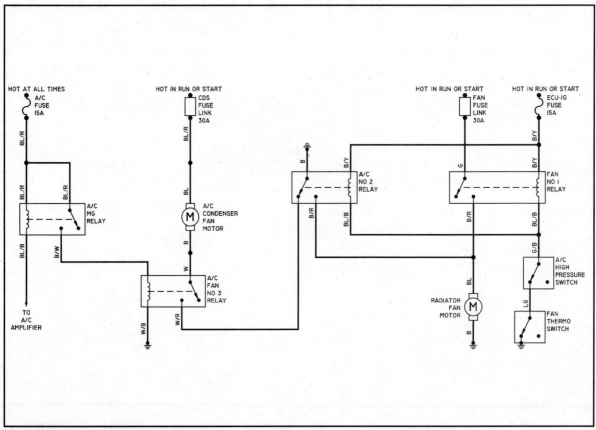

DIA. 66 - 1995-99 Geo/Chevrolet PRISM 1.6L/1.8L (W/ A/C)

79229W33

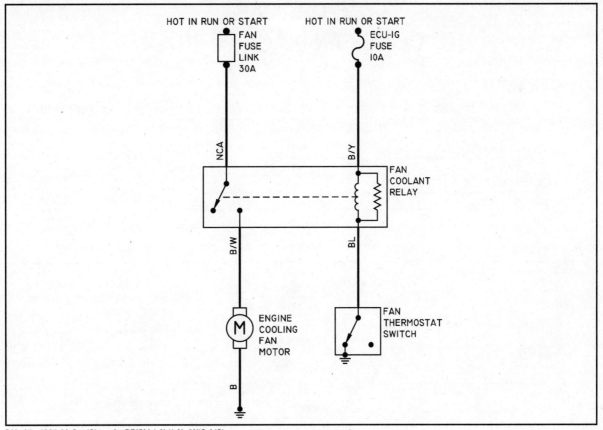

DIA. 67 - 1995-99 Geo/Chevrolet PRISM 1.6L/1.8L (W/O A/C)

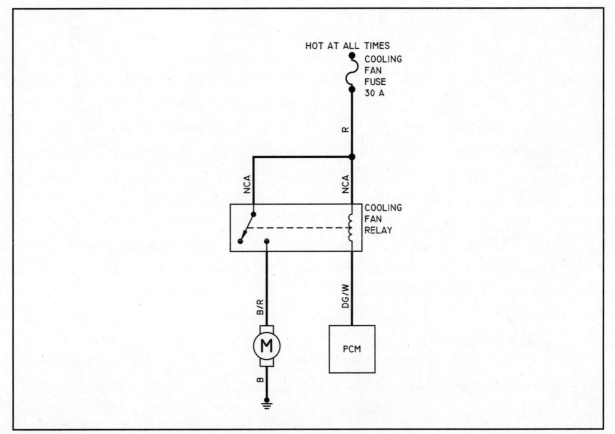

DIA. 68 - 1995-99 Saturn 1.9L

IMPORT CAR
COOLING FAN DIAGRAM INDEX

MANUFACTURER
Model and Engine   Diagram

Acura

Model and Engine	Condition	Diagram
Integra, GSR 1.8L		1
2.5TL		2
Legend Coupe 3.2L		3
3.0L		4
3.5RL		5
2.2CL		6
NSX 3.0L / 3.2L		7

Audi

Model and Engine	Condition	Diagram
90 2.8L	1995 with automatic A/C and A/T	11
	1995 with automatic A/C and M/T	10
	1995 with manual A/C and M/T	8
	1995 with manual A/C and A/T	9
A4 1.8L		12
A4 2.8L		13
A6 2.8L	1995 with automatic A/C and A/T	11
	1995 with automatic A/C and M/T	10
	1995 with manual A/C and M/T	8
	1995 with manual A/C and A/T	9
	1996-99	13
Cabriolet 2.8L	1995-96	9
	1997-99	13
S6 2.8L		13

BMW

Model and Engine	Condition	Diagram
318i 1.8L / 2.5L	1995	14
318is-c, 320i, 325i-c, M3 2.8L	1996-99	15
525i 2.5L / 3.0L / 4.0L		16
318is-c, 320i, 325i-c, M3 1.9L / 2.8L	1996-99	17
318ts 1.9L / 2.8L	1996-99	18

Chrysler Imports

Model and Engine	Diagram
Summit Wagon, Expo 1.8L / 2.4L	19
Summit 1.5L	20
Colt, Summit 1.8L	21

Honda

Model and Engine	Condition	Diagram
Accord 2.2L / 2.7L		24
Civic 1.5L / 1.6L	1995	22
	1996-99	23
Del Sol 1.6L		23
Prelude 2.2L / 2.3L	1995	25
	1996	26
	1997-99	27

Hyundai

Model and Engine	Condition	Diagram
Accent 1.5L	1995	28
	1996-99	31

91261C04

IMPORT CAR
COOLING FAN DIAGRAM INDEX

MANUFACTURER

Model and Engine		Diagram
Hyundai (cont.)		
Elantra 1.5L / 1.6L / 1.8L/2.0L	1995	29
	1996-99	32
Sonata 2.0L / 3.0L	1995	30
	1996-99	33
Tiburon 1.8L / 2.0L		34
Infiniti		
G20 2.0L		35
J30 3.0L		36
I30 3.0L		37
Q45 (US) 4.5L	1995-96 U.S.	38
	1995-96 Canada	39
	1997-99	40
Jaguar		
XJ6, XJ12, XJR 4.0L		41
XK8, XJ8 4.0L		42
XJX 6.0L		43
Kia		
Sephia 1.6L / 1.8L		44
Lexus		
ES300 3.0L		45
GS300, SC300 3.0L		46
SC400 4.0L		47
LS400 4.0L		48
Mazda		
323, Protégé 1.5L / 1.8L		49
MX3 1.6L		50
626, MX6 2.0L	1995 M/T	51
	1995 A/T	52
	1996-99	51
626, MX6 2.5L		52
929 3.0L		53
Miata 1.8L		54
Millenia 2.3L		55
Millenia 2.5L		56
RX7 1.3L		57
Mercedes-Benz		
C220, C280 2.2L / 2.8L		58
E320, E420 3.2L / 4.2L		59
S320, S420, S500 3.2L / 4.2L / 5.0L		60
Mitsubishi		
3000GT 3.0L		61
Diamante 3.0L/3.5L	1995 except Wagon	62
	1995 Wagon	63
	1996-99	64

91261C05

IMPORT CAR
COOLING FAN DIAGRAM INDEX

MANUFACTURER

Model and Engine		Diagram
Mitsubishi (cont.)		
Eclipse 2.0L Non-turbo		65
Eclipse 2.0L Turbo M/T		65
Eclipse 2.0L Turbo A/T		66
Eclipse 2.4L M/T		65
Eclipse 2.4L A/T		66
Galant 2.4L		67
Mirage 1,5L/1.8L	1995-96 1.5L	68
	1995-96 1.8L	69
	1997-99	70
Nissan		
200SX, Sentra 1.6L / 2.0L M/T		71
200SX, Sentra 1.6L / 2.0L A/T		72
240SX 2.4L		73
300ZX 3.0L		74
Altima 2.4L		75
Maxima 3.0L		76
Porsche		
928 GTS		77
968		77
911 Carrera, Carrera 4, Turbo		78
Boxster		79
Saab		
900 2.0L / 2.3L / 2.5L		80
9000 2.3L		81
9000, 9-5 3.0L		82
Subaru		
Impreza 1.8L / 2.2L	1995-96	83
Impreza 2.2L / 2.5L	1997-99	84
Legacy 2.0L	1995	85
Legacy 2.2L / 2.5L	1995-97	86
	1998-99	87
Legacy Brighton, Outback 2.2L / 2.5L		88
SVX 3.3L		89
Suzuki		
Esteem 1.6L		90
Swift 1.0L / 1.3L		91
Toyota		
Avalon 3.0L		92
Camry 2.2L/3.0L		
	1995-96 2.2L	93
	1995-96 3.0L	94
	1997-99	95
Celica 1.8L / 2.2L		96
Corolla 1.8L		97

91261C06

IMPORT CAR
COOLING FAN DIAGRAM INDEX

MANUFACTURER

Model and Engine	Diagram
Toyota (cont.)	
Paseo, Tercel 1.5L	98
Supra 3.0L	99
Volkswagen	
Beetle, Cabrio, Golf, Jetta 2.0L	100
Beetle, Golf, Jetta 1.9L Turbo Diesel	101
GTI 2.0L	100
GTI, Jetta 2.8L	102
Passat 1.9L Turbo Diesel	103
Passat 2.0L	104
Passat 2.8L	105
Volvo	
850, C70, V70 2.3L Turbo / 2.4L Diesel	106
940 2.3L	107
960, S90, V90 2.9L	108

91261C07

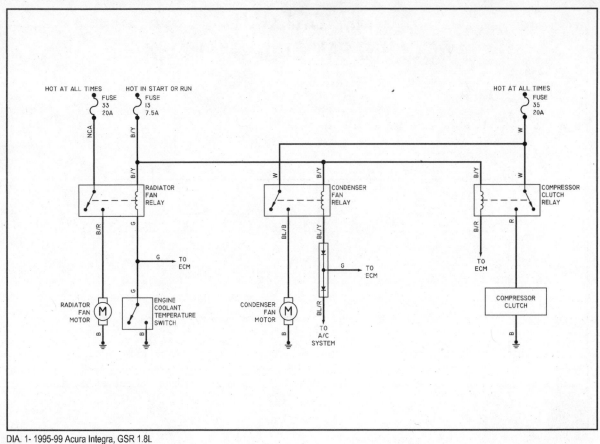

DIA. 1- 1995-99 Acura Integra, GSR 1.8L

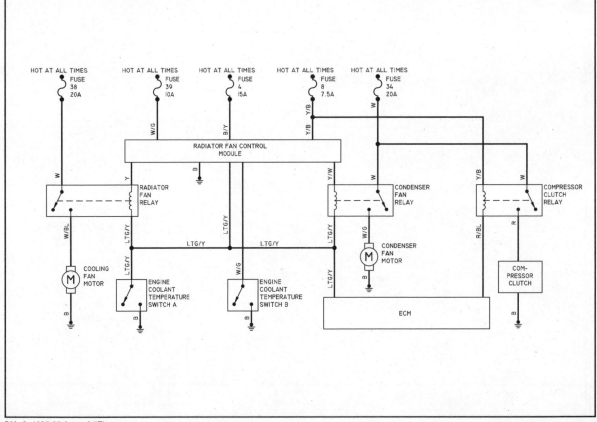

DIA. 2- 1995-97 Acura 2.5TL

79239W01

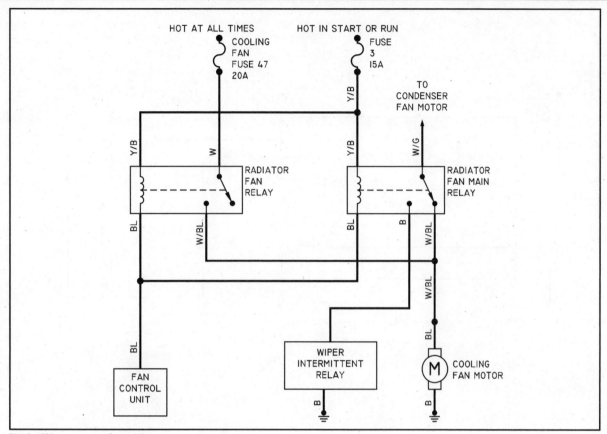

DIA. 3- 1995 Acura Legend Coupe 3.2L

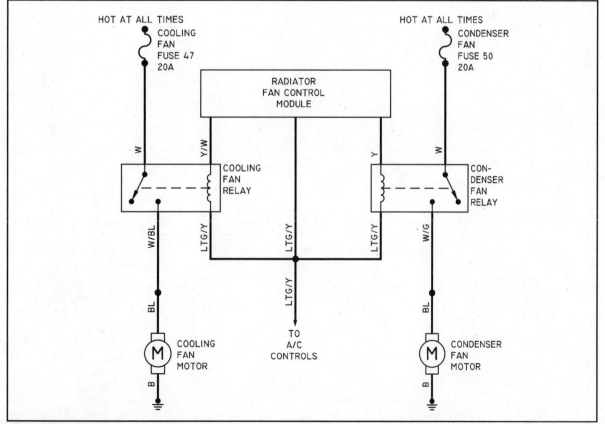

DIA. 4- 1995-97 Acura 3.0L

79239W02

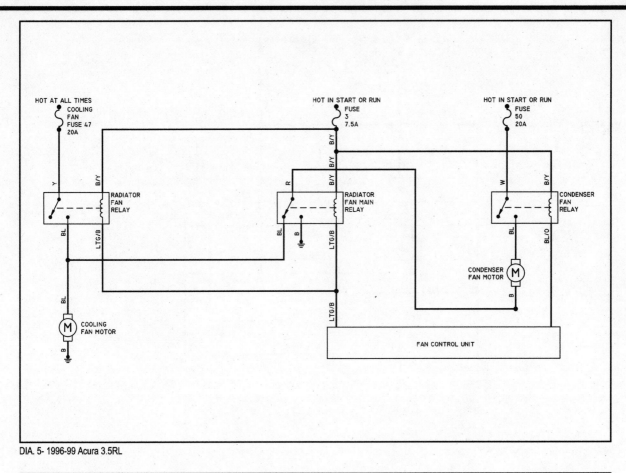

DIA. 5- 1996-99 Acura 3.5RL

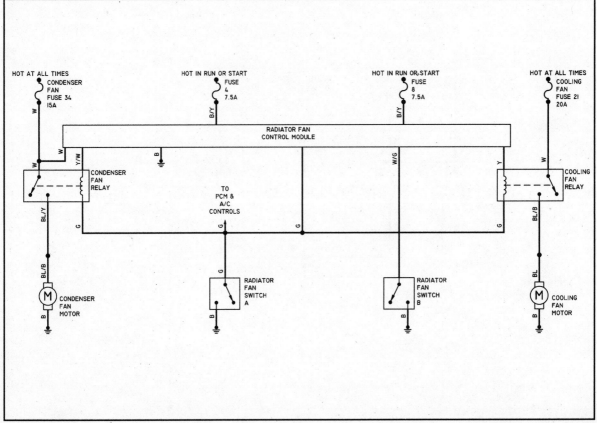

DIA. 6- 1997 Acura 2.2CL

79239W03

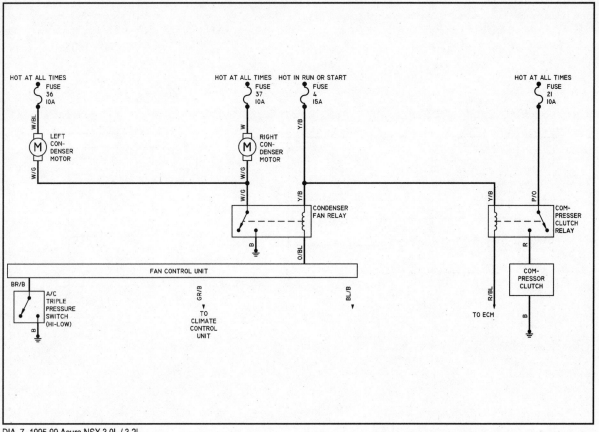

DIA. 7- 1995-99 Acura NSX 3.0L / 3.2L

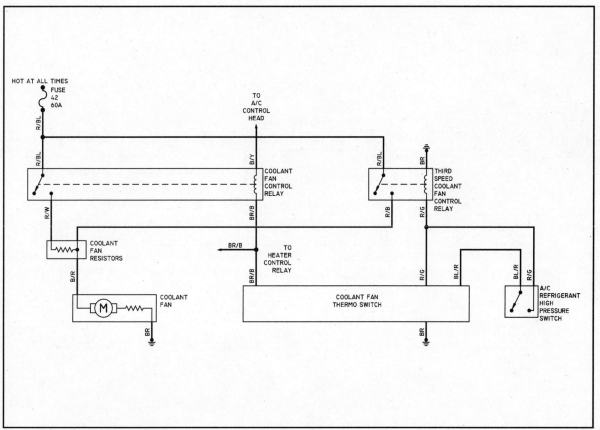

DIA. 8- 1995 Audi 90, A6 Manual A/C w/ M/T 2.8L

79239W04

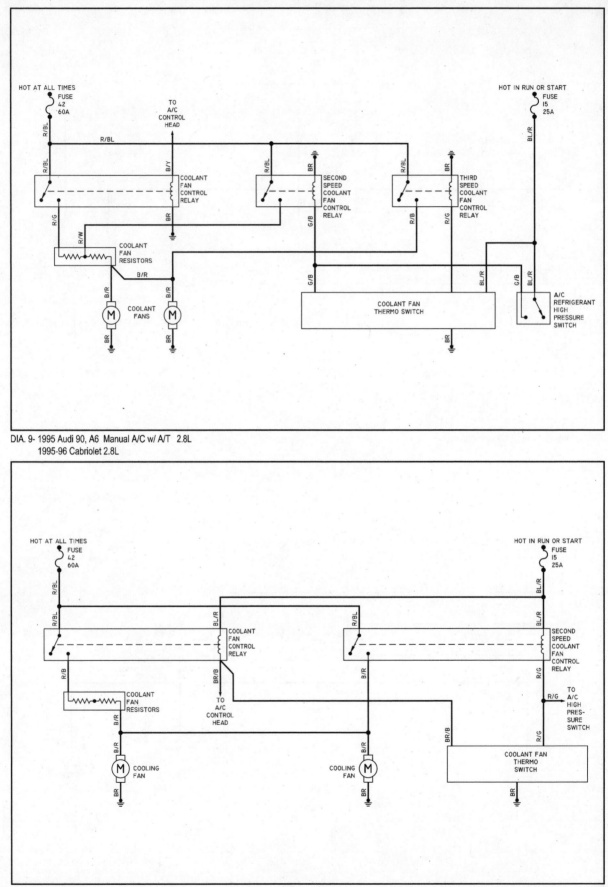

DIA. 9- 1995 Audi 90, A6 Manual A/C w/ A/T 2.8L
         1995-96 Cabriolet 2.8L

DIA. 10- 1995 Audi 90, A6 Auto A/C w/ M/T 2.8L

79239W05

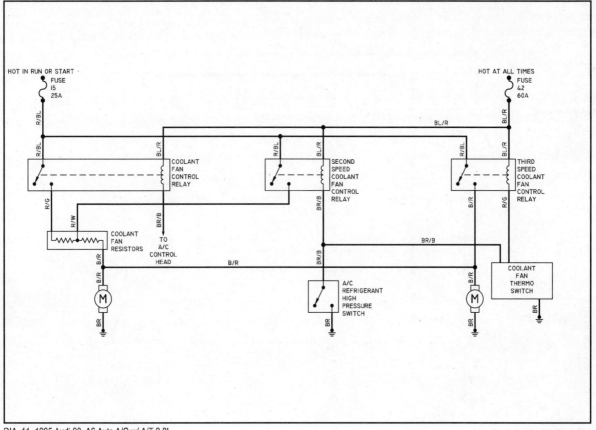

DIA. 11- 1995 Audi 90, A6 Auto A/C w/ A/T 2.8L

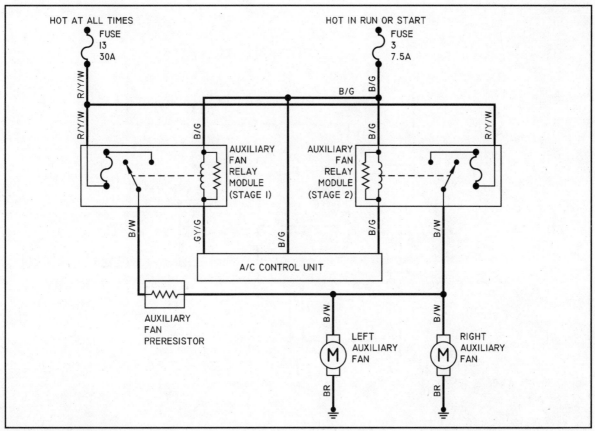

DIA. 12- 1996-99 Audi A4 1.8L

79239W06

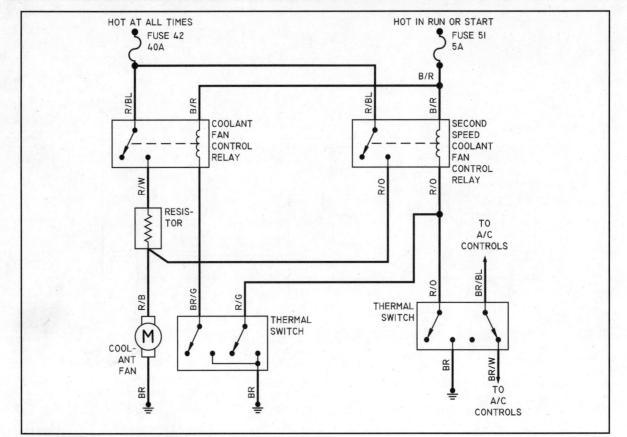

DIA. 13- 1996-99 Audi A4, A6, S6 2.8L, 97-99 Cabriolet

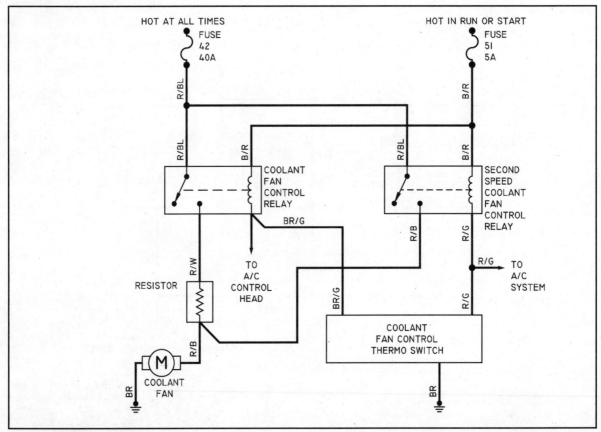

DIA. 14- 1995 BMW 318i 1.8L / 2.5L

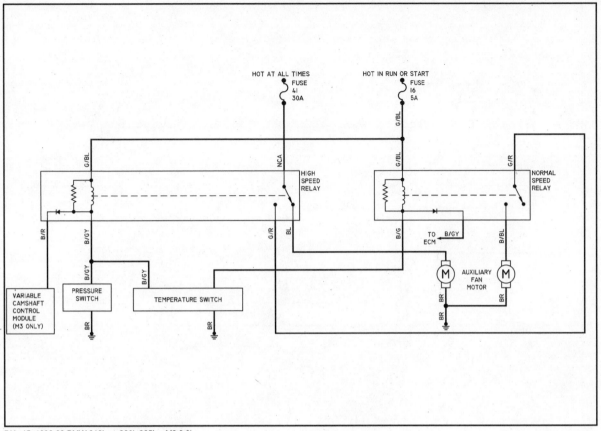

DIA. 15- 1996-99 BMW 318is-c, 320i, 325i-c, M3 2.8L

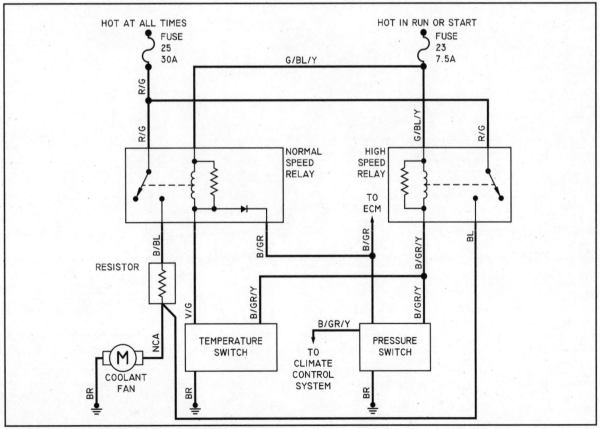

DIA. 16- 1995-99 BMW 525i 2.5L / 3.0L / 4.0L

79239W08

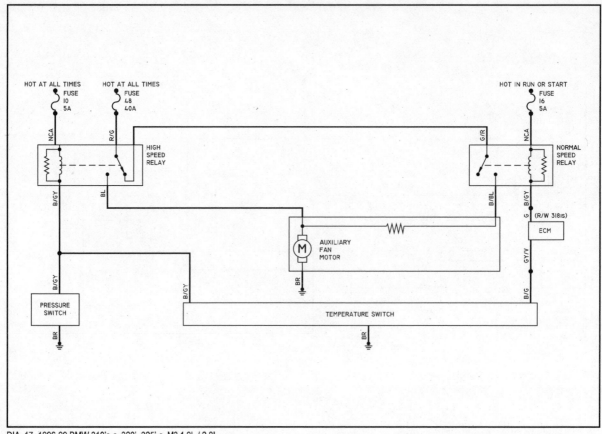

DIA. 17- 1996-99 BMW 318is-c, 320i, 325i-c, M3 1.9L / 2.8L

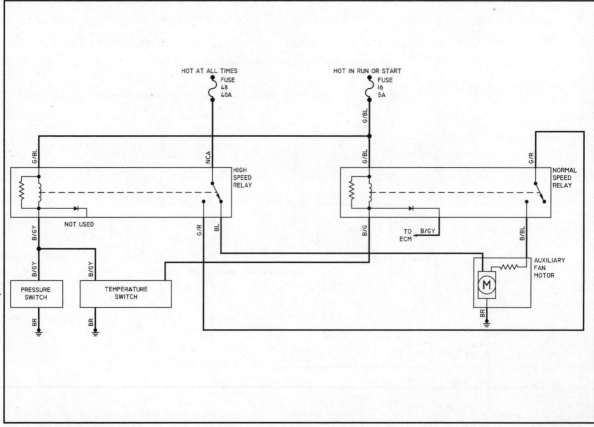

DIA. 18- 1996-99 BMW 318ts 1.9L / 2.8L

79239W09

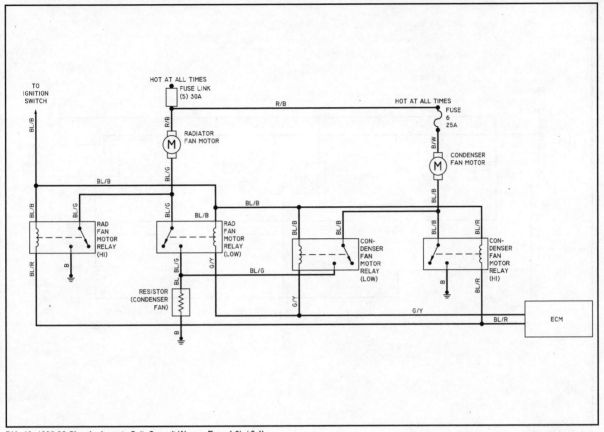

DIA. 19- 1995-96 Chrysler Imports Colt, Summit Wagon, Expo 1.8L / 2.4L

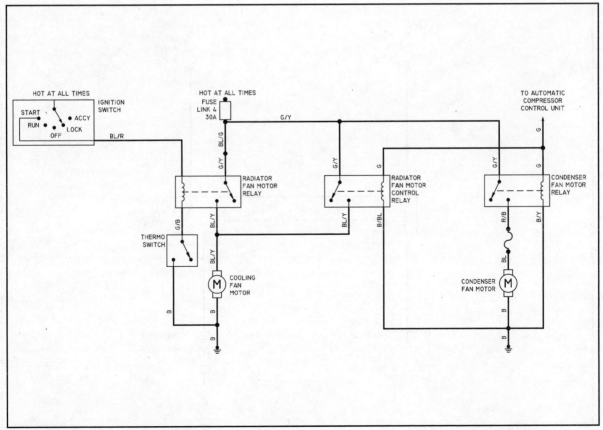

DIA. 20- 1995-96 Chrysler Imports Colt, Summit 1.5L

79239W10

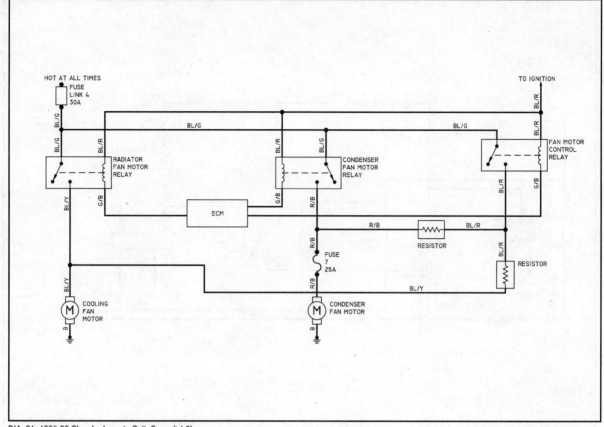

DIA. 21- 1995-96 Chrysler Imports Colt, Summit 1.8L

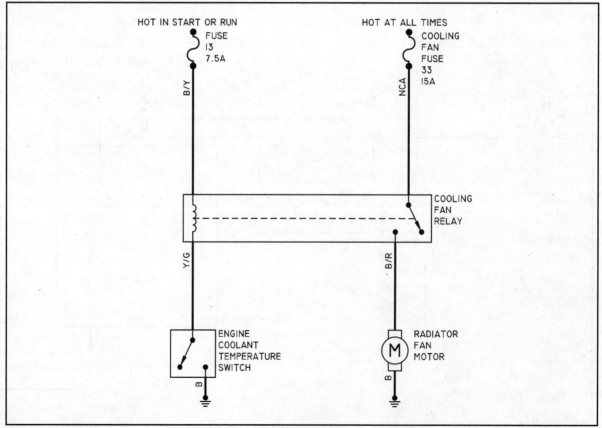

DIA. 22- 1995 Honda Civic 1.5L / 1.6L

79239W11

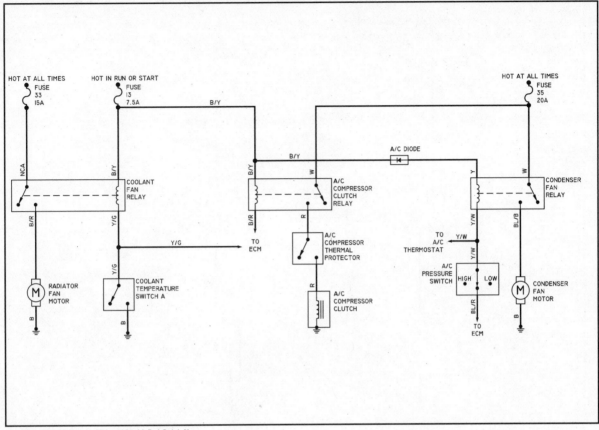

DIA. 23- 1996-99 Honda Civic, 1995-99 Del Sol 1.6L

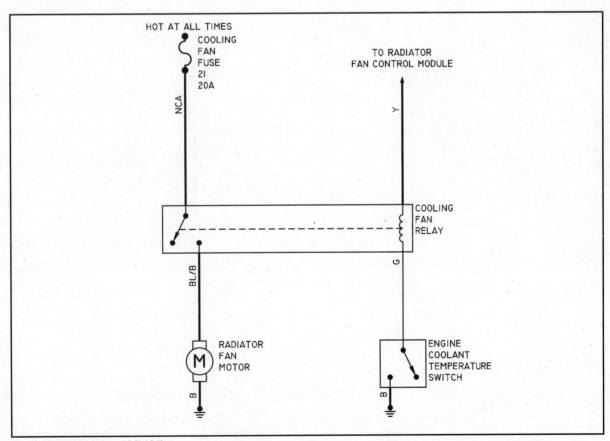

DIA. 24- 1995-99 Honda Accord 2.2L / 2.7L

79239W12

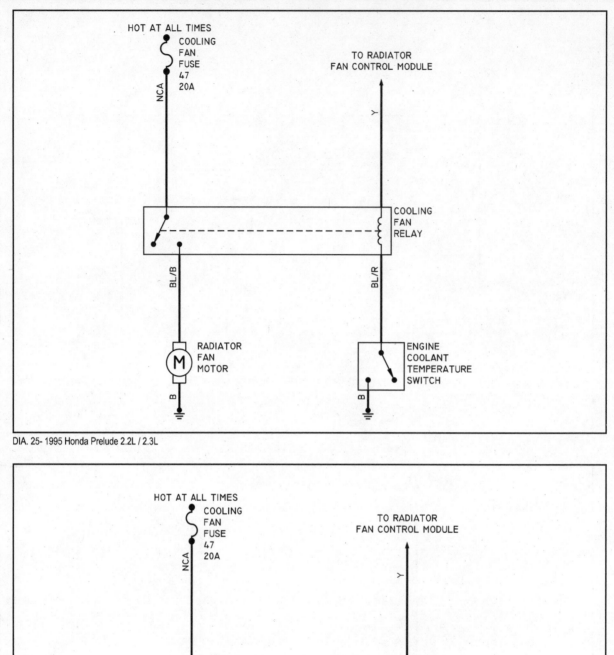

DIA. 25- 1995 Honda Prelude 2.2L / 2.3L

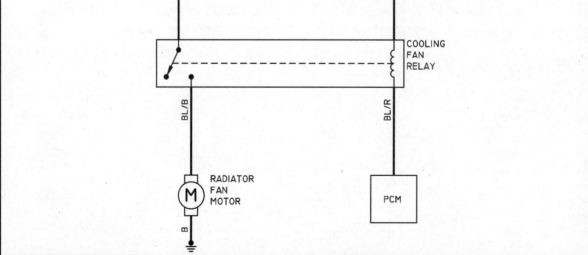

DIA. 26- 1996 Honda Prelude 2.2L / 2.3L

79239W13

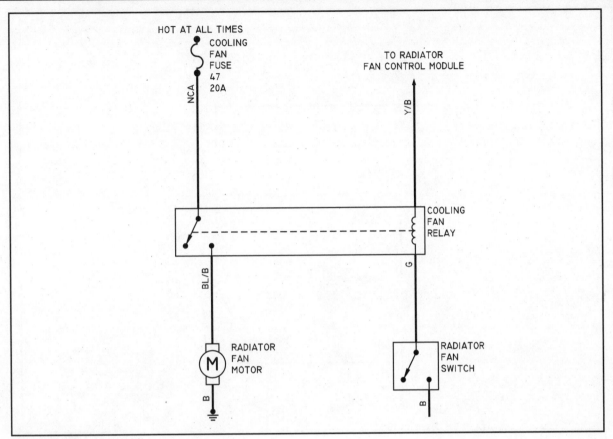

DIA. 27- 1997-99 Honda Prelude 2.2L

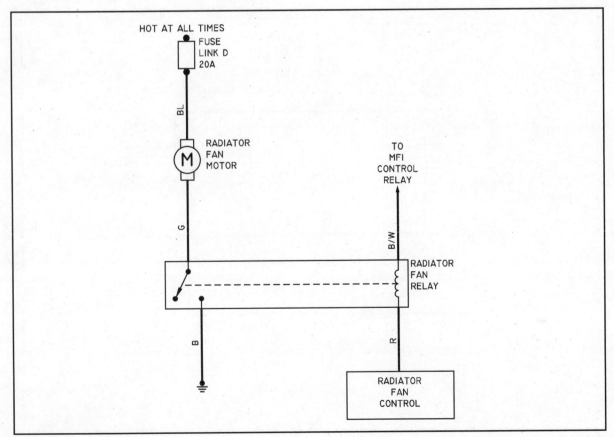

DIA. 28- 1995 Hyundai Accent 1.5L

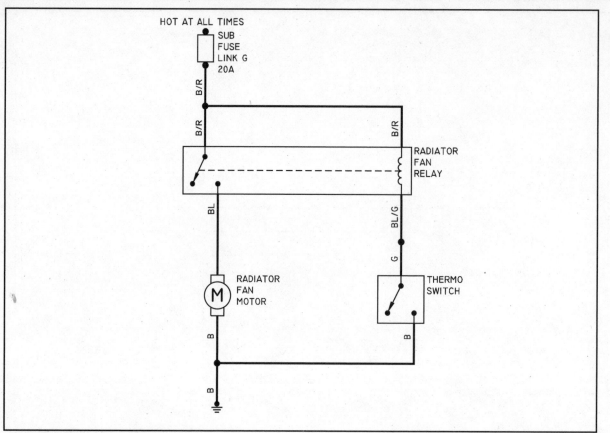

DIA. 29- 1995 Hyundai Elantra 1.5L / 1.6L / 1.8L

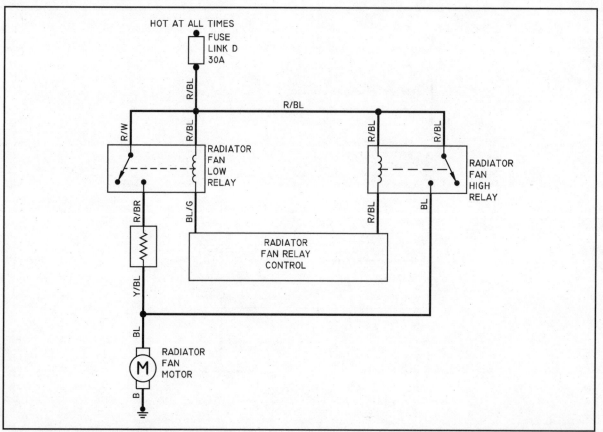

DIA. 30- 1995 Hyundai Sonata 2.0L / 3.0L

79239W15

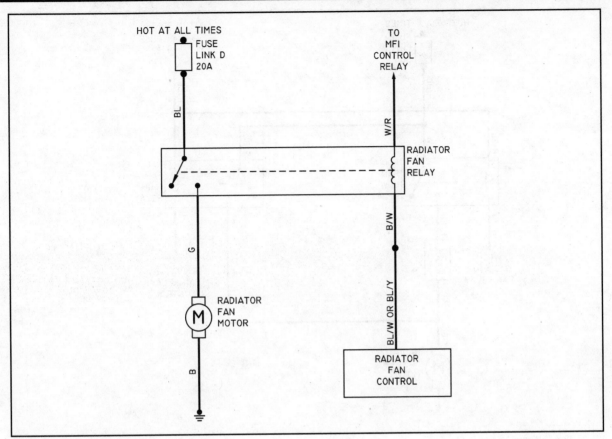

DIA. 31- 1996-99 Hyundai Accent 1.5L

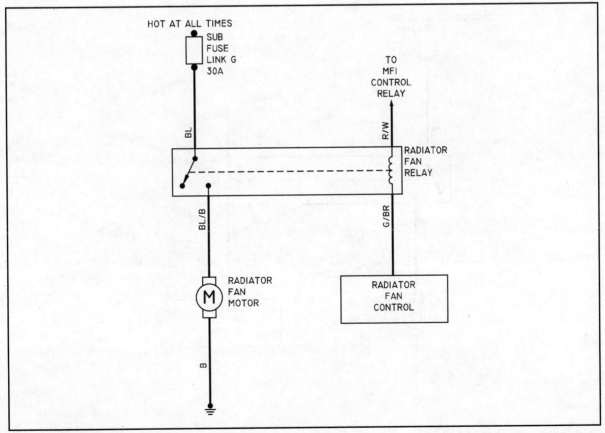

DIA. 32- 1996-99 Hyundai Elantra 1.8L / 2.0L

79239W16

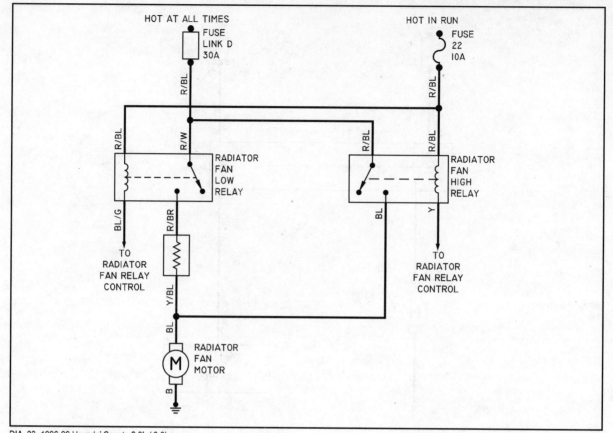

DIA. 33- 1996-99 Hyundai Sonata 2.0L / 3.0L

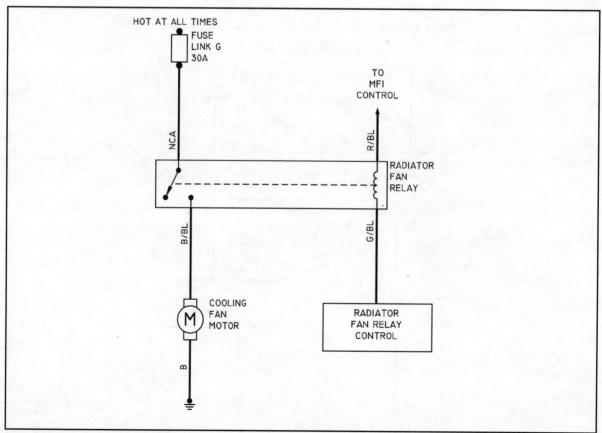

DIA. 34- 1997-99 Hyundai Tiburon 1.8L / 2.0L

79239W17

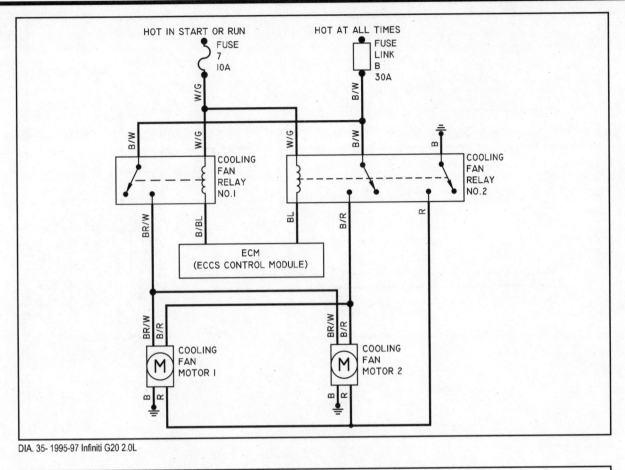

DIA. 35- 1995-97 Infiniti G20 2.0L

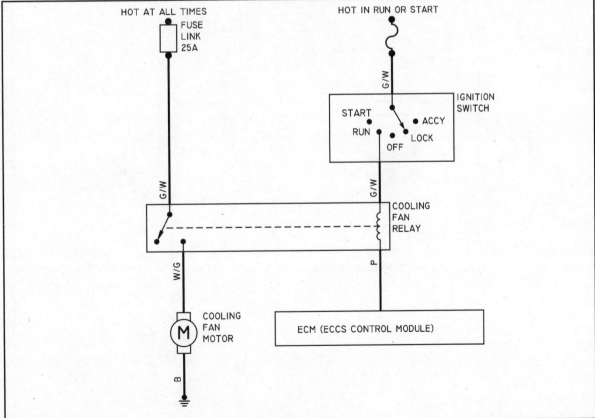

DIA. 36- 1995-97 Infiniti J30 3.0L

79239W18

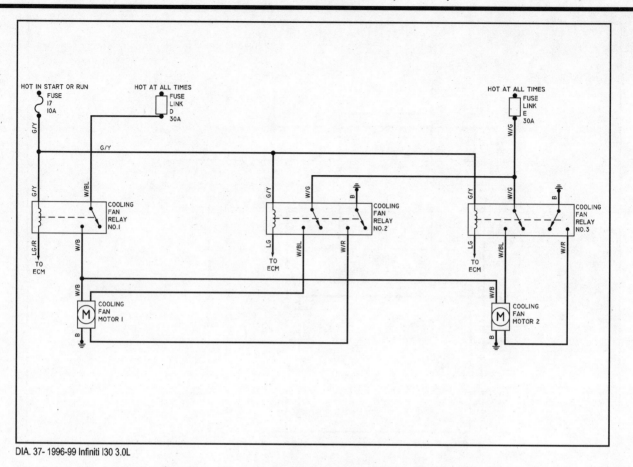

DIA. 37- 1996-99 Infiniti I30 3.0L

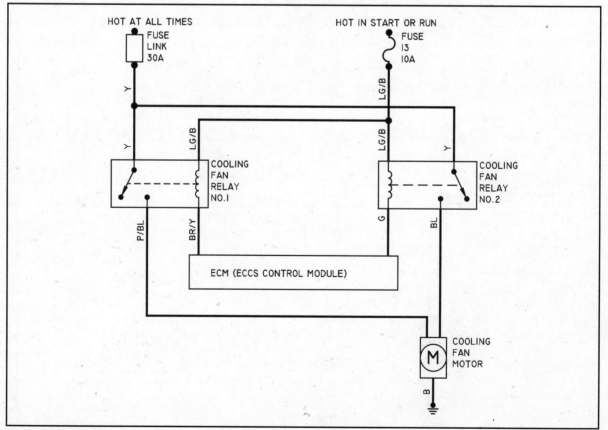

DIA. 38- 1995-96 Q45 (US)

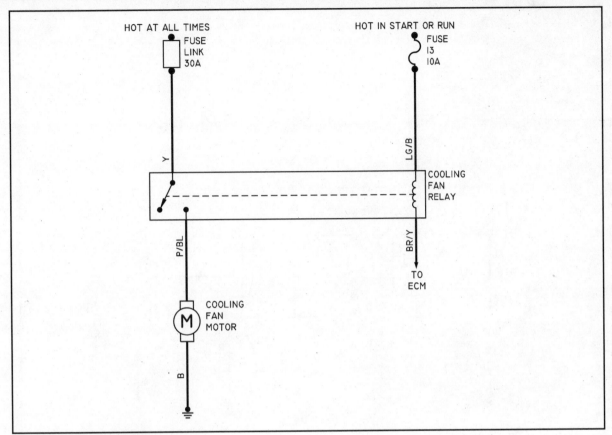

DIA. 39- 1995-96 Infiniti Q45 (CANADA) 4.5L

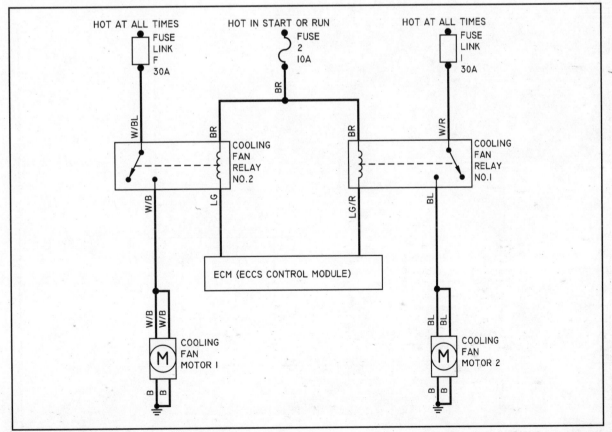

DIA. 40- 1997-99 Infiniti Q45 4.1L

79239W20

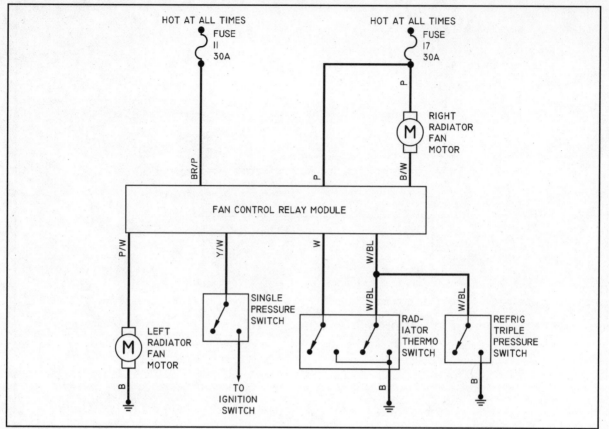

DIA. 41- 1995-99 Jaguar XJ6, XJ12, XJR 4.0L

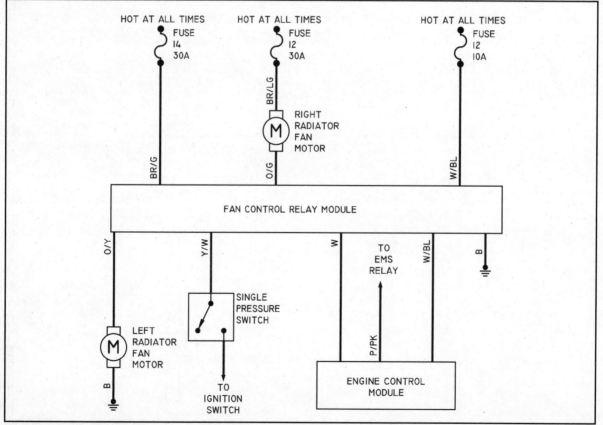

DIA. 42- 1997-99 Jaguar XK8, XJ8, 4.0L

79239W21

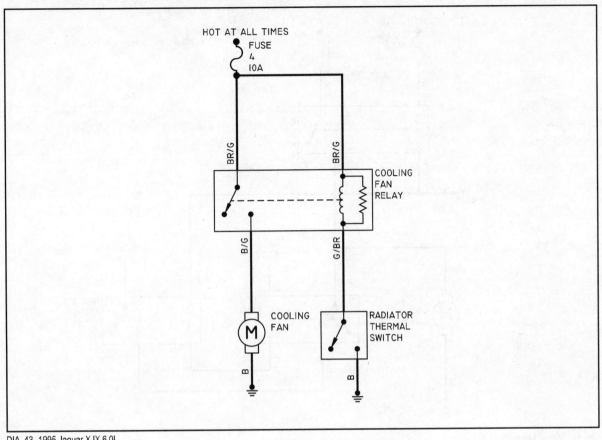

DIA. 43- 1995 Jaguar XJX 6.0L

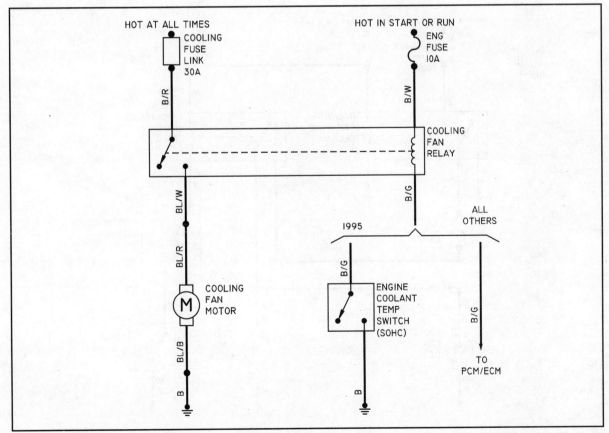

DIA. 44- 1995-99 Kia Sephia 1.6L / 1.8L

79239W22

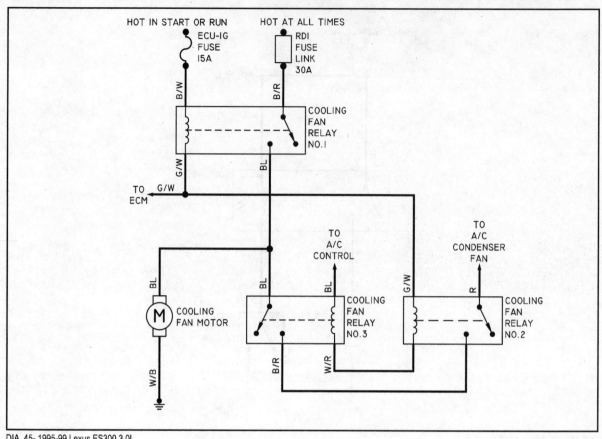

DIA. 45- 1995-99 Lexus ES300 3.0L

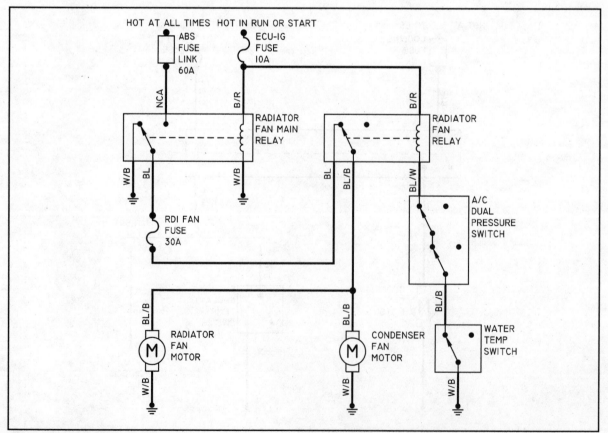

DIA. 46- 1995-99 Lexus GS 300, SC 300 3.0L

79239W23

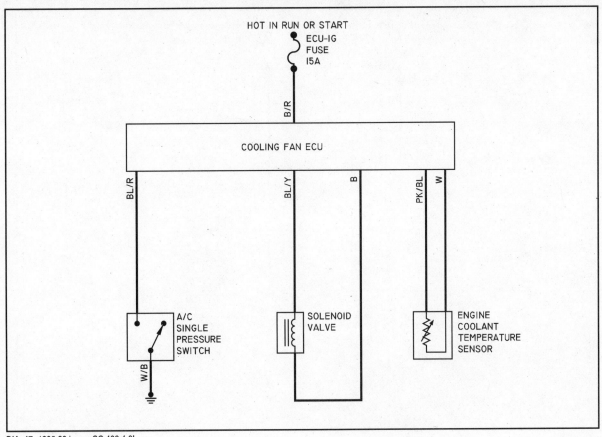

DIA. 47- 1995-99 Lexus SC 400 4.0L

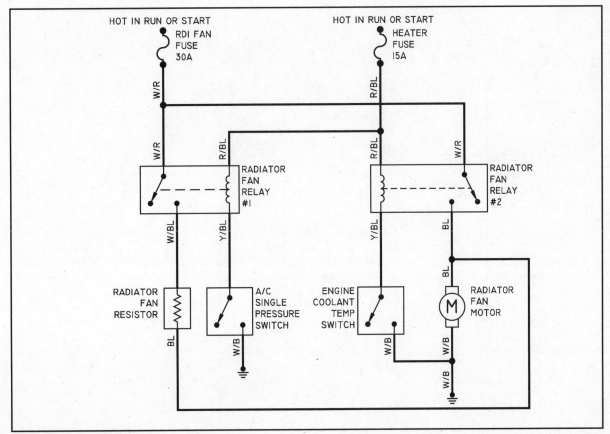

DIA. 48- 1995-99 Lexus LS 400 4.0L

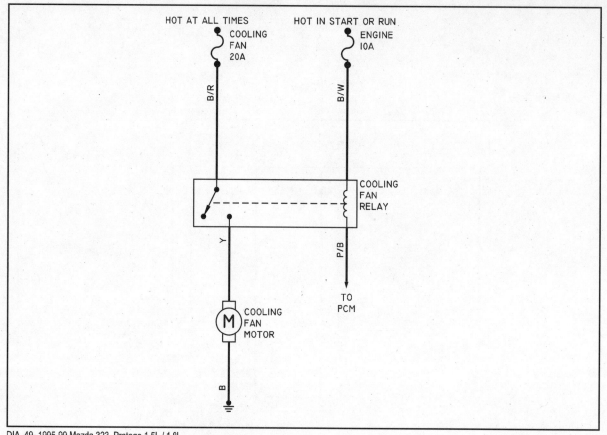

DIA. 49- 1995-99 Mazda 323, Protege 1.5L / 1.8L

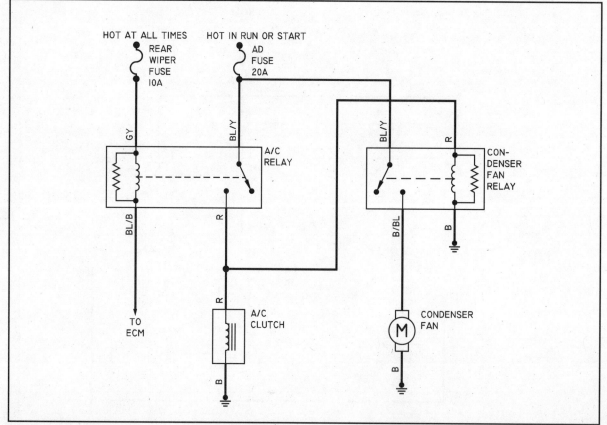

DIA. 50- 1995 Mazda MX3 1.6L

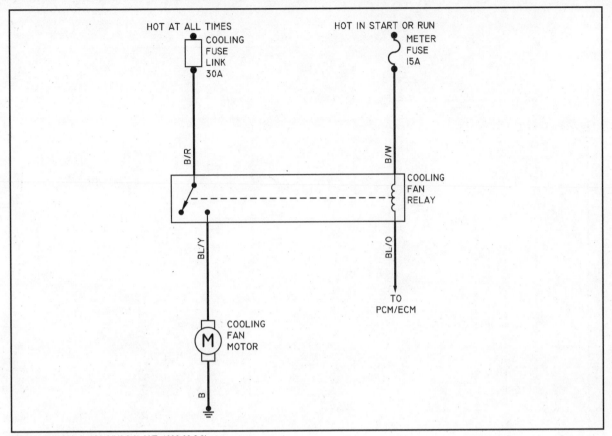

DIA. 51- 1995 Mazda 626, MX6 2.0L M/T, 1996-99 2.0L

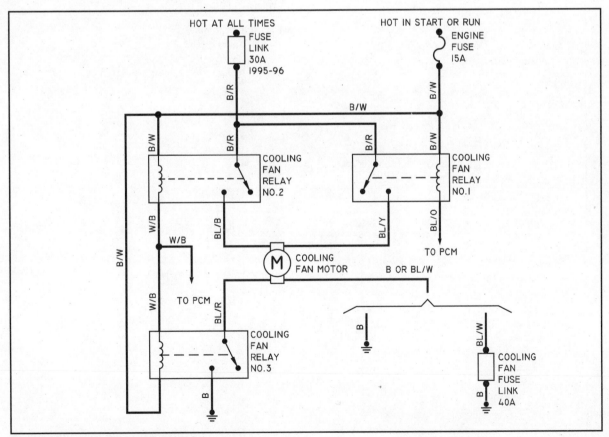

DIA. 52- 1995 Mazda 626, MX6 2.0L A/T, 1996-99 2.5L

79239W26

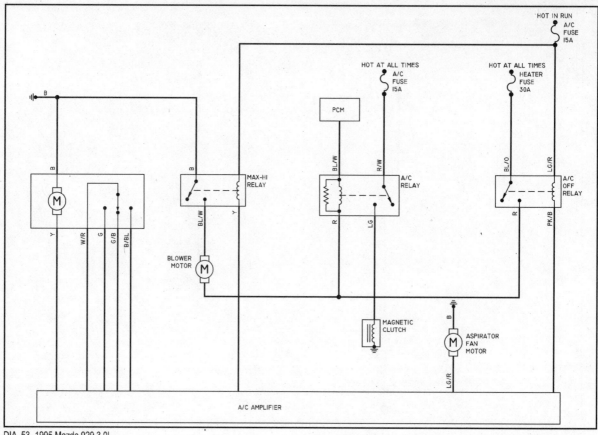

DIA. 53- 1995 Mazda 929 3.0L

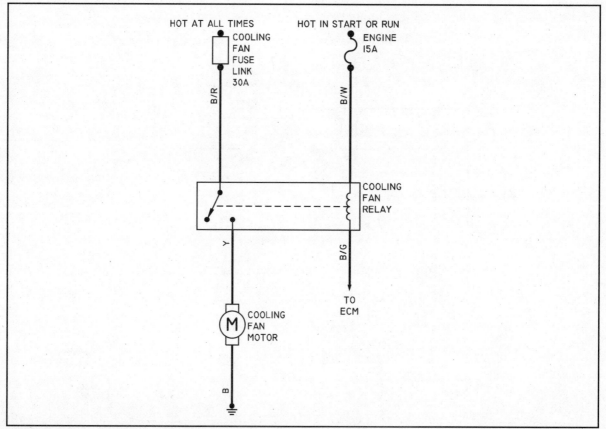

DIA. 54- 1995-99 Mazda Miata 1.8L

79239W27

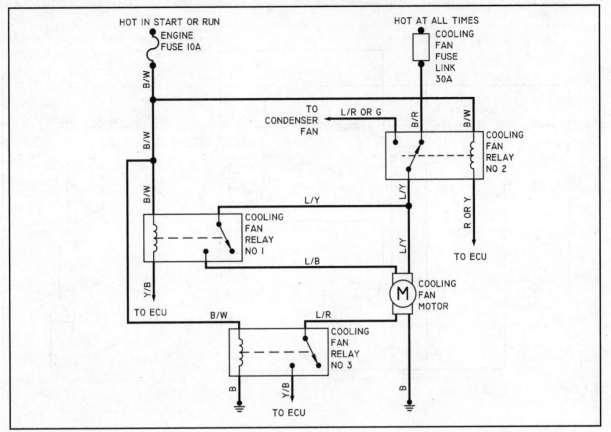

DIA. 55- 1995-99 Mazda Millenia 2.3L

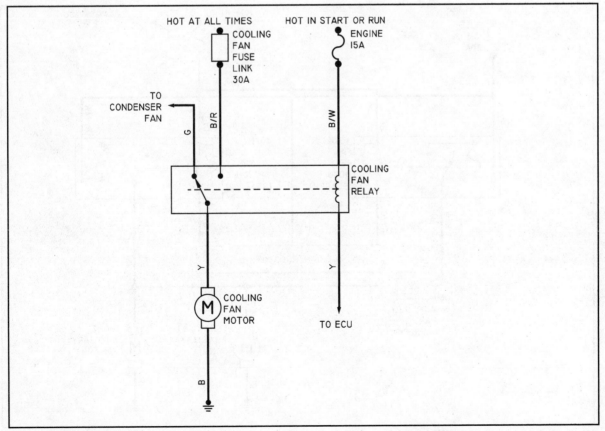

DIA. 56- 1995-99 Mazda Millenia 2.5L

79239W28

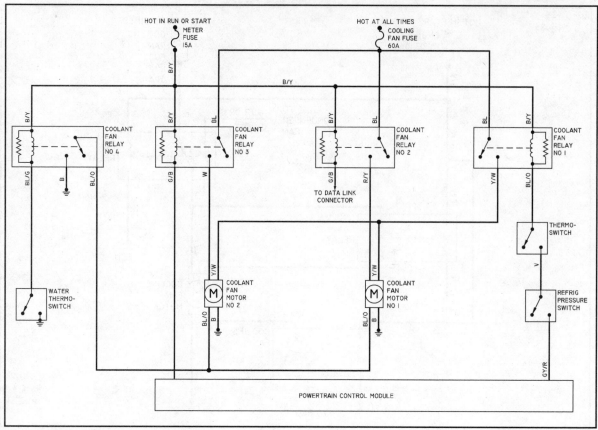

DIA. 57- 1995 Mazda RX7 1.3L

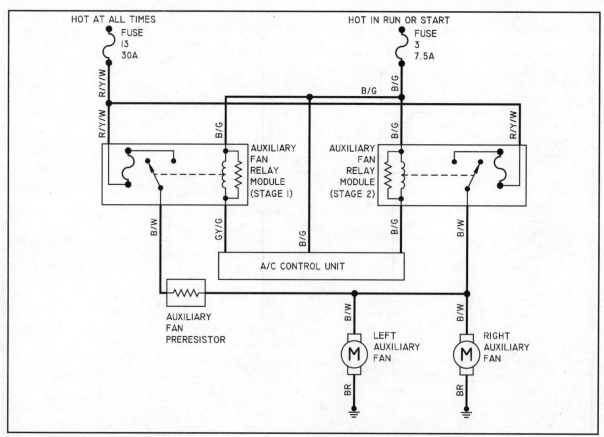

DIA. 58- 1995-99 Mercedes-Benz C220, C280 2.2L / 2.8L

79239W29

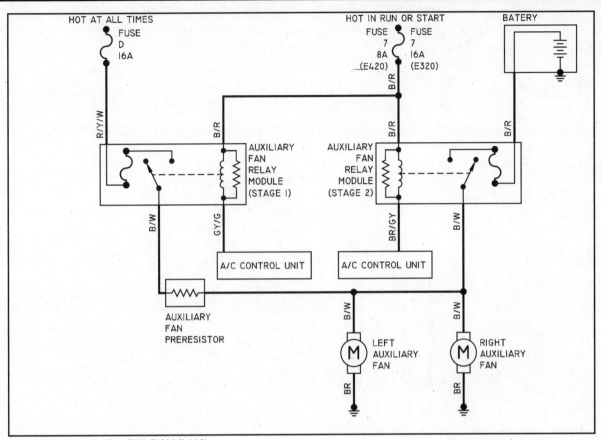

DIA. 59- 1995-99 Mercedes-Benz E320, E420 3.2L / 4.2L

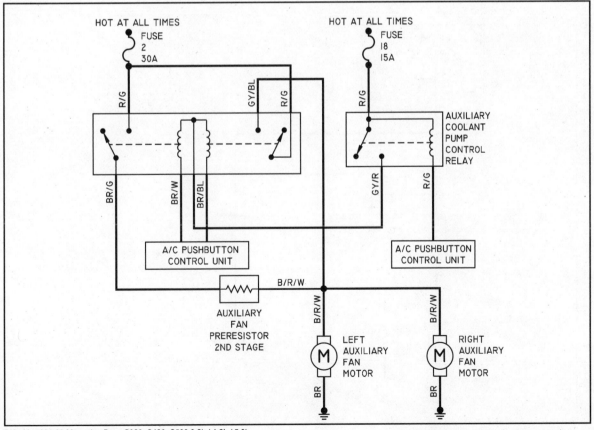

DIA. 60- 1995-99 Mercedes-Benz S320, S420, S500 3.2L / 4.2L / 5.0L

79239W30

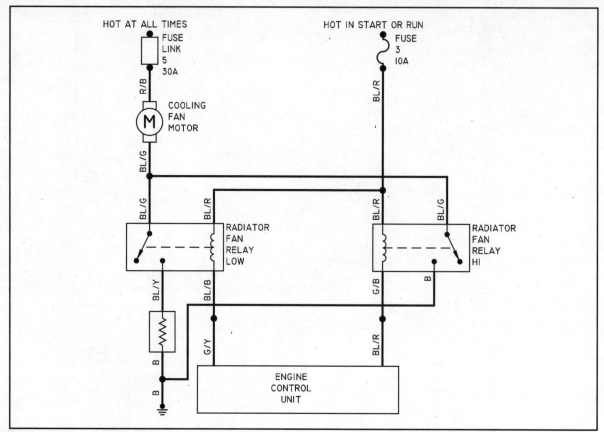

DIA. 61- 1995-99 Mitsubishi 3000GT 3.0L

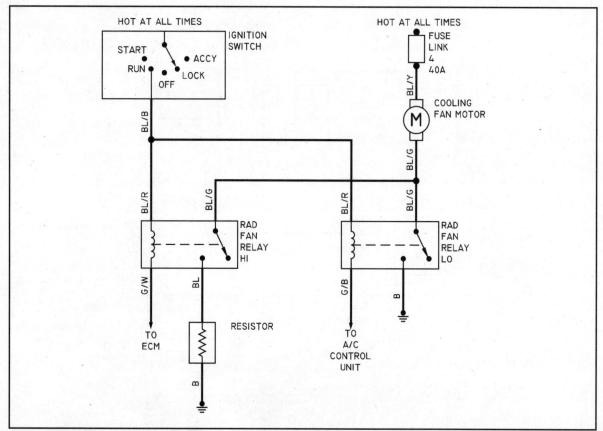

DIA. 62- 1995 Mitsubishi Diamante 3.0L (Except wagon)

79239W31

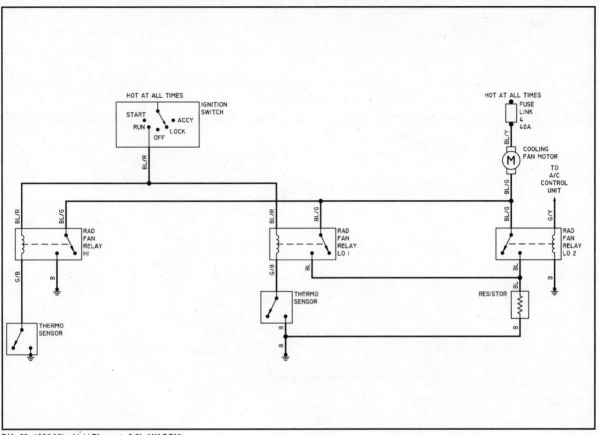

DIA. 63- 1995 Mitsubishi Diamante 3.0L (WAGON)

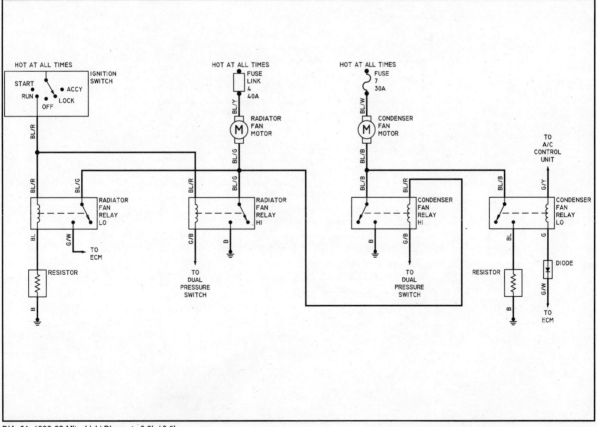

DIA. 64- 1996-99 Mitsubishi Diamante 3.0L / 3.5L

79239W32

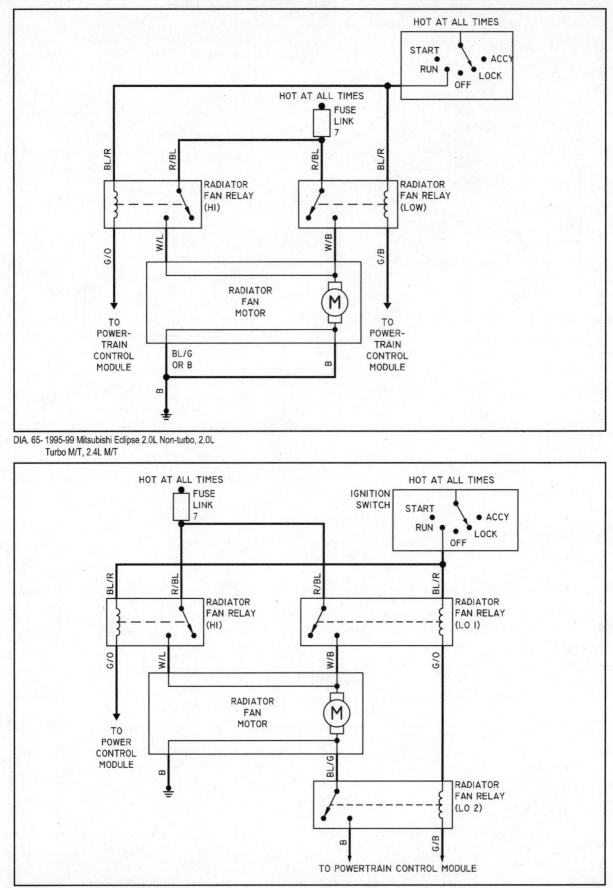

DIA. 65- 1995-99 Mitsubishi Eclipse 2.0L Non-turbo, 2.0L
Turbo M/T, 2.4L M/T

DIA. 66- 1996-99 Mitsubishi Eclipse 2.0L Turbo A/T, 2.4L A/T

79239W33

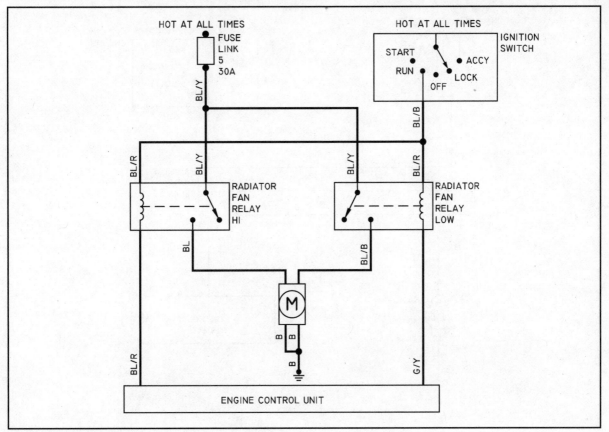

DIA. 67- 1995-99 Mitsubishi Galant 2.4L

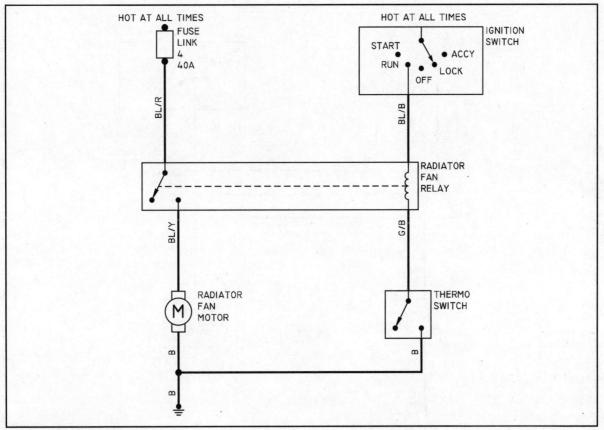

DIA. 68- 1995-96 Mitsubishi Mirage 1.5L

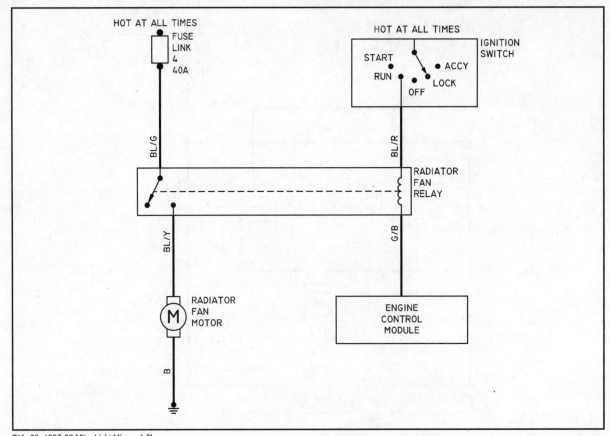

DIA. 69- 1995-96 Mitsubishi Mirage 1.8L

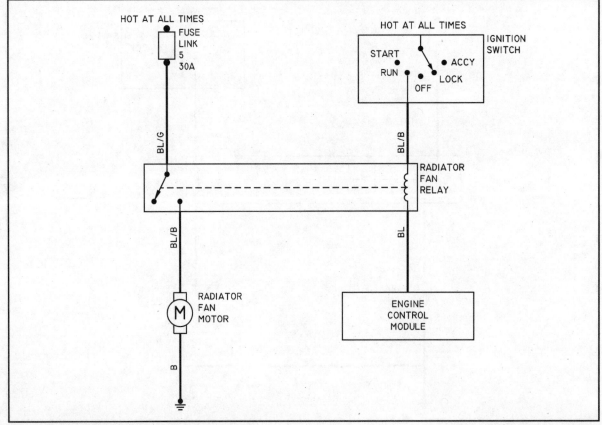

DIA. 70- 1997-99 Mitsubishi Mirage 1.5L/1.8L

79239W35

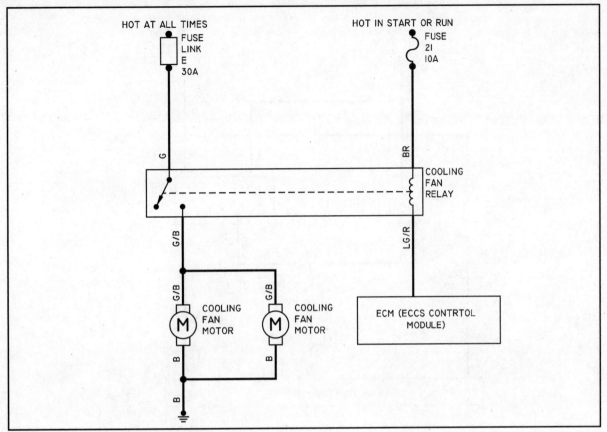

DIA. 71- 1995-99 Nissan 200SX, Sentra 1.6L / 2.0L M/T

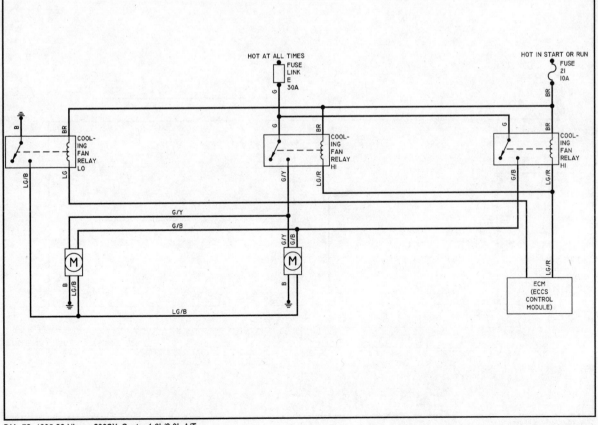

DIA. 72- 1995-99 Nissan 200SX, Sentra 1.6L/2.0L A/T

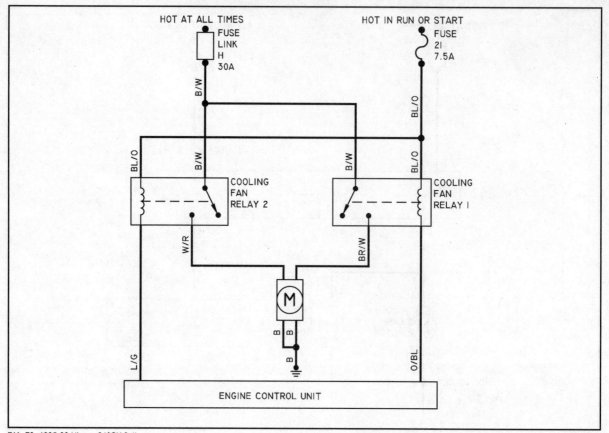

DIA. 73- 1995-99 Nissan 240SX 2.4L

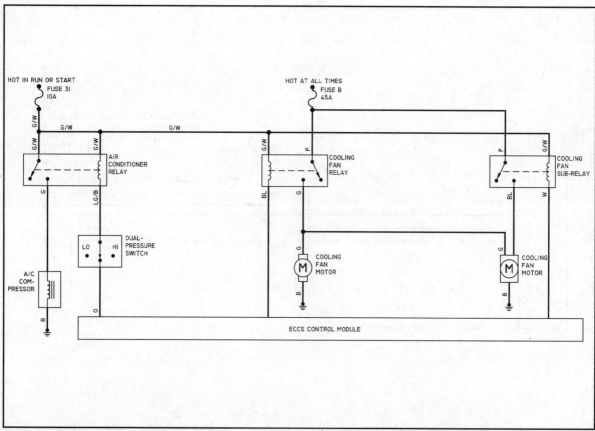

DIA. 74- 1995-96 Nissan 300ZX 3.0L

79239W37

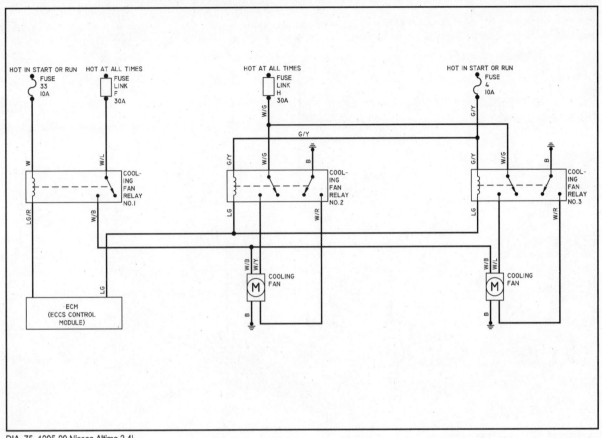

DIA. 75- 1995-99 Nissan Altima 2.4L

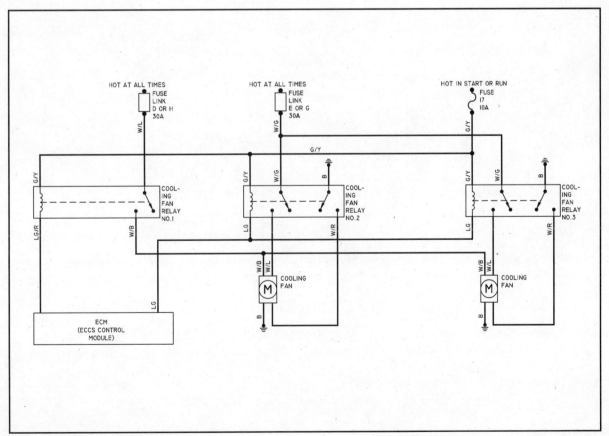

DIA. 76- 1995-99 Nissan Maxima 3.0L

79239W38

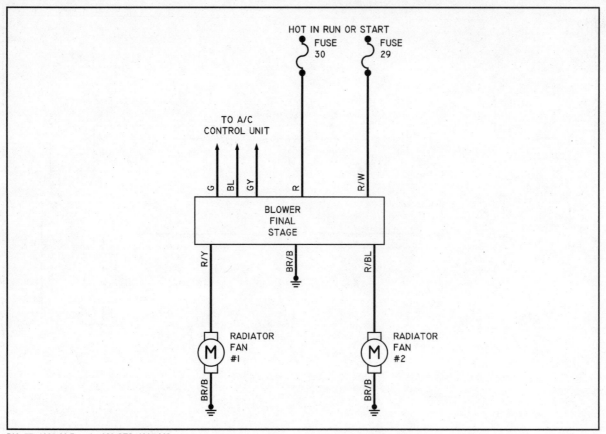

DIA. 77- 1995-99 Porsche 928 GTS, 1995 968

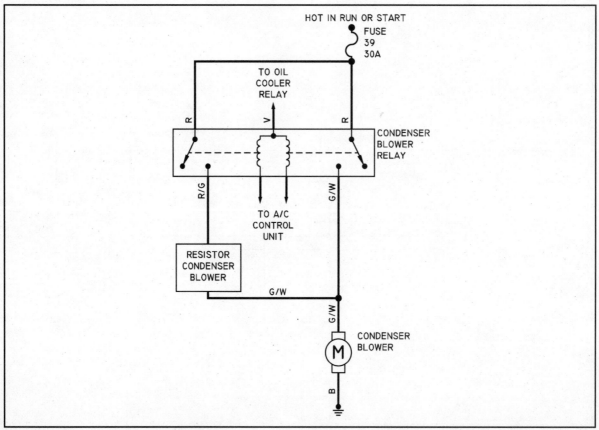

DIA. 78- 1995-99 Porsche 911 Carrera, Carrera 4, Turbo

79239W39

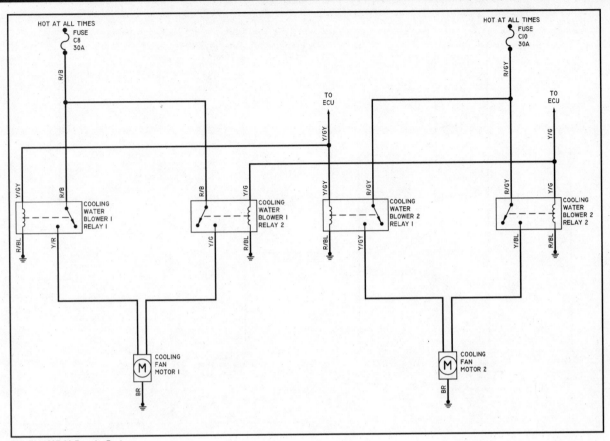

DIA. 79- 1997-99 Porsche Boxter

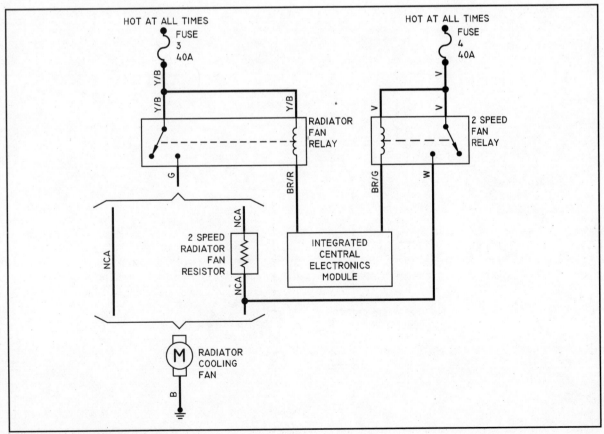

DIA. 80- 1995-99 SAAB 900 2.0L / 2.3L / 2.5L

79239W40

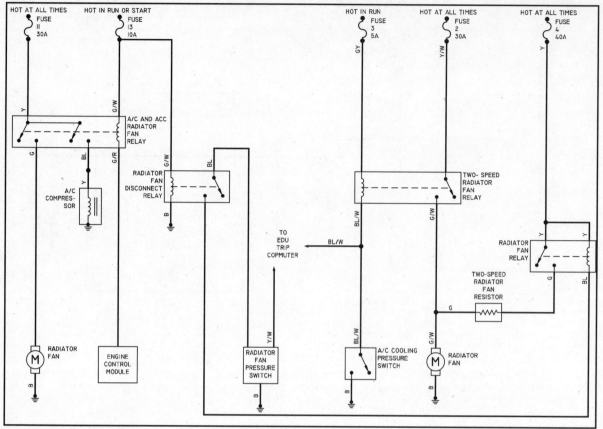

DIA. 81- 1995-99 SAAB 9000 2.3L

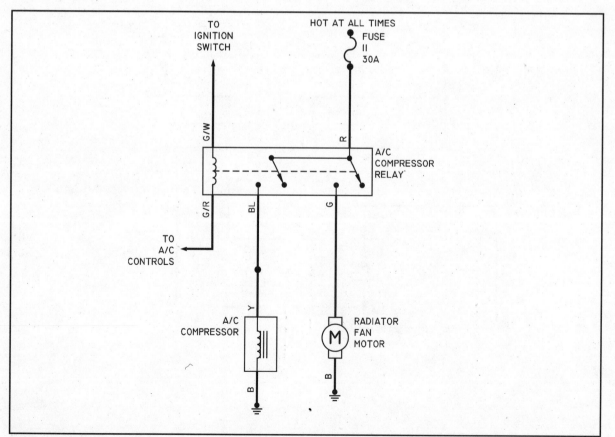

DIA. 82- 1996-99 SAAB 9000, 9-5 3.0L

79239W41

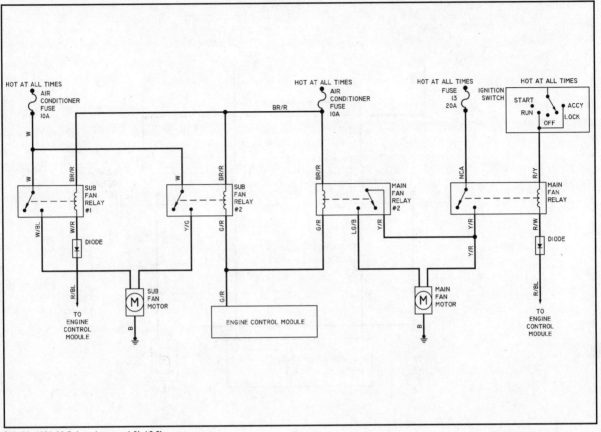

DIA. 83- 1995-96 Subaru Impreza 1.8L / 2.2L

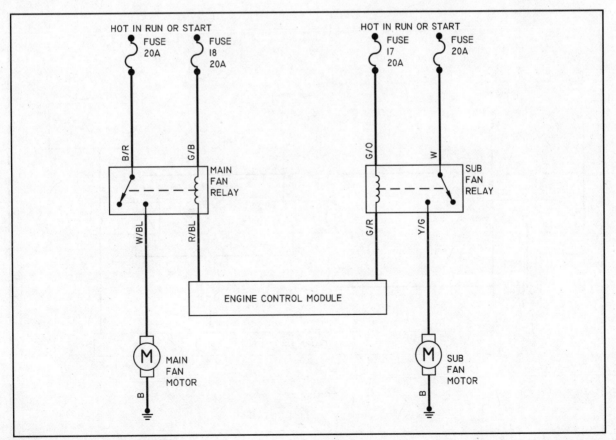

DIA. 84- 1997-99 Subaru Impreza 2.2L / 2.5L

79239W42

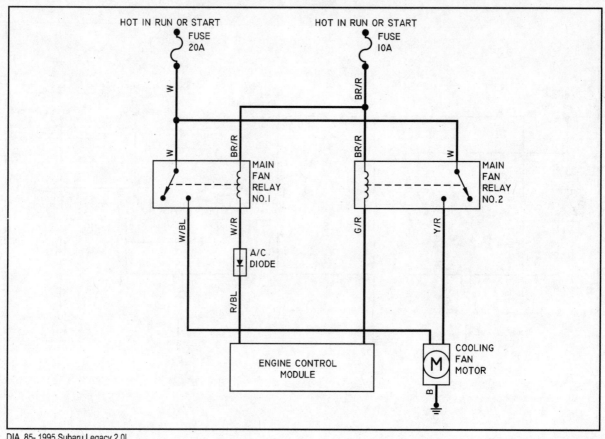

DIA. 85- 1995 Subaru Legacy 2.0L

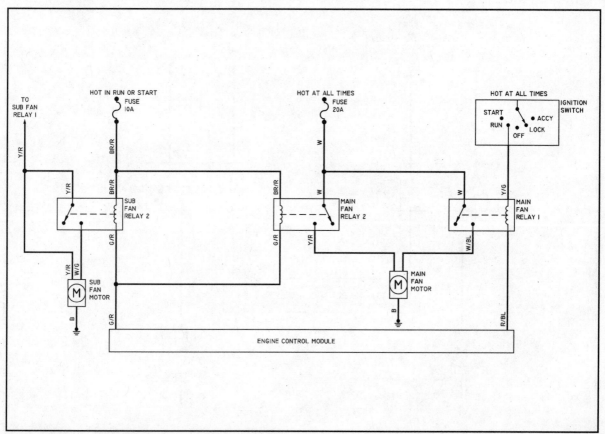

DIA. 86- 1995-97 Subaru Legacy 2.2L / 2.5L

79239W43

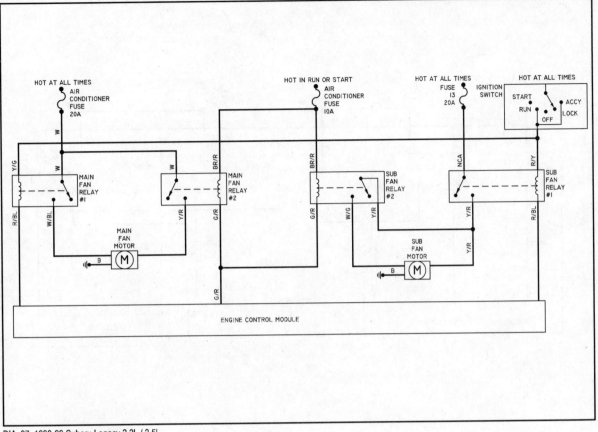

DIA. 87- 1998-99 Subaru Legacy 2.2L / 2.5L

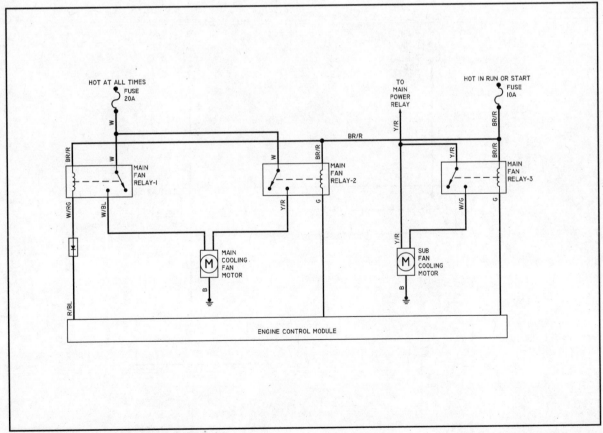

DIA. 88- 1995-99 Subaru Legacy Brighton, Outback 2.2L, 2.5L

79239W44

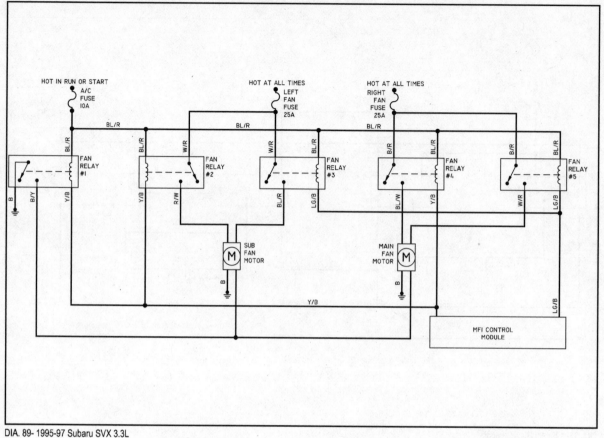

DIA. 89- 1995-97 Subaru SVX 3.3L

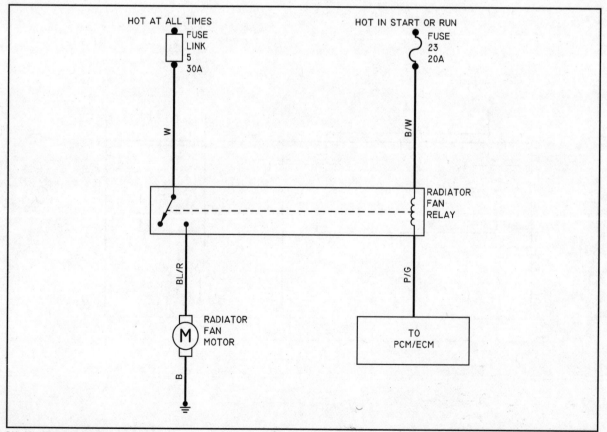

DIA. 90- 1995-99 Suzuki Esteem 1.6L

79239W45

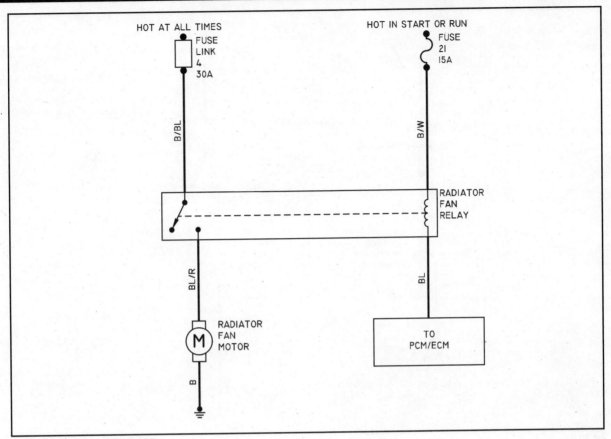

DIA. 91- 1995-99 Suzuki Swift 1.0L / 1.3L

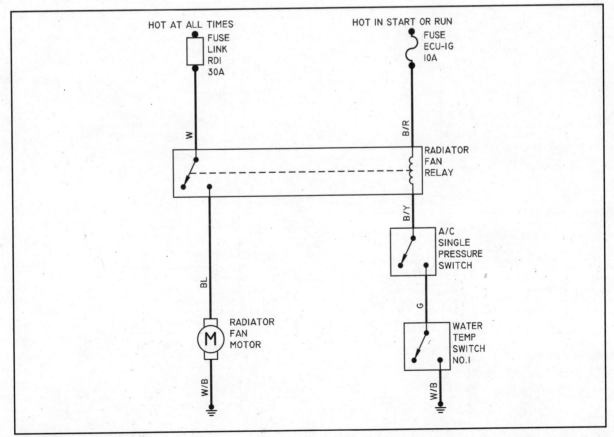

DIA. 92- 1995-99 Toyota Avalon 3.0L

79239W46

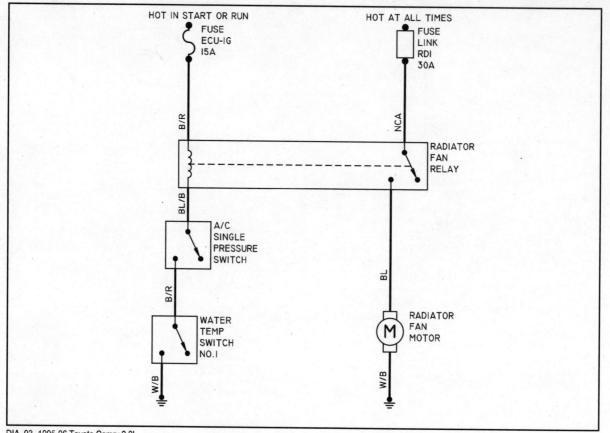

DIA. 93- 1995-96 Toyota Camry 2.2L

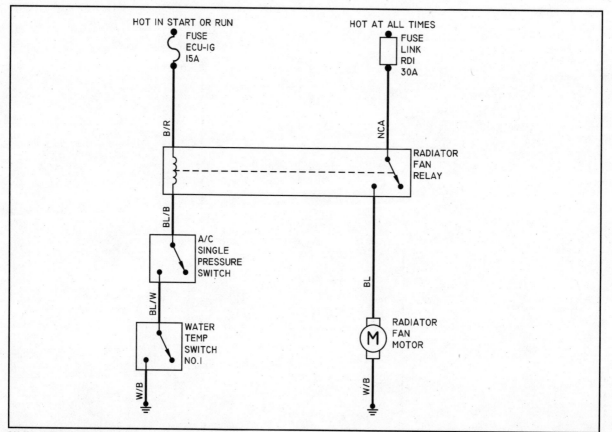

DIA. 94- 1995-96 Toyota Camry 3.0L

79239W47

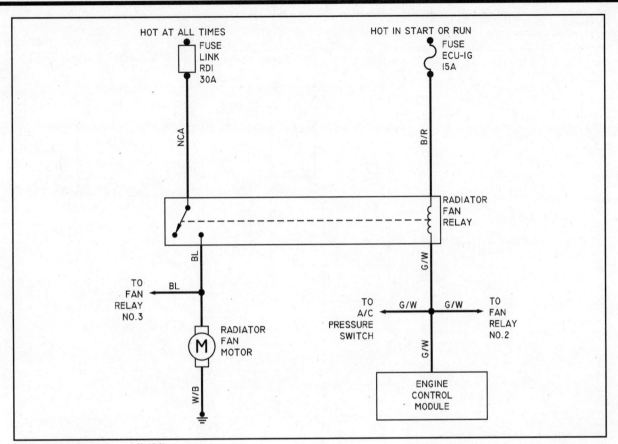

DIA. 95- 1997-99 Toyota Camry 2.2L / 3.0L

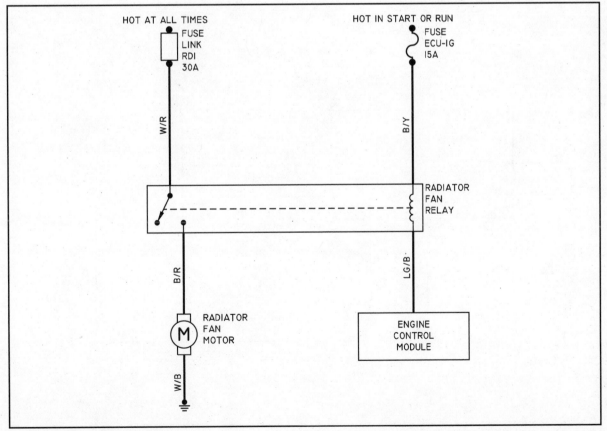

DIA. 96- 1995-99 Toyota Celica 1.8L / 2.2L

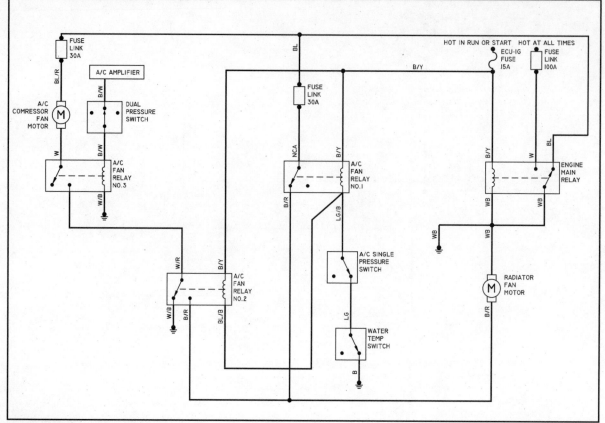

DIA. 97- 1995-99 Toyota Corolla 1.6L

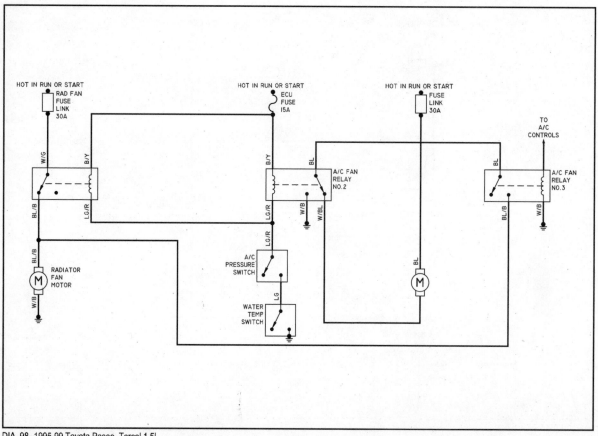

DIA. 98- 1995-99 Toyota Paseo, Tercel 1.5L

79239W49

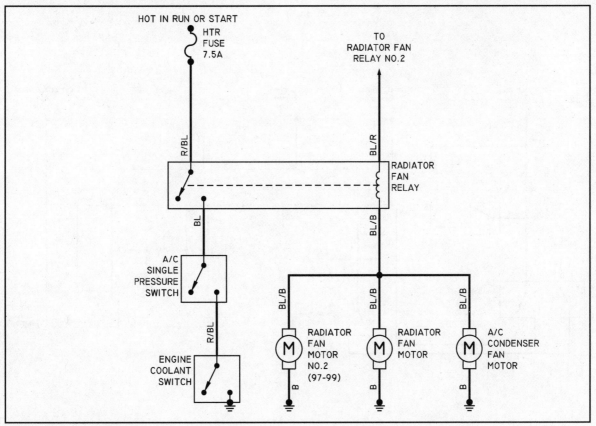

DIA. 99- 1995-99 Toyota Supra 3.0L

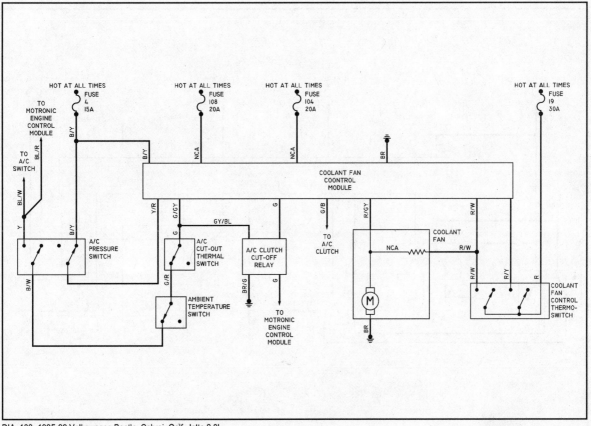

DIA. 100- 1995-99 Volkswagen Beetle, Cabroi, Golf, Jetta 2.0L
1997-99 GTI 2.0L

79239W50

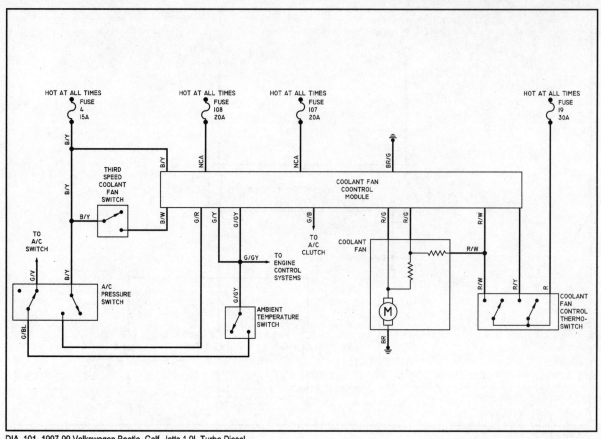

DIA. 101- 1997-99 Volkswagen Beetle, Golf, Jetta 1.9L Turbo Diesel

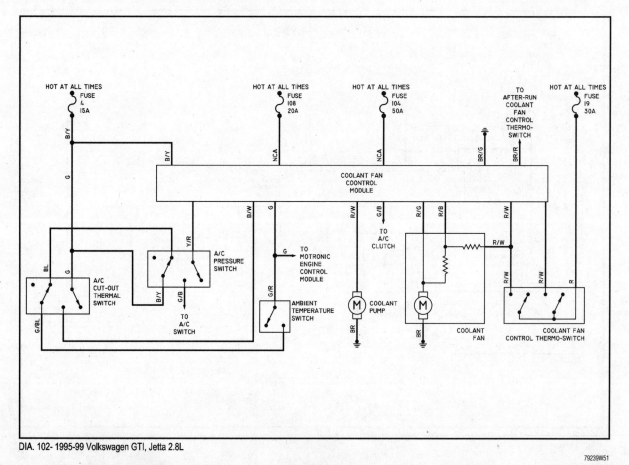

DIA. 102- 1995-99 Volkswagen GTI, Jetta 2.8L

79239W51

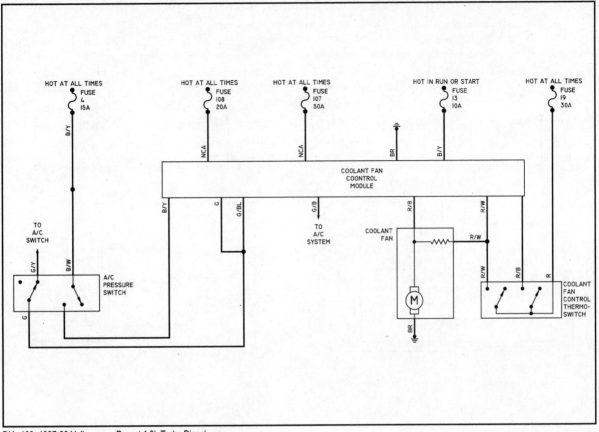

DIA. 103- 1997-99 Volkswagen Passat 1.9L Turbo Diesel

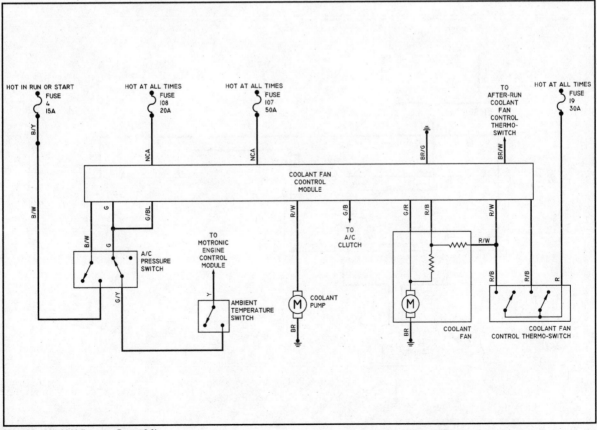

DIA. 104- 1995-99 Volkswagen Passat 2.0L

79239W52

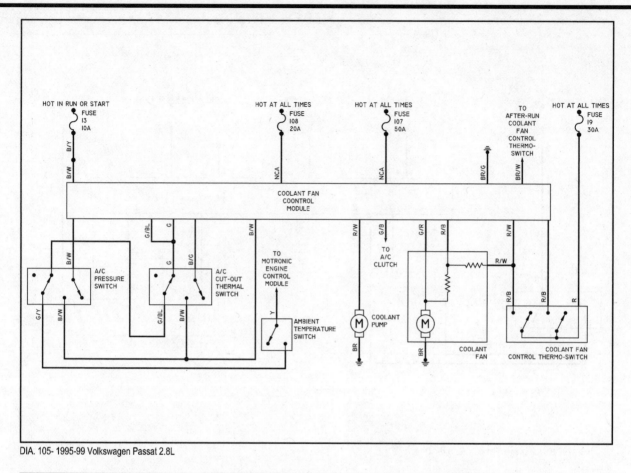

DIA. 105- 1995-99 Volkswagen Passat 2.8L

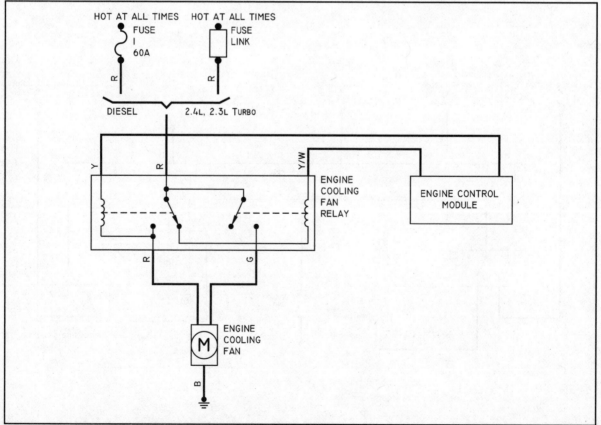

DIA. 106- 1995-99 Volvo 850, C70, S70, V70 2.3L Turbo / 2.4L Diesel

79239W53

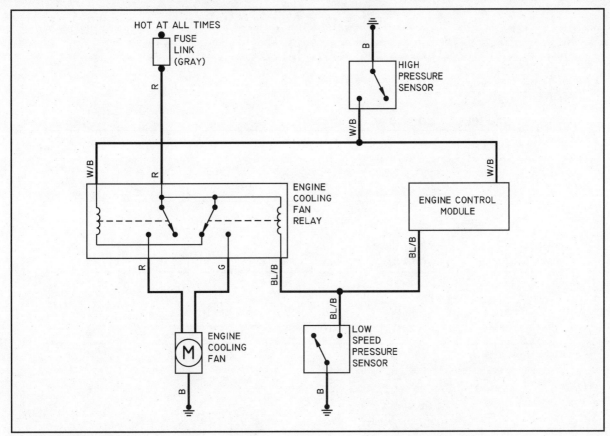

DIA. 107- 1995-99 Volvo 940 2.3L

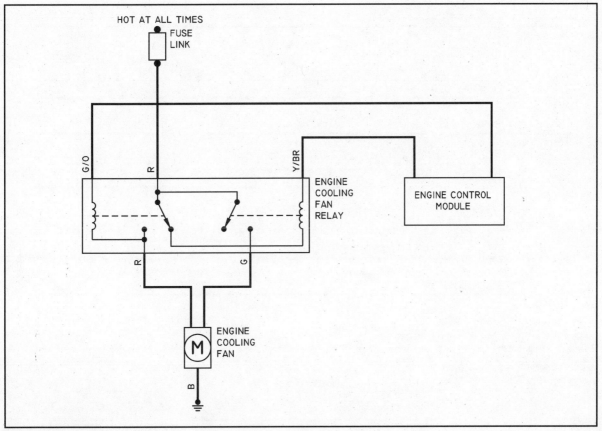

DIA. 108- 1995-99 Volvo 960, S90, V90 2.9L

TRUCK AND VAN
COOLING FAN DIAGRAM INDEX

MANUFACTURER

Model and Engine	Diagram
Chrysler	
Town & Country/Caravan/Voyager 2.4L/2.5L/3.0L/3.3L/3.8L	
1995 Models	1
1996-99 Models	2
Dodge	
Dakota 2.5L	
1995-96 Models	3
1997-99 Models	4
Ford	
Windstar 3.0L/3.8L	
1995-96 Models	5
1997-99 Models	6
General Motors	
Lumina APV/Silhouette/Trans Sport 3.1L	7
Lumina APV/Silhouette/Trans Sport/Venture 3.4L	
1996 Models	9
1997-99 Models	10
Lumina APV/Silhouette/Trans Sport 3.8L	8
Geo/Chevrolet	
Tracker 1.6L	11
Honda	
CR-V 2.0L	15
Oasis 2.2L	17
Odyssey 2.0L/2.2L	16
Kia	
Sportage 2.0L	18
Land Rover	
Discovery 3.9L	19
Range Rover 4.0L	20
Mazda	
MPV 3.0L	
1995 Models	21
1996-99 Models	22
Mercury	
Villager 3.0L	23
Mitsubishi	
Montero 3.0L/3.5L	24
Montero Sport 2.4L/3.0L	25
Nissan	
Quest 3.0L	
1995 Models	26
1996-97 Models	27
1998-99 Models	28

91261C08

TRUCK AND VAN
COOLING FAN DIAGRAM INDEX

MANUFACTURER

Model and Engine	Diagram
Suzuki	
Sidekick 1.6L	
1995-96 Models	12
1997-98 Models	13
X90 1.6L, Sport 1.8L	14
Toyota	
RAV4 2.0L	29

91261C09

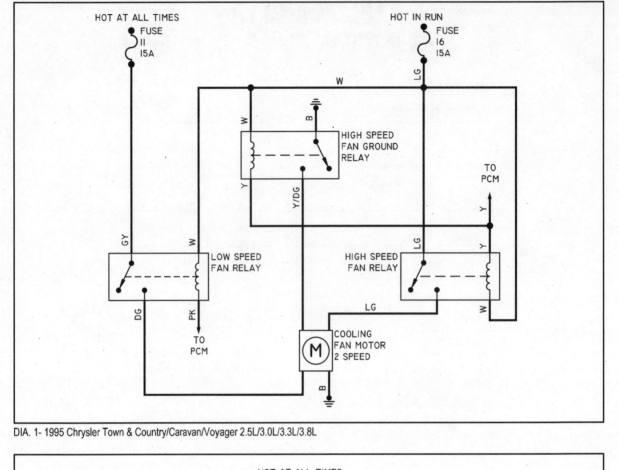

DIA. 1- 1995 Chrysler Town & Country/Caravan/Voyager 2.5L/3.0L/3.3L/3.8L

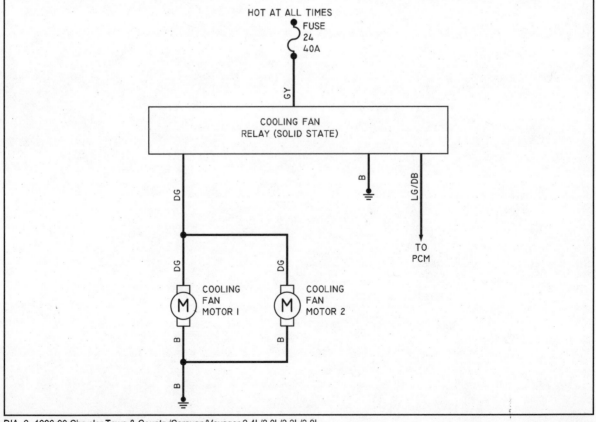

DIA. 2- 1996-99 Chrysler Town & Country/Caravan/Voyager 2.4L/3.0L/3.3L/3.8L

79249W01

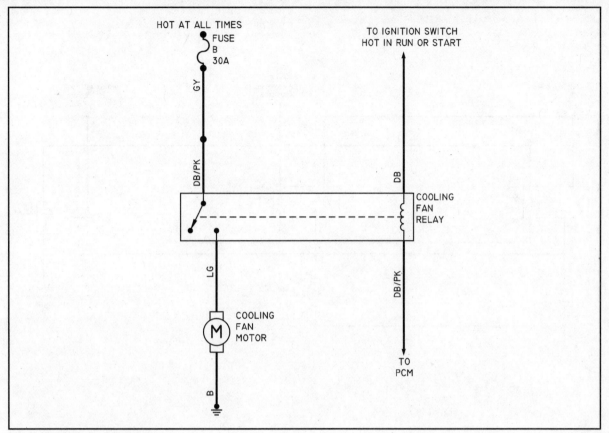

DIA. 3- 1995-96 Dodge Dakota 2.5L

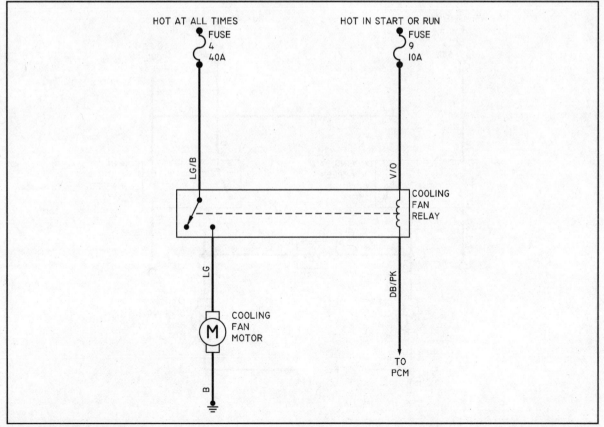

DIA. 4- 1997-99 Dodge Dakota 2.5L

79249W02

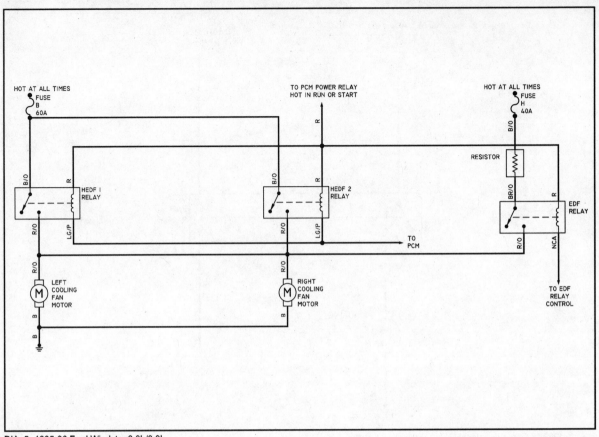

DIA. 5- 1995-96 Ford Windstar 3.0L/3.8L

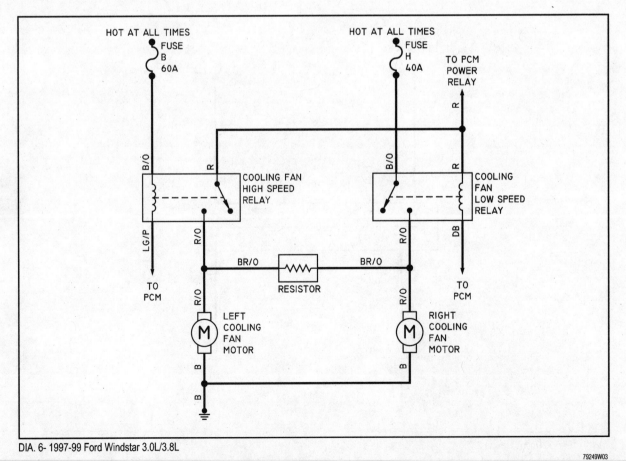

DIA. 6- 1997-99 Ford Windstar 3.0L/3.8L

79249W03

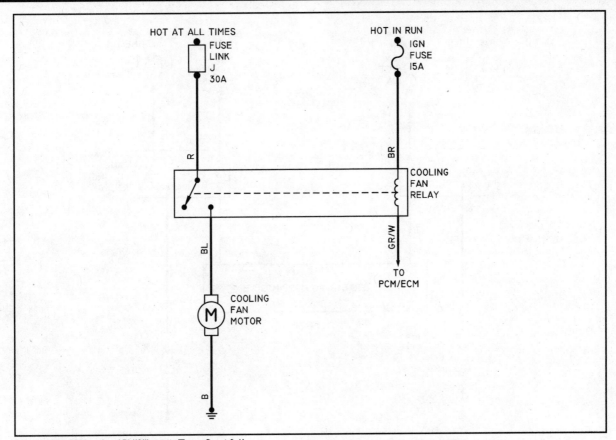

DIA. 7- 1995 GM Lumina APV/Silhouette/Trans Sport 3.1L

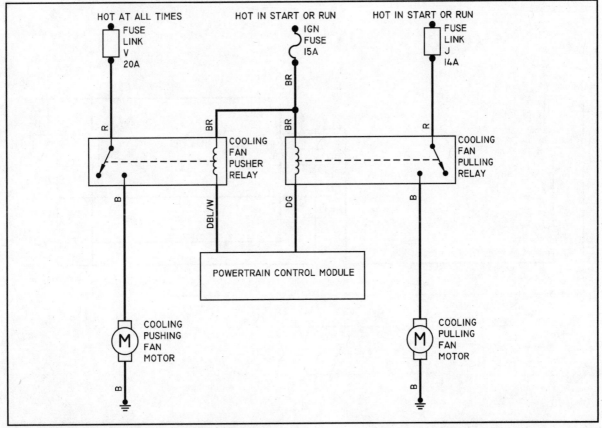

DIA. 8- 1995 GM Lumina APV/Silhouette/Trans Sport 3.8L

79249W04

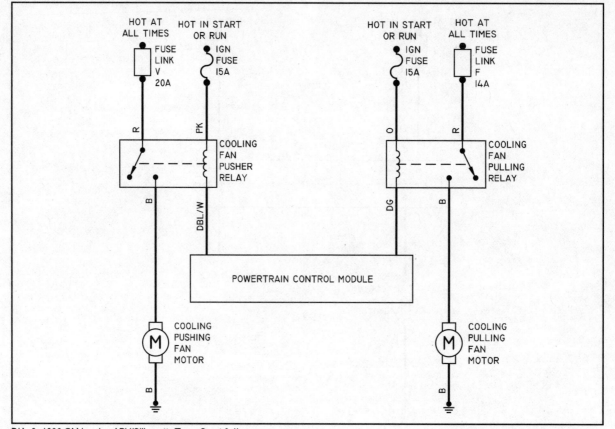

DIA. 9- 1996 GM Lumina APV/Silhouette/Trans Sport 3.4L

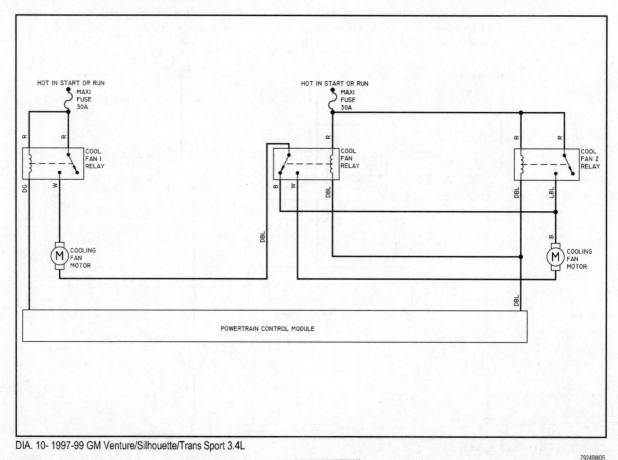

DIA. 10- 1997-99 GM Venture/Silhouette/Trans Sport 3.4L

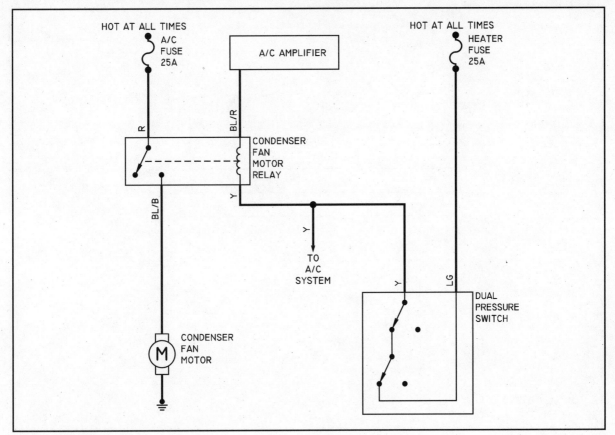

DIA. 11- 1995-98 Geo Tracker 1.6L

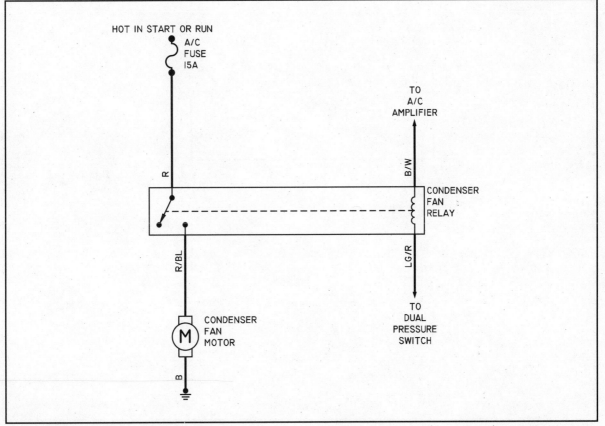

DIA. 12- 1995-96 Suzuki Sidekick 1.6L

79249W06

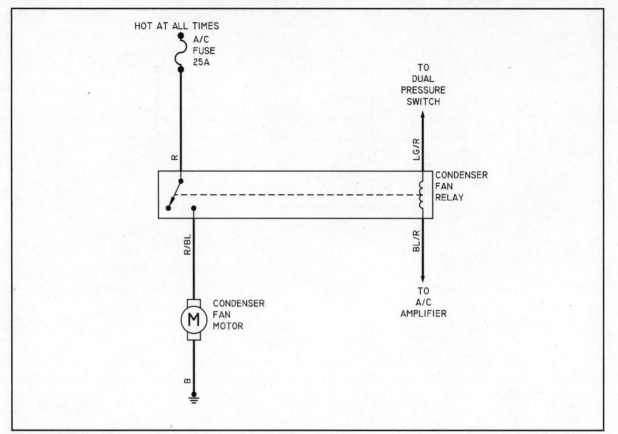

DIA. 13- 1997-98 Sukuki Sidekick 1.6L

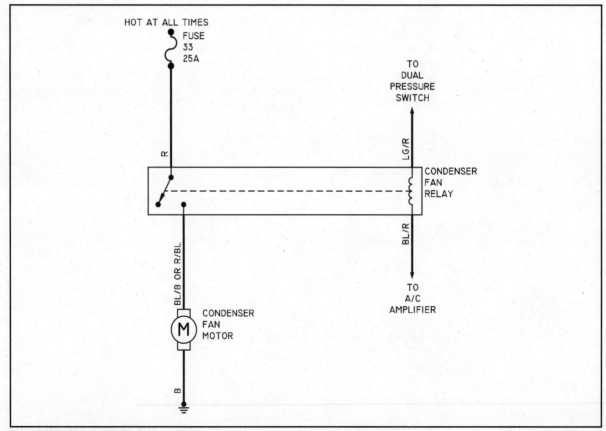

DIA. 14- 1996-98 Suzuki X90 1.6L, Sport 1.8L

79249W07

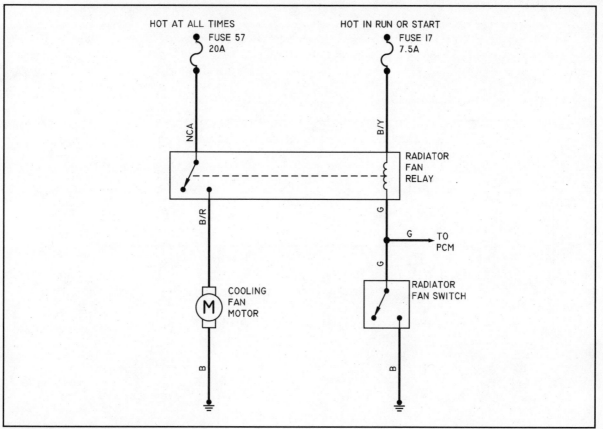

DIA. 15- 1997-99 Honda CR-V 2.0L

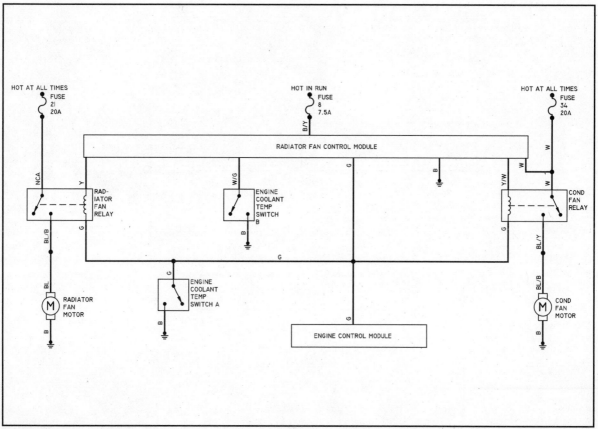

DIA. 16- 1995-99 Honda Odyssey 2.0L/2.2L

79249W08

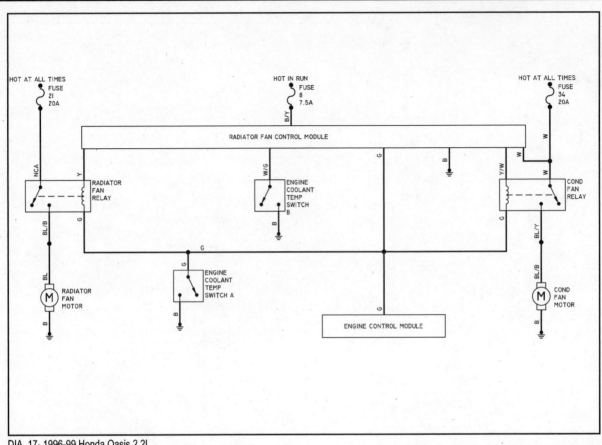

DIA. 17- 1996-99 Honda Oasis 2.2L

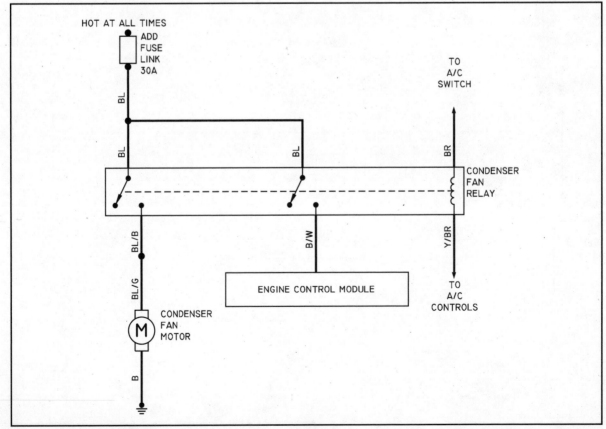

DIA. 18- 1995-99 Kia Sportage 2.0L

79249W09

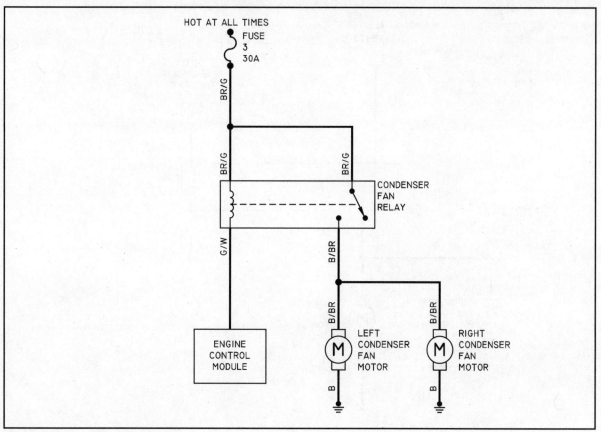

DIA. 19- 1995-99 Land Rover Discovery 3.9L

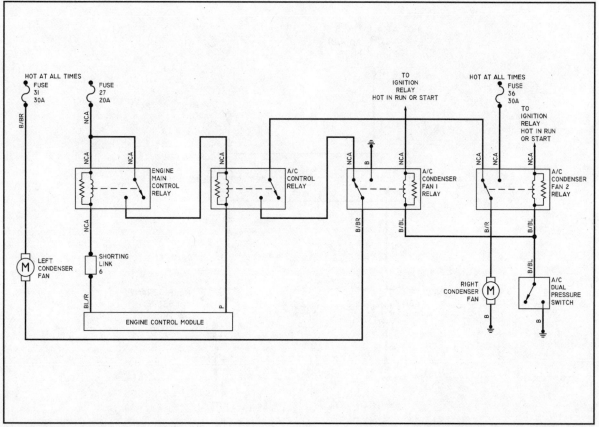

DIA. 20- 1995-99 Land Rover Range Rover 4.0L

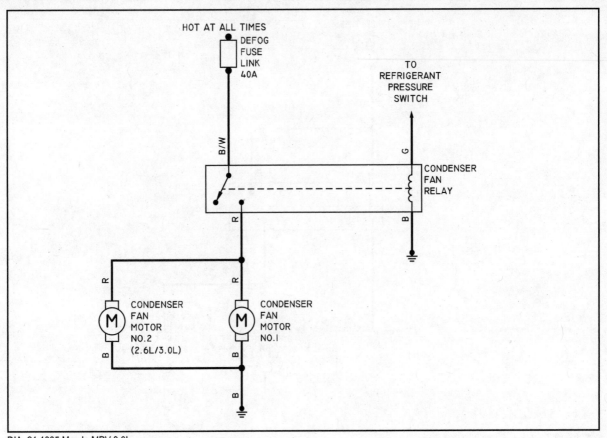

DIA. 21 1995 Mazda MPV 3.0L

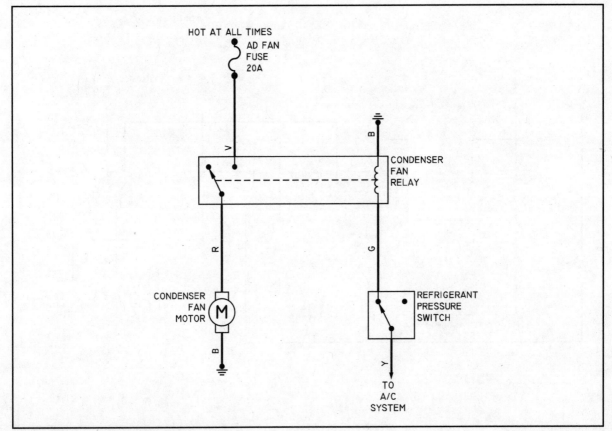

DIA. 22- 1996-99 Mazda MPV 3.0L

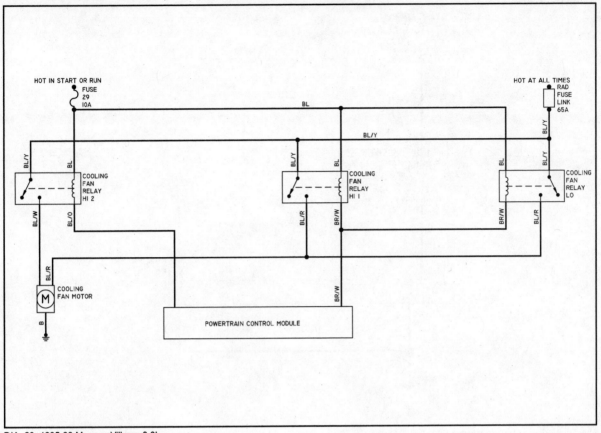

DIA. 23- 1995-99 Mercury Villager 3.0L

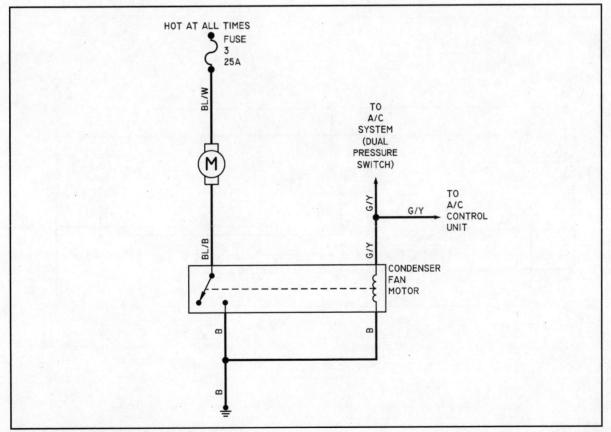

DIA. 24- 1995-99 Mitsubishi Montero 3.0L/3.5L

79249W12

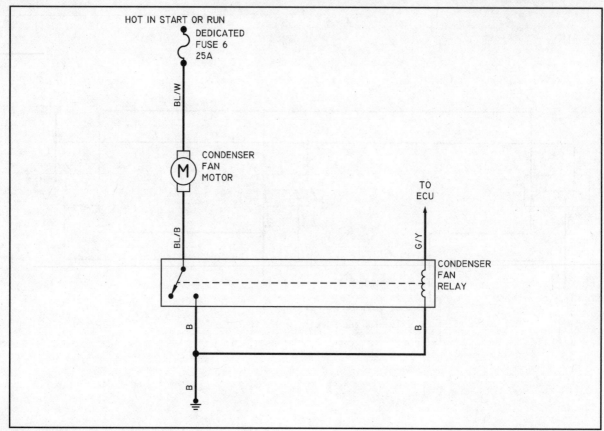

DIA. 25- 1997-99 Mitsubishi Montero Sport 2.4L/3.0L

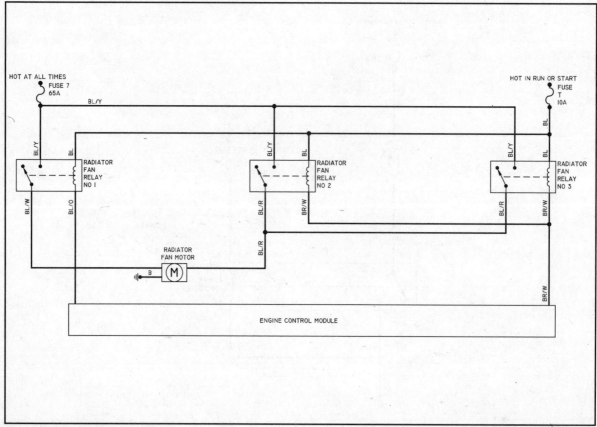

DIA. 26- 1995 Nissan Quest 3.0L

79249W13

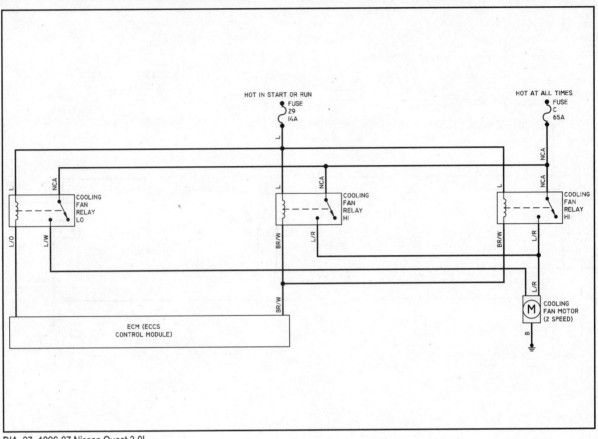

DIA. 27- 1996-97 Nissan Quest 3.0L

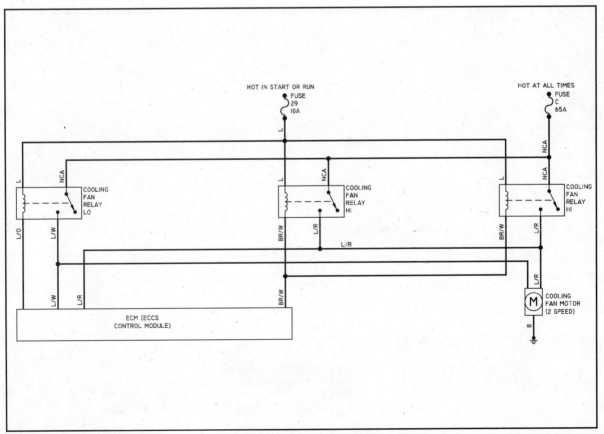

DIA. 28- 1998-99 Nissan Quest 3.0L

79249W14

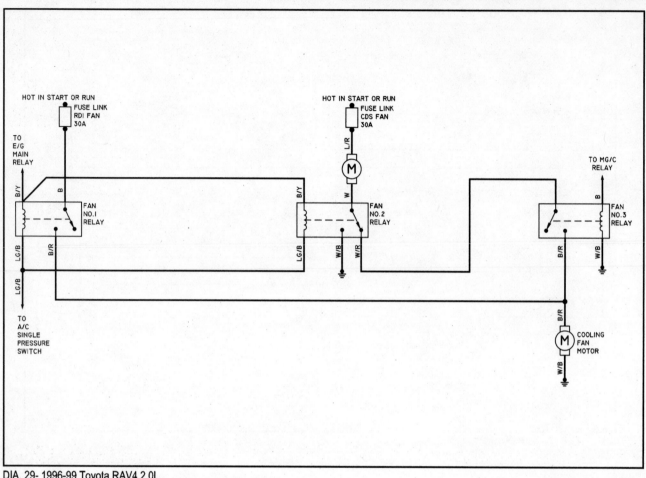

DIA. 29- 1996-99 Toyota RAV4 2.0L

79249W14

ACCESSORY DRIVE BELTS

2

GENERAL INFORMATION

Accessory drive belts are usually divided into two basic types: V-belts (conventional, cogged, and flat multi-ribbed) and serpentine (multi-ribbed) belts. The flat multi-ribbed V-belt actually resembles a serpentine belt, however, unlike a serpentine belt, only the inner surface of the belt makes contact with the components' pulleys. (Rarely, the back of multi-ribbed V-belts may ride against an idler or tensioner pulley, however.) V-belts ride in pulleys with V-shaped groove(s) to rotate various accessories, such as the power steering pump, air conditioner compressor, alternator/generator, water pump, and air pump. Only the inside of a V-belt is used, unlike a serpentine belt which utilizes both sides. V-belts typically operate one or two accessories per belt, whereas a single serpentine belt can drive all of the accessories. V-belts and a few serpentine belts require periodic adjustment because the belts are under tension and stretch over time. Most serpentine belts utilize an automatic belt tensioner that constantly provides the proper tension to the belt.

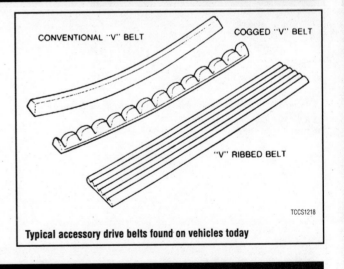

Typical accessory drive belts found on vehicles today

V-BELTS

Inspection

Although different maintenance intervals are given by each manufacturer, it is a good rule of thumb to inspect the drive belts every 15,000 miles (24,000 km) or 12 months (whichever occurs first). Determine the belt tension at a point half-way between the pulleys by pressing on the belt with moderate thumb pressure. The belt should deflect about ¼ – ½ in. (6–13mm) at this point. Note that "deflection" is not play, but the ability of the belt, under actual tension, to stretch slightly and give.

Inspect the belts for the following signs of damage or wear: glazing, cracking, fraying, crumbling or missing chunks. A glazed belt will be perfectly smooth from slippage, while a good belt will have a slight texture of fabric visible. Cracks will usually start at the inner edge of the belt and run outward. A belt that is fraying will have the fabric backing de-laminating itself from the belt. A belt that is crumbling or missing chunks will have voids in the cross-section of the belt, some times the section missing chunks will be in the pulley groove and not easily seen. All worn or damaged drive belts should be replaced immediately. It is best to

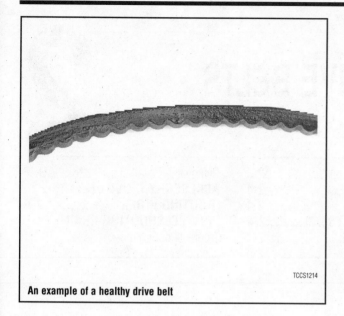

An example of a healthy drive belt

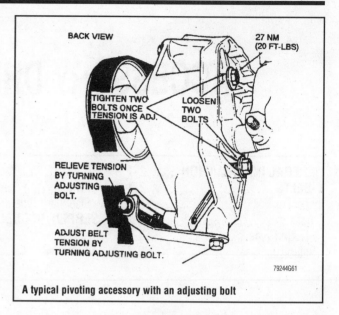

A typical pivoting accessory with an adjusting bolt

Deep cracks in this belt will cause flex, building up heat that will eventually lead to belt failure

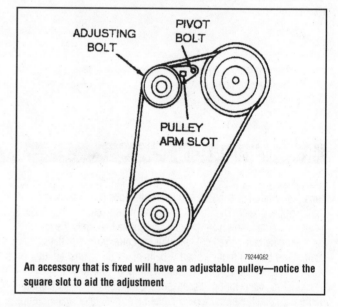

An accessory that is fixed will have an adjustable pulley—notice the square slot to aid the adjustment

replace all drive belts at one time, as a preventive maintenance measure.

Although it is generally easier on the component to have the belt too loose than too tight, a very loose belt may place a high impact load on a bearing due to the whipping or snapping action of the belt. A belt that is slightly loose may slip, especially when component loads are high. This slippage may be hard to identify. For example, the generator belt may run okay during the day, then slip at night when headlights are turned on. Slipping belts wear quickly not only due to the direct effect of slippage but also because of the heat the slippage generates. . Extreme slippage may even cause a belt to burn. A very smooth, glazed appearance on the belt's sides, as opposed to the obvious pattern of a fabric cover, indicates that the belt has been slipping.

Adjustment

✳✳ CAUTION

On vehicles with an electric cooling fan, disable the power to the fan by disengaging the fan motor wiring con-

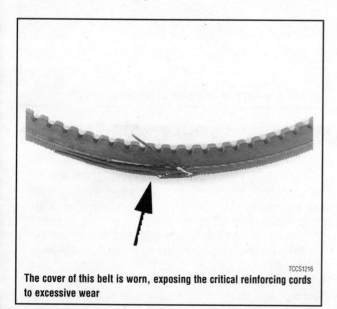

The cover of this belt is worn, exposing the critical reinforcing cords to excessive wear

nector or removing the negative battery cable before replacing or adjusting the drive belts. Otherwise, the fan may engage even though the ignition is OFF.

Belt tension can be checked by pressing on the belt at the center point of its longest straight span. The belt should give approximately ¼ – ½ in. (6–13mm). If the belt is loose it will slip, whereas if the belt is too tight it will damage the bearings in the driven unit.

For the purposes of V-belt tensioning, there are generally three types of mounting for the various components driven by the drive belt. The first method, referred to as pivoting type without adjuster, is designed so that the component is secured by at least 2 bolts. One of the bolts is a pivoting bolt and the other is the lockbolt. When both bolts are loosened so that the component may move, the component pivots on the pivoting bolt. The lockbolt passes through the component and a slotted bracket, so that when the lockbolt's nut is tightened, the component is held in that position. There are not automatic adjusting mechanisms used with this type of mounting.

The second method of component mounting, referred to as pivoting type with adjuster, is almost identical except for the addition of an adjuster of some sort. Usually the adjuster is composed of a bracket attached to the component and a threaded adjusting bolt. After loosening the pivoting and lockbolts, the adjusting bolt can be tightened or loosened to increase or decrease the drive belt's tension. With this type of mounting, you do not have to hold the component in a tensioned position and tighten the pivoting and lockbolts; the adjusting bolt does the job for you.

Some versions of this method of mounting use an adjuster which is built into one of the components mounting braces. The brace attaches the component to the engine and incorpor-ates a threaded adjuster in its mid-span, so that when the threaded adjuster is turned the brace shortens or lengthens. This in turn increases or decreases the amount of tension on the component.

The third type of mounting, referred to as stationary type, is designed so that the component is mounted on its brackets. There are no pivoting or lockbolts, and the component is not designed to be moved. Rather, this type of mounting uses an extra tensioner idler pulley assembly. The drive belt is tensioned by adjusting the position of the idler pulley, usually accomplished by turning the adjuster bolt on the idler mechanism.

PIVOTING TYPE

Without Adjuster

1. Disconnect the negative battery cable.
2. Loosen the component's lockbolt and pivoting bolt only enough for the component to move.
3. Using a strong wooden, plastic or metal prytool, move the component either closer to, or farther away from, the engine to provide the correct tension on the belt.

✳✳ WARNING

If using a metal prytool, always wrap the end with a rag or towel to prevent accidentally damaging the component from undue stress.

4. Once the proper amount of tension is applied to the drive belt, hold the prytool with one hand while tightening the lockbolt securely with the other hand.
5. Release the pressure from the prytool and tighten the pivoting bolt securely.

6. Double check the drive belt's tension, in case the component moved slightly while tightening the bolts.
7. Connect the negative battery cable.

With Adjuster

This type of drive belt is tensioned by a tensioner, which makes precise tension adjustment easy.
1. Disconnect the negative battery cable.
2. Loosen the component's pivot and lockbolts.
3. Inspect the tensioner assembly on the component; the tensioner adjusting bolt may use a locknut or screw to prevent it from loosening over time. On the type of adjuster with a threaded mounting brace, there may be two jam nuts used on either side of the threaded coupling. If such locking fasteners are found, loosen them.
4. Turn the tensioner adjusting bolt or threaded coupling to increase or decrease the amount of tension on the drive belt, as necessary.
5. When the belt tension is correct, tighten the lockbolt and the pivot bolt.
6. If equipped, tighten the tension adjusting bolt locknut or screw to prevent the adjuster from slowly loosening over time. If equipped, tighten the two jam nuts.
7. Connect the negative battery cable.

STATIONARY TYPE

Idler Pulley With Adjusting Bolt

1. Loosen the idler bracket pivot bolt and locking bolts.
2. Adjust the belt tension by inserting the proper size ratchet in the square slot of the idler bracket and rotating the bracket until tension is applied.
3. While holding the tension on the belt with the ratchet, tighten the locking bolts, then the pivot bolt.

Idler Pulley Without Adjusting Bolt

1. Loosen the mounting/pivot bolt behind the idler pulley.
2. Swivel the idler pulley with a pair of pliers or a wrench on the bearing mounting until the proper tension is achieved.
3. While holding the idler pulley, at the proper tension, tighten the mounting/pivot bolt.

Removal & Installation

If a belt must be replaced, the driven unit or idler pulley must be loosened and moved to its extreme loosest position, generally by moving it toward the center of the motor. After removing the old belt, check the pulleys for dirt or built-up material which could affect belt contact. Carefully install the new belt, remembering that it is new and unused; it may appear to be just a little too small to fit over the pulley flanges. Fit the belt over the largest pulley (usually the crankshaft pulley at the bottom center of the motor) first, then work on the smaller one(s). Gentle pressure in the direction of rotation is helpful. Some belts run around a third, or idler pulley, which acts as an additional pivot in the belt's path. It may be possible to loosen the idler pulley as well as the main component, making your job much easier. Depending on which belt(s) you are changing, it may be necessary to loosen or remove other interfering belts to get at the one(s) you want.

When buying replacement belts, remember that the fit is critical according to the length of the belt ("diameter"), the width of the belt, the depth of the belt and the angle or profile of the V shape or the ribs. The belt shape should match the shape of the pulley exactly; belts that are not an exact match can cause noise, slippage and premature failure.

After the new belt is installed, draw tension on it by moving the driven unit or idler pulley away from the motor and tighten its mounting bolts. This is sometimes a three or four-handed job; you may find an assistant helpful. Be sure that all the bolts you loosened get retightened and that any other loosened belts also have the correct tension. A new belt can be expected to stretch a bit after installation so be prepared to readjust your new belt, if needed, within the first two hundred miles of use.

PIVOTING TYPE

✳✳ CAUTION

On vehicles with an electric cooling fan, disable the power to the fan by disengaging the fan motor wiring connector or removing the negative battery cable before replacing or adjusting the drive belts. Otherwise, the fan may engage even though the ignition is OFF.

Without Adjuster

1. Disconnect the negative battery cable.
2. Loosen the accessory's slotted adjusting bracket bolt. If the hinge bolt is excessively tight, it too will have to be loosened.
3. Push the component toward the engine to provide enough slack in the belt so that it will slide over one of the accessory drive pulleys. Remove the drive belt from the accessory drive pulleys and from the vehicle.
   To install:
4. Position the new drive belt over the component pulleys. Be sure that it is routed correctly.
5. Adjust the tension of the belt, as described earlier in this section.
6. Connect the negative battery cable.

With Adjuster

1. Disconnect the negative battery cable.
2. Loosen the component's pivot and lockbolts.
3. Inspect the tensioner assembly on the component; the tensioner adjusting bolt may use a locknut or screw to prevent it from loosening over time. On the type of adjuster with a threaded mounting brace, there may be two jam nuts used on either side of the threaded coupling. If such locking fasteners are found, loosen them.

4. Turn the tensioner adjusting bolt or threaded coupling to relieve all tension from the drive belt until the most possible slack is gained from the component.
5. Slip the belt off of the accessory pulley, then remove it from the other pulleys. Remove the belt from the vehicle.
   To install:
6. Route the new belt on the component pulleys. Make certain that it is routed correctly; incorrect routing could cause a components to spin backward, possibly damaging it.
7. Once the belt is correctly positioned on all of the pulleys, adjust the tension as described earlier in this section.
8. Connect the negative battery cable.

STATIONARY TYPE

Idler Pulley With Adjusting Bolt

1. Disconnect the negative battery cable.
2. Loosen the idler bracket pivot bolt and locking bolts.
3. Move the idler pulley until the most amount of slack is gained.
4. Remove the drive belt from the accessory pulley, then from the other applicable pulleys.
   To install:
5. Position the new belt over the crankshaft pulley, the idler pulley and the accessory pulley. Make certain that it is correctly routed, otherwise it could cause the accessory to be rotated backwards. This could cause damage to the accessory.
6. Adjust the belt tension, as described earlier in this section.
7. While holding the tension on the belt with the ratchet, tighten the locking bolts, then the pivot bolt.
8. Connect the negative battery cable.

Idler Pulley Without Adjusting Bolt

1. Disconnect the negative battery cable.
2. Loosen the mounting/pivot bolt behind the idler pulley.
3. Remove the drive belt from the accessory pulley, then from the other applicable pulleys.
   To install:
4. Position the new belt over the crankshaft pulley, the idler pulley and the accessory pulley. Make certain that it is correctly routed, otherwise it could cause the accessory to be rotated backwards. This could cause damage to the accessory.
5. Swivel the idler pulley with a pair of pliers or a wrench on the bearing mounting until the proper tension is achieved.
6. While holding the idler pulley, at the proper tension, tighten the mounting/pivot bolt.
7. Connect the negative battery cable.

SERPENTINE BELTS

Inspection

Although many manufacturers recommend that the drive belt(s) be inspected every 30,000 miles (48,000 km) or more, it is really a good idea to check them at least once a year, or at every major fluid change. Whichever interval you choose, the belts should be checked for wear or damage. Obviously, a damaged drive belt can cause problems should it give way while the vehicle is in operation. But, improper length belts (too short or long), as well as excessively worn belts, can also cause problems. Loose accessory drive belts can lead to poor engine cooling and diminished output from the alternator, air conditioning compressor or power steering pump. A belt that is too tight places a severe strain on the driven unit and can wear out bearings quickly.

Serpentine drive belts should be inspected for rib chunking (pieces of the ribs breaking off), severe glazing, frayed cords or other visible damage. Any belt which is missing sections of 2 or more adjacent ribs which are ½ in. (13mm) or longer must be replaced. You might want to note that serpentine belts do tend to form small cracks across the backing. If the only wear you find is in the form of one or more cracks are across the backing and NOT parallel to the ribs, the belt is still good and does not need to be replaced.

Troubleshooting the Serpentine Drive Belt

Problem	Cause	Solution
Tension sheeting fabric failure (woven fabric on outside circumference of belt has cracked or separated from body of belt)	• Grooved or backside idler pulley diameters are less than minimum recommended • Tension sheeting contacting (rubbing) stationary object • Excessive heat causing woven fabric to age • Tension sheeting splice has fractured	• Replace pulley(s) not conforming to specification • Correct rubbing condition • Replace belt • Replace belt
Noise (objectional squeal, squeak, or rumble is heard or felt while drive belt is in operation)	• Belt slippage • Bearing noise • Belt misalignment • Belt-to-pulley mismatch • Driven component inducing vibration • System resonant frequency inducing vibration	• Adjust belt • Locate and repair • Align belt/pulley(s) • Install correct belt • Locate defective driven component and repair • Vary belt tension within specifications. Replace belt.
Rib chunking (one or more ribs has separated from belt body)	• Foreign objects imbedded in pulley grooves • Installation damage • Drive loads in excess of design specifications • Insufficient internal belt adhesion	• Remove foreign objects from pulley grooves • Replace belt • Adjust belt tension • Replace belt
Rib or belt wear (belt ribs contact bottom of pulley grooves)	• Pulley(s) misaligned • Mismatch of belt and pulley groove widths • Abrasive environment • Rusted pulley(s) • Sharp or jagged pulley groove tips • Rubber deteriorated	• Align pulley(s) • Replace belt • Replace belt • Clean rust from pulley(s) • Replace pulley • Replace belt
Longitudinal belt cracking (cracks between two ribs)	• Belt has mistracked from pulley groove • Pulley groove tip has worn away rubber-to-tensile member	• Replace belt • Replace belt
Belt slips	• Belt slipping because of insufficient tension • Belt or pulley subjected to substance (belt dressing, oil, ethylene glycol) that has reduced friction • Driven component bearing failure • Belt glazed and hardened from heat and excessive slippage	• Adjust tension • Replace belt and clean pulleys • Replace faulty component bearing • Replace belt
"Groove jumping" (belt does not maintain correct position on pulley, or turns over and/or runs off pulleys)	• Insufficient belt tension • Pulley(s) not within design tolerance • Foreign object(s) in grooves	• Adjust belt tension • Replace pulley(s) • Remove foreign objects from grooves

TCCS3C09

Troubleshooting the Serpentine Drive Belt

Problem	Cause	Solution
"Groove jumping" (belt does not maintain correct position on pulley, or turns over and/or runs off pulleys)	• Excessive belt speed • Pulley misalignment • Belt-to-pulley profile mismatched • Belt cordline is distorted	• Avoid excessive engine acceleration • Align pulley(s) • Install correct belt • Replace belt
Belt broken (Note: identify and correct problem before replacement belt is installed)	• Excessive tension • Tensile members damaged during belt installation • Belt turnover • Severe pulley misalignment • Bracket, pulley, or bearing failure	• Replace belt and adjust tension to specification • Replace belt • Replace belt • Align pulley(s) • Replace defective component and belt
Cord edge failure (tensile member exposed at edges of belt or separated from belt body)	• Excessive tension • Drive pulley misalignment • Belt contacting stationary object • Pulley irregularities • Improper pulley construction • Insufficient adhesion between tensile member and rubber matrix	• Adjust belt tension • Align pulley • Correct as necessary • Replace pulley • Replace pulley • Replace belt and adjust tension to specifications
Sporadic rib cracking (multiple cracks in belt ribs at random intervals)	• Ribbed pulley(s) diameter less than minimum specification • Backside bend flat pulley(s) diameter less than minimum • Excessive heat condition causing rubber to harden • Excessive belt thickness • Belt overcured • Excessive tension	• Replace pulley(s) • Replace pulley(s) • Correct heat condition as necessary • Replace belt • Replace belt • Adjust belt tension

TCCS3C10

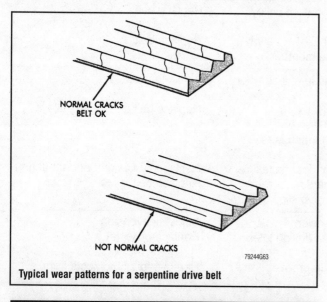

NORMAL CRACKS BELT OK

NOT NORMAL CRACKS

79244G63

Typical wear patterns for a serpentine drive belt

Adjustment

Periodic drive belt tensioning is not necessary, because an automatic spring-loaded tensioner is used with these belts to maintain proper adjustment at all times. The tensioner is also useful as a wear indicator. When the belt is properly installed, the arrow on the tensioner housing must point within the acceptable range lines on the tensioner's face. If the arrow falls outside the range, either an improper belt has been installed or the belt is worn beyond its useful life span.

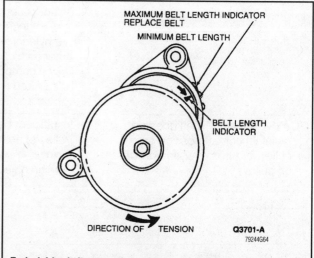

MAXIMUM BELT LENGTH INDICATOR REPLACE BELT

MINIMUM BELT LENGTH

BELT LENGTH INDICATOR

DIRECTION OF TENSION

Q3701-A
79244G64

Typical drive belt automatic tensioner wear indicator

In either case, a new belt must be installed immediately to assure proper engine operation and to prevent possible accessory damage.

Removal & Installation

Because serpentine belts use a spring loaded tensioner for adjustment, belt replacement tends to be somewhat easier than it used to be on engines where accessories were pivoted and bolted in place for tension adjustment. Basically, all belt replacement involves is to pivot the tensioner to loosen the belt, then slide the belt off of the pulleys. The two most important points are to pay CLOSE attention to the proper belt routing (since serpentine belts tend to be "snaked" all different ways through the pulleys) and to be sure the V-ribs are properly seated in all the pulleys.

Although belt routing diagrams have been included in this section, the first places you should check for proper belt routing are the labels in your engine compartment. These should include a belt routing diagram which may reflect changes made during a production run.

1. Disconnect the negative battery cable for safety. This will help assure that no one mistakenly cranks the engine over with your hands between the pulleys, and that the cooling fan cannot activate while servicing the belt(s).

➥**Take a good look at the installed belt and make a note of the routing. Before removing the belt, be sure the routing matches that of the belt routing label or one of the diagrams in this book. If for some reason a diagram does not match (you may not have the original engine or it may have been modified), carefully note the changes on a piece of paper.**

2. For tensioners equipped with a ½ in. (13mm) square hole, insert the drive end of a large breaker bar into the hole. Use the breaker bar to pivot the tensioner away from the drive belt. For tensioners not equipped with this hole, use the proper-sized socket and breaker bar (or a large handled wrench) on the tensioner idler pulley center bolt to pivot the tensioner away from the belt. This will loosen the belt sufficiently that it can be pulled off of one or more of the pulleys. It is usually easiest to carefully pull the belt out from underneath the tensioner pulley itself.

3. Once the belt is off one of the pulleys, gently pivot the tensioner back into position. DO NOT allow the tensioner to snap back, as this could damage the tensioner's internal parts.

4. Now finish removing the belt from the other pulleys and remove it from the engine.

To install:

5. While referring to the proper routing diagram (which you identified earlier), begin to route the belt over the pulleys, leaving whichever pulley you first released it from for last.

6. Once the belt is mostly in place, carefully pivot the tensioner and position the belt over the final pulley. As you begin to allow the tensioner back into contact with the belt, run your hand around the pulleys and be sure the belt is properly seated in the ribs. If not, release the tension and seat the belt.

7. Once the belt is installed, take another look at all the pulleys to double check your installation.

8. Connect the negative battery cable, then start and run the engine to check belt operation.

9. Once the engine has reached normal operating temperature, turn the ignition **OFF** and check that the belt tensioner arrow is within the proper adjustment range.

Relieve the belt tension by pivoting the automatic tensioner away from the belt, then remove the belt

Often the underhood label will display the Serpentine belt routing

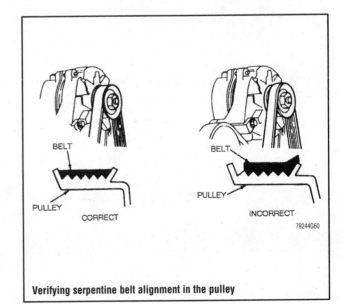
Verifying serpentine belt alignment in the pulley

ACCESSORY DRIVE BELT ROUTING INDEX

MANUFACTURER

Model and Engine	Description	Figure
Acura Cars		
1.8L engines	V-belt routing	1
2.2L engines	V-belt routing	2
2.5L engines	V-belt routing	3
3.0L engines except NSX	V-belt routing	4
3.0L NSX engines	V-belt routing	5
3.2L and 3.5L engines	V-belt routing	6
Acura SLX		
3.2L engines	V-belt routing	7
3.5L engines	Serpentine belt routing	8
Audi		
AAN engines without A/C	Serpentine belt routing	9
AAN engines with A/C	Serpentine belt routing	10
A4 AFC engines	Serpentine belt routing	11
AFC, AAH engines	Serpentine belt routing	12
Chrysler Domestic Cars		
2.0L (VIN C, Y) Neon engine	V-belt routing	13
2.0L (VIN Y) except Neon engine	V-belt routing	14
2.0L (VIN F) turbo engine	V-belt routing	15
2.4L engine	V-belt routing	16
2.5L engine	V-belt routing	17
2.7L engine	V-belt routing	18
3.2L engine	V-belt routing	19
3.3L engine	V-belt routing	20
3.5L engine	V-belt routing	21
Chrysler Import Cars		
1.5L and 2.4L engines	V-belt routing	22
1.8L engines	V-belt routing	23
Chrysler Mini-Vans		
2.4L engines	V-belt routing	24
2.5L engines	V-belt routing	25
3.0L engines	V-belt routing	26
3.3L engines	V-belt routing	27
3.8L engines	V-belt routing	27
Dodge Trucks		
2.5L engines		
1995 Models	V-belt routing	28
1996-99 Models with A/C	Serpentine belt routing	29
1996-99 Models without A/C	Serpentine belt routing	30
3.9L, 5.2L and 5.9L LDC gasoline engines	Serpentine belt routing	31
5.9L Diesel engines		
Models with A/C	Serpentine belt routing	32
Models without A/C	Serpentine belt routing	33

91262C01

ACCESSORY DRIVE BELT ROUTING INDEX

MANUFACTURER Model and Engine	Description	Figure
Dodge (cont.)		
5.9L HDC gasoline engines		
Models with A/C	Serpentine belt routing	34
Models without A/C	Serpentine belt routing	35
8.0L engines with A/C		
Models with A/C	Serpentine belt routing	34
Models without A/C	Serpentine belt routing	35
Ford Cars		
1.3L Aspire engine	V-belt routing	36
1.8L engine	V-belt routing	37
1.9L engine	Serpentine belt routing	38
2.0L Probe engine	V-belt routing	39
2.0L (VIN 3) engine	V-belt routing	40
2.5L (VIN B) Probe engine without A/C	Alternator drive belt routing	41
2.5L (VIN B) Probe engine with A/C	Alternator drive belt routing	42
2.5L (VIN B) Probe engine	Power steering/water pump belt	43
2.5L (VIN L) engine	Water pump drive belt routing	44
2.5L (VIN L) engine with A/C	V-belt routing	45
2.5L (VIN L) engine without A/C	V-belt routing	46
3.0L SHO engine	Serpentine belt routing	47
3.0L (VIN S) engines	Serpentine belt routing	49
3.0L (VIN U and 1) engine	Serpentine belt routing	48
3.2L SHO engine	Serpentine belt routing	50
3.4L SHO engine	Serpentine belt routing	51
3.4L SHO engine	Water pump drive belt routing	52
3.8L Taurus/Sable engine	Serpentine belt routing	53
3.8L (except Taurus/Sable) engine	V-belt routing	54
3.8L supercharged engine	V-belt routing	55
4.6L (except Continental) engine	V-belt routing	56
4.6L Continental engine	Serpentine belt routing	57
5.0L engine	V-belt routing	58
Ford Trucks		
2.3L Ranger engines	Serpentine belt routing	59
3.0L Aerostar engines		
Models with A/C	Serpentine belt routing	60
Models without A/C	Serpentine belt routing	61
3.0L Ranger engines	Serpentine belt routing	62
3.0L Windstar engines	Serpentine belt routing	63
3.8L Windstar engines	Serpentine belt routing	64
4.0L Ranger engines	Serpentine belt routing	65
4.0L SOHC engines	Serpentine belt routing	66

ACCESSORY DRIVE BELT ROUTING INDEX

MANUFACTURER

Model and Engine	Description	Figure
Ford Trucks (cont.)		
4.2L engines		
Models with A/C	Serpentine belt routing	67
Models without A/C	Serpentine belt routing	68
4.6L engines		
Models with A/C	Serpentine belt routing	69
Models without A/C	Serpentine belt routing	70
4.9L engines	Serpentine belt routing	71
5.0L (except Explorer) engines	Serpentine belt routing	72
5.0L Explorer engines	Serpentine belt routing	73
5.4L engines		
Models with A/C	Serpentine belt routing	69
Models without A/C	Serpentine belt routing	70
5.8L engines	Serpentine belt routing	72
6.8L engines		
Models with A/C	Serpentine belt routing	69
Models without A/C	Serpentine belt routing	70
7.3L turbo Diesel engines	Serpentine belt routing	74
7.5L (except Motorhome) engines	Serpentine belt routing	75
7.5L F-Super Duty Motorhome engines	Serpentine belt routing	76
General Motors Cars		
2.2L engines		
A body models	Serpentine belt routing	77
Except A body models		
1995-97 engines	Serpentine belt routing	78
1998-99 engines with A/C	Serpentine belt routing	79
1998-99 engines without A/C	Serpentine belt routing	80
2.3L engines	Serpentine belt routing	81
2.4L engines	Serpentine belt routing	81
3.0L engine	Serpentine belt routing	82
3.1L engines		
A body models	Serpentine belt routing	83
Except A and L/N body models	Serpentine belt routing	84
L/N body models	Serpentine belt routing	85
3.4L (except F body) engines	Serpentine belt routing	86
3.4L F body engines	Serpentine belt routing	87
3.8L engines		
1995 C & H bodies 3.8L (VIN L/K) engines	Serpentine belt routing	88
1996-99 C & H bodies 3.8L (VIN K) engines	Serpentine belt routing	89
C & H bodies 3.8L (VIN 1) engines	Serpentine belt routing	90
F body models	Serpentine belt routing	91
G body 3.8L (VIN 1) engines	Serpentine belt routing	92
G body 3.8L (VIN K) engines	Serpentine belt routing	93

91262C03

ACCESSORY DRIVE BELT ROUTING INDEX

MANUFACTURER Model and Engine	Description	Figure
General Motors Cars (cont.)		
3.8L engines (cont.)		
W body models	Serpentine belt routing	93
4.0L engines	Serpentine belt routing	94
4.3L engines	Serpentine belt routing	95
4.6L engines	Serpentine belt routing	96
4.9L engines	Serpentine belt routing	97
5.7L engines		
5.7L (VIN P) except B or F body engines	Serpentine belt routing	98
5.7L (VIN J) except B or F body engines	Serpentine belt routing	99
5.7L (VIN G) except B or F body engines	Serpentine belt routing	100
5.7L B body engines	Serpentine belt routing	96
5.7L 1995-97 F body engines	Serpentine belt routing	101
5.7L 1998-99 F body engines	Serpentine belt routing	102
5.7L 1998-99 F body engines	A/C drive belt routing	103
General Motors Trucks		
2.2L engines with A/C	Serpentine belt routing	104
2.2L engines without A/C	Serpentine belt routing	106
3.1L and 3.4L engines	Serpentine belt routing	106
3.8L engines	Serpentine belt routing	88
4.3L (except Vin W, X and Full-size) engines		
1996-99 Models with A/C	Serpentine belt routing	109
1996-99 Models without A/C	Serpentine belt routing	110
4.3L (VIN W and X) engines		
1995 Models	Serpentine belt routing	107
1996-99 Models	Serpentine belt routing	108
4.3L Full-size Truck engines		
1995 Models with A.I.R.	Serpentine belt routing	111
1995 Models without A.I.R.	Serpentine belt routing	112
5.0L engines		
1995 Models with A.I.R.	Serpentine belt routing	111
1995 Models without A.I.R.	Serpentine belt routing	112
1996-99 Models with A/C	Serpentine belt routing	111
1996-99 Models without A/C	Serpentine belt routing	112
5.7L engines		
1995 Models with A.I.R.	Serpentine belt routing	111
1995 Models without A.I.R.	Serpentine belt routing	112
1996-99 Models with A/C	Serpentine belt routing	109
1996-99 Models without A/C	Serpentine belt routing	110
7.4L engines		
1995 Models	Serpentine belt routing	113
1996-99 Models	Serpentine belt routing	114

ACCESSORY DRIVE BELT ROUTING INDEX

MANUFACTURER

Model and Engine	Description	Figure
General Motors Trucks (cont.)		
Diesel engines		
1995 Models	Serpentine belt routing	115
1996-99 Models	Serpentine belt routing	116
Geo/Chevrolet		
All engines	V-belt routing	117
Honda Cars		
4-cylinder engines (except Civic/Del Sol) without A/C except C	V-belt routing	118
4-cylinder engines (except Civic/Del Sol) with A/C except Civic	V-belt routing	119
Civic/Del Sol 4-cylinder engines	V-belt routing	120
V6 engines	V-belt routing	121
Honda Trucks		
2.0L, 2.2L, 2.3L, and 2.6L engines	V-belt routing	122
3.2L engines		
1995-97 Models	V-belt routing	123
1998-99 Models	Serpentine belt routing	124
Hyundai		
Accent 1.5L engines	V-belt routing	125
1.5L (except Accent) engines	V-belt routing	126
1995 1.8L and 2.0L engines	V-belt routing	127
1996-99 1.8L and 2.0L engines	V-belt routing	128
3.0L engines	V-belt routing	129
Infiniti		
2.0L engines	V-belt routing	130
3.0L (VG30DE) engines	V-belt routing	131
3.0L (VQ30DE) engines	V-belt routing	132
4.1L engines	V-belt routing	133
4.5L engines	V-belt routing	134
Isuzu		
2.2L (except Hombre) and 2.6L engines	V-belt routing	122
2.2L Hombre engines		
Models with A/C	Serpentine belt routing	135
Models without A/C	Serpentine belt routing	136
3.2L Trooper engines	V-belt routing	137
3.5L Trooper engines	Serpentine belt routing	138
3.2L Amigo and Rodeo engines		
1995-97 Models	V-belt routing	139
1998-99 Models	Serpentine belt routing	140
Jaguar		
4.0L non-supercharged engines	Serpentine belt routing	141
4.0L supercharged engines	Serpentine belt routing	142
6.0L engines	Serpentine belt routing	143

91262C05

ACCESSORY DRIVE BELT ROUTING INDEX

MANUFACTURER

Model and Engine	Description	Figure
Jeep		
2.5L Cherokee engines		
1995 Models with A/C	Serpentine belt routing	144
1995 Models without A/C	Serpentine belt routing	145
1996-99 Models with A/C	Serpentine belt routing	146
1996-99 Models without A/C	Serpentine belt routing	147
2.5L Wrangler engines		
1997-99 Models with A/C	Serpentine belt routing	148
1997-99 Models without A/C	Serpentine belt routing	149
4.0L Cherokee (right-hand drive) engines		
1995 Models	Serpentine belt routing	152
1996-99 Models with A/C	Serpentine belt routing	150
1996-99 Models without A/C	Serpentine belt routing	151
4.0L Cherokee engines		
1995 Models with A/C	Serpentine belt routing	153
1995 Models without A/C	Serpentine belt routing	154
1996-99 Models with A/C	Serpentine belt routing	155
1996-99 Models without A/C	Serpentine belt routing	156
4.0L Grand Cherokee engines		
1995 Models	Serpentine belt routing	157
1996-99 Models	Serpentine belt routing	158
4.0L Wrangler engines		
1995 Models with A/C	Serpentine belt routing	159
1995 Models without A/C	Serpentine belt routing	161
1997-99 Models with A/C	Serpentine belt routing	160
1997-99 Models without A/C	Serpentine belt routing	162
5.2L Grand Cherokee engines		
1995 Models	Serpentine belt routing	163
1996-99 Models	Serpentine belt routing	164
KIA		
1.6L and 1.8L engines	V-belt routing	165
Land Rover		
3.9L engines		
Models with A/C	Serpentine belt routing	166
Models without A/C	Serpentine belt routing	170
4.0L engines		
Models with A/C	Serpentine belt routing	166
Models without A/C	Serpentine belt routing	167
4.6L engines		
Models with A/C	Serpentine belt routing	166
Models without A/C	Serpentine belt routing	167

ACCESSORY DRIVE BELT ROUTING INDEX

MANUFACTURER

Model and Engine	Description	Figure
Lexus Cars		
3.0L (2JZ-GE) engines	Serpentine belt routing	168
3.0L (1MZ-FE) engines	V-belt routing	169
4.0L engines	Serpentine belt routing	170
Lexus LX450		
4.5L engines	V-belt routing	171
Mazda Cars		
4-cylinder engines	Serpentine belt routing	172
2.3L (KJ) engines	Serpentine belt routing	173
V-6 engines except 2.3L (KJ) engines	Serpentine belt routing	174
Mazda Trucks		
2.3L B-series truck engines	Serpentine belt routing	175
3.0L B-series truck engines	Serpentine belt routing	176
3.0L MPV engines	Alternator belt routing	177
3.0L MPV engines	Power steering belt routing	178
3.0L MPV engines	A/C compressor belt routing	179
4.0L B-series truck engines	Serpentine belt routing	180
Mercedes-Benz Cars		
Except 2.2L, 2.3L and Diesel engines	Serpentine belt routing	37
2.2L, 2.3L (non-supercharged) and Diesel engines	Serpentine belt routing	38
2.3L supercharged engines	Serpentine belt routing	39
Mercedes-Benz Truck		
3.2L engines	Serpentine belt routing	74
Mercury Trucks		
3.0L engines	V-belt routing	181
5.0L engines	Serpentine belt routing	182
Mitsubishi Cars		
1.5L, 2.0L (turbo) and 2.4L engines	V-belt routing	183
1.8L and 2.0L (non-turbo) engines	V-belt routing	184
3.0L engines	V-belt routing	185
3.5L engines	V-belt routing	186
Mitsubishi Trucks		
2.4L engines	Serpentine belt routing	187
3.0L engines	Serpentine belt routing	188
3.5L engines	Serpentine belt routing	189
Nissan Cars		
1.6L engines	V-belt routing	190
2.0L engines	V-belt routing	191
2.4L (except 240 SX) engines	V-belt routing	192
240 SX 2.4L engines	V-belt routing	193
3.0L engines	V-belt routing	194

91262C07

ACCESSORY DRIVE BELT ROUTING INDEX

MANUFACTURER

Model and Engine	Description	Figure
Nissan Trucks		
2.4L (KA24E) engines	V-belt routing	195
3.0L (except Quest) engines	V-belt routing	196
3.0L Quest engines	V-belt routing	198
3.3L (VG33E) engines	V-belt routing	197
Porsche		
3.6L engines	V-belt routing	199
4.5L engines	V-belt routing	200
Saab		
4-cylinder engines	Serpentine belt routing	201
6-cylinder engines	Serpentine belt routing	202
Saturn		
All engines	Serpentine belt routing	203
Subaru		
4-cylinder engines	V-belt routing	204
6-cylinder engines	V-belt routing	205
Suzuki		
All engines	V-belt routing	206
Toyota Cars		
1.5L (5EFE) engines	V-belt routing	207
Toyota Cars (cont.)		
1.8L (1ZZFE) engines	Serpentine belt routing	208
1.8L (7AFE) and 1.6L (4AFE) engines	V-belt routing	209
2.2L (5SFE) engines	V-belt routing	210
3.0L (2JZGTE and 2JZGE) engines	Serpentine belt routing	211
3.0L (1MZFE) engines	V-belt routing	212
Toyota Trucks		
2.0L (3SFE) engines	V-belt routing	213
2.4L (2RZFE) engines	V-belt routing	214
2.4L (2TZFZE) engines	V-belt routing	215
2.7L engines	V-belt routing	216
3.4L engines	V-belt routing	217
4.5L engines	V-belt routing	218
Volkswagen		
2.8L engines	V-belt routing	219
all engines (except 2.8L) without A/C	V-belt routing	220
all engines (except 2.8L) with A/C	V-belt routing	221
Volvo		
All engines	Serpentine belt routing	222

91262C08

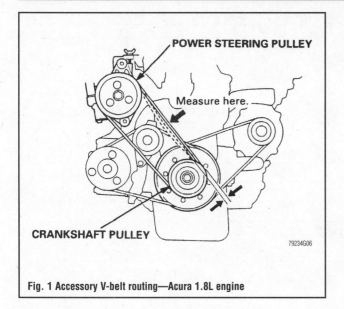

Fig. 1 Accessory V-belt routing—Acura 1.8L engine

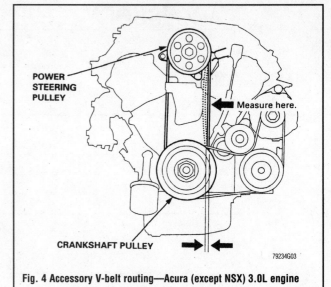

Fig. 4 Accessory V-belt routing—Acura (except NSX) 3.0L engine

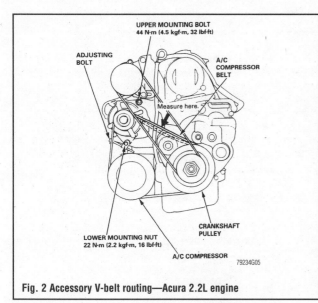

Fig. 2 Accessory V-belt routing—Acura 2.2L engine

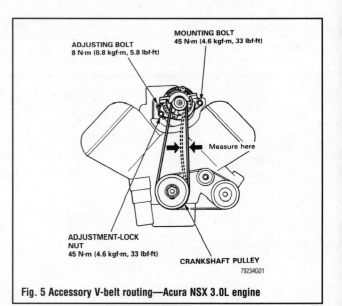

Fig. 5 Accessory V-belt routing—Acura NSX 3.0L engine

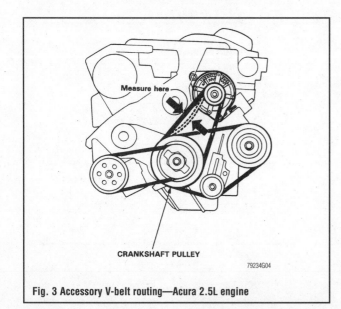

Fig. 3 Accessory V-belt routing—Acura 2.5L engine

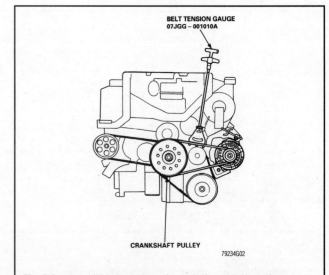

Fig. 6 Accessory V-belt routing—Acura 3.2L and 3.5L engines

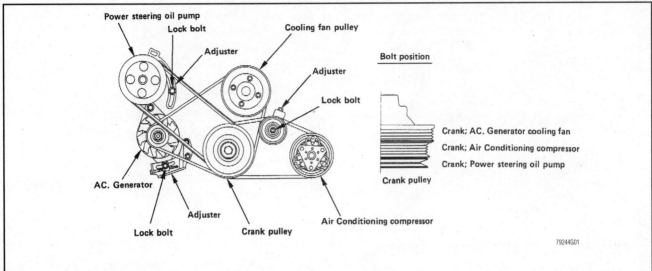

Fig. 7 Accessory V-belt routing—Acura 3.2L engines

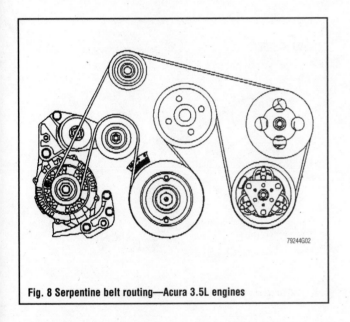

Fig. 8 Serpentine belt routing—Acura 3.5L engines

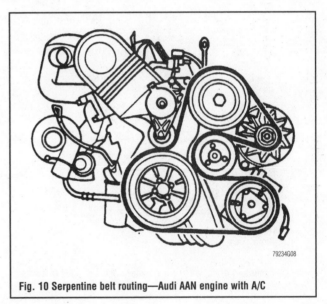

Fig. 10 Serpentine belt routing—Audi AAN engine with A/C

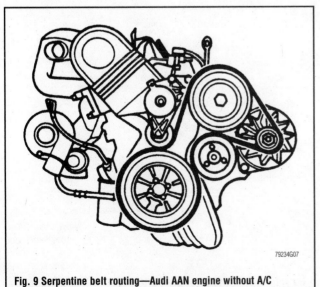

Fig. 9 Serpentine belt routing—Audi AAN engine without A/C

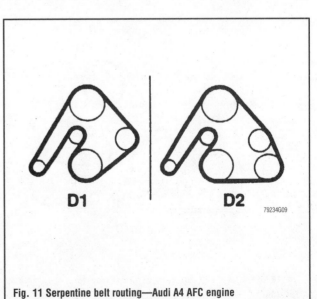

Fig. 11 Serpentine belt routing—Audi A4 AFC engine

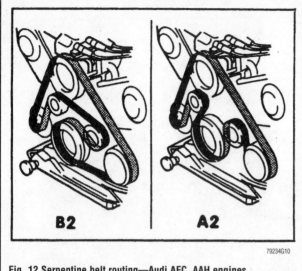

Fig. 12 Serpentine belt routing—Audi AFC, AAH engines

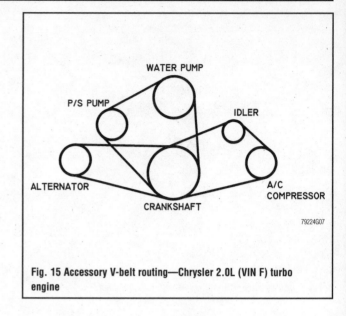

Fig. 15 Accessory V-belt routing—Chrysler 2.0L (VIN F) turbo engine

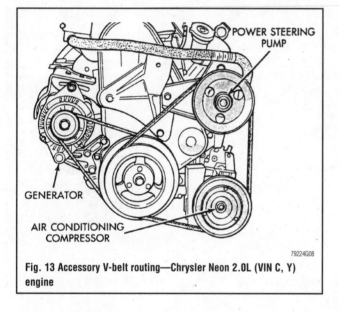

Fig. 13 Accessory V-belt routing—Chrysler Neon 2.0L (VIN C, Y) engine

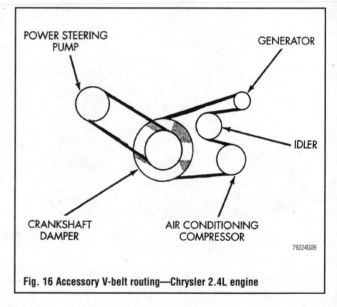

Fig. 16 Accessory V-belt routing—Chrysler 2.4L engine

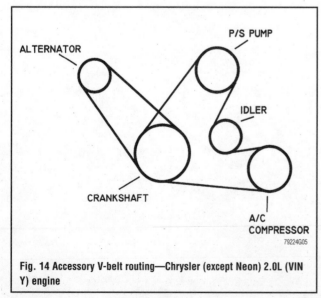

Fig. 14 Accessory V-belt routing—Chrysler (except Neon) 2.0L (VIN Y) engine

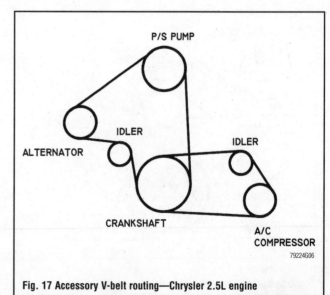

Fig. 17 Accessory V-belt routing—Chrysler 2.5L engine

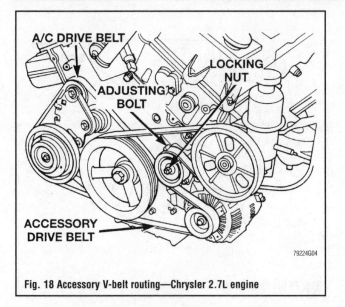

Fig. 18 Accessory V-belt routing—Chrysler 2.7L engine

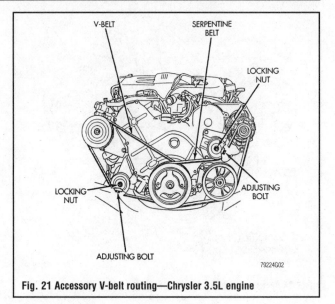

Fig. 21 Accessory V-belt routing—Chrysler 3.5L engine

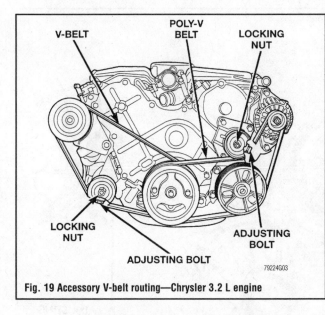

Fig. 19 Accessory V-belt routing—Chrysler 3.2 L engine

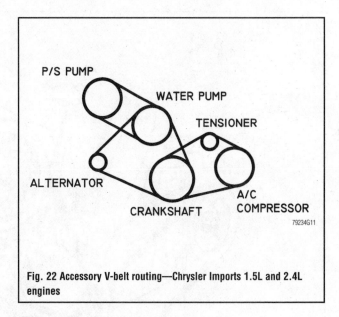

Fig. 22 Accessory V-belt routing—Chrysler Imports 1.5L and 2.4L engines

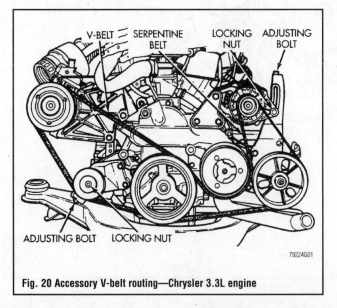

Fig. 20 Accessory V-belt routing—Chrysler 3.3L engine

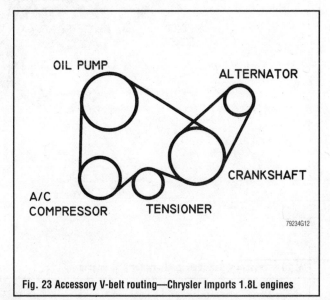

Fig. 23 Accessory V-belt routing—Chrysler Imports 1.8L engines

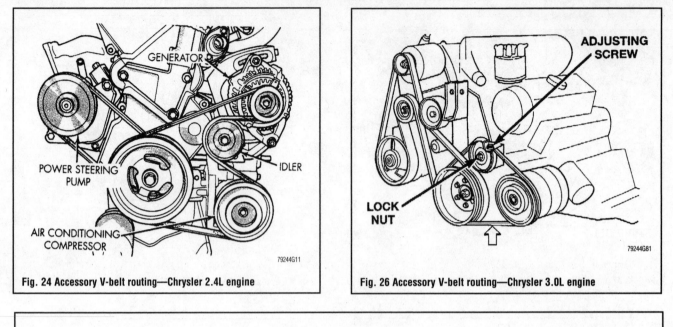

Fig. 24 Accessory V-belt routing—Chrysler 2.4L engine

Fig. 26 Accessory V-belt routing—Chrysler 3.0L engine

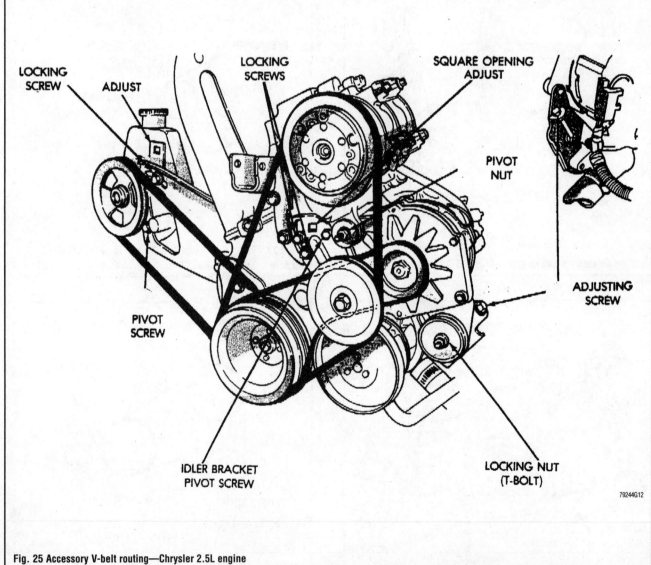

Fig. 25 Accessory V-belt routing—Chrysler 2.5L engine

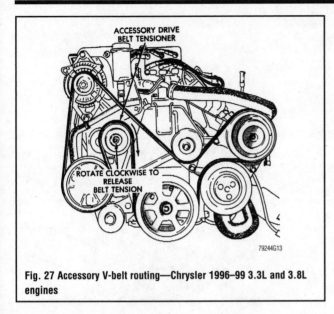

Fig. 27 Accessory V-belt routing—Chrysler 1996–99 3.3L and 3.8L engines

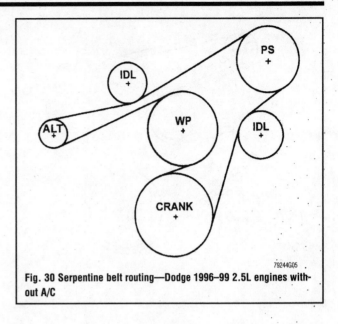

Fig. 30 Serpentine belt routing—Dodge 1996–99 2.5L engines without A/C

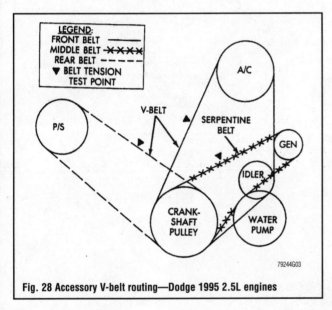

Fig. 28 Accessory V-belt routing—Dodge 1995 2.5L engines

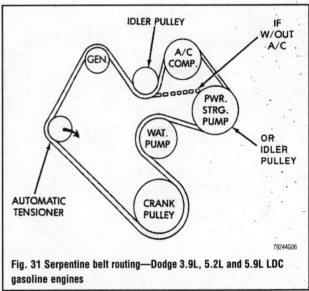

Fig. 31 Serpentine belt routing—Dodge 3.9L, 5.2L and 5.9L LDC gasoline engines

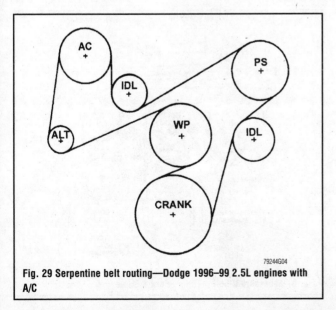

Fig. 29 Serpentine belt routing—Dodge 1996–99 2.5L engines with A/C

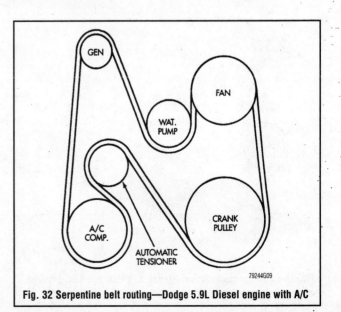

Fig. 32 Serpentine belt routing—Dodge 5.9L Diesel engine with A/C

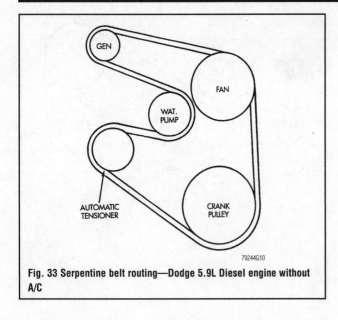

Fig. 33 Serpentine belt routing—Dodge 5.9L Diesel engine without A/C

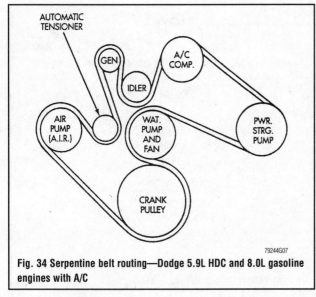

Fig. 34 Serpentine belt routing—Dodge 5.9L HDC and 8.0L gasoline engines with A/C

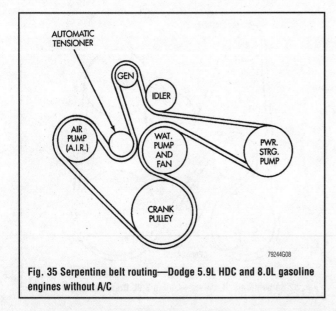

Fig. 35 Serpentine belt routing—Dodge 5.9L HDC and 8.0L gasoline engines without A/C

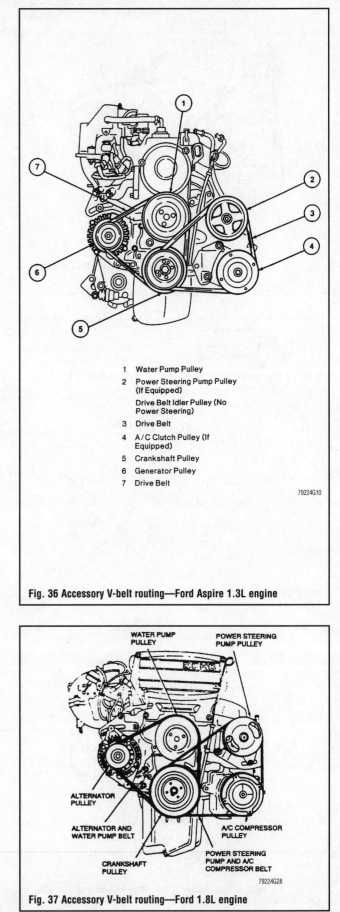

1 Water Pump Pulley
2 Power Steering Pump Pulley (If Equipped)
   Drive Belt Idler Pulley (No Power Steering)
3 Drive Belt
4 A/C Clutch Pulley (If Equipped)
5 Crankshaft Pulley
6 Generator Pulley
7 Drive Belt

Fig. 36 Accessory V-belt routing—Ford Aspire 1.3L engine

Fig. 37 Accessory V-belt routing—Ford 1.8L engine

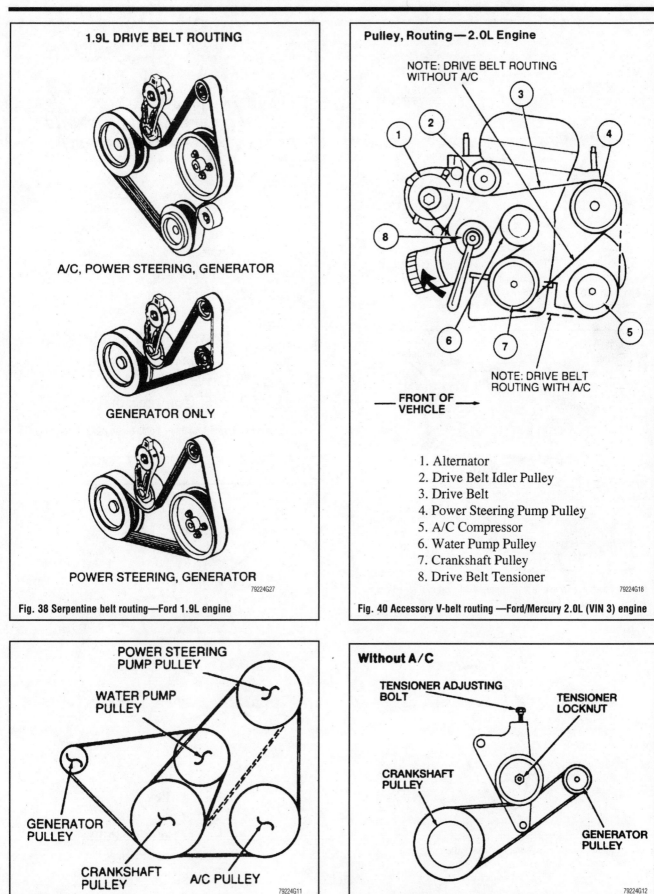

1.9L DRIVE BELT ROUTING

A/C, POWER STEERING, GENERATOR

GENERATOR ONLY

POWER STEERING, GENERATOR

79224G27

Fig. 38 Serpentine belt routing—Ford 1.9L engine

Pulley, Routing—2.0L Engine

NOTE: DRIVE BELT ROUTING WITHOUT A/C

NOTE: DRIVE BELT ROUTING WITH A/C

FRONT OF VEHICLE

1. Alternator
2. Drive Belt Idler Pulley
3. Drive Belt
4. Power Steering Pump Pulley
5. A/C Compressor
6. Water Pump Pulley
7. Crankshaft Pulley
8. Drive Belt Tensioner

79224G18

Fig. 40 Accessory V-belt routing —Ford/Mercury 2.0L (VIN 3) engine

POWER STEERING PUMP PULLEY

WATER PUMP PULLEY

GENERATOR PULLEY

CRANKSHAFT PULLEY

A/C PULLEY

79224G11

Fig. 39 Accessory V-belt routing—Ford Probe 2.0L engine

Without A/C

TENSIONER ADJUSTING BOLT

TENSIONER LOCKNUT

CRANKSHAFT PULLEY

GENERATOR PULLEY

79224G12

Fig. 41 Alternator drive belt routing—Ford Probe 2.5L (VIN B) engine without A/C

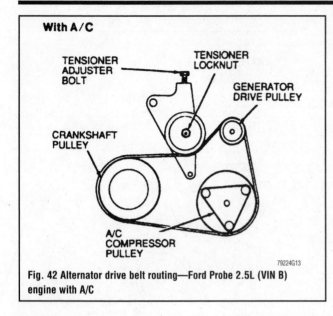

Fig. 42 Alternator drive belt routing—Ford Probe 2.5L (VIN B) engine with A/C

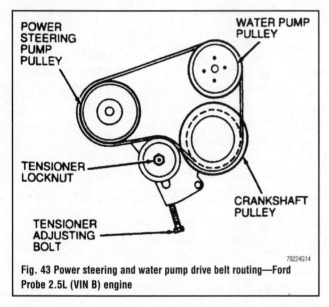

Fig. 43 Power steering and water pump drive belt routing—Ford Probe 2.5L (VIN B) engine

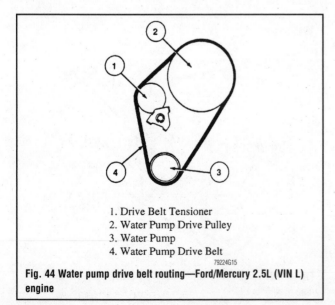

1. Drive Belt Tensioner
2. Water Pump Drive Pulley
3. Water Pump
4. Water Pump Drive Belt

Fig. 44 Water pump drive belt routing—Ford/Mercury 2.5L (VIN L) engine

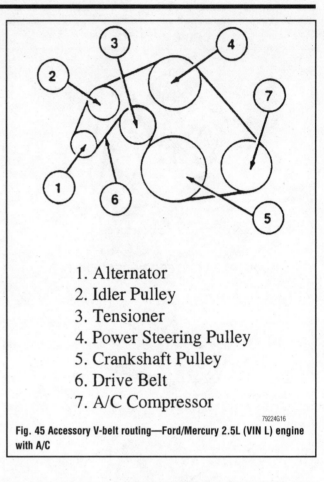

1. Alternator
2. Idler Pulley
3. Tensioner
4. Power Steering Pulley
5. Crankshaft Pulley
6. Drive Belt
7. A/C Compressor

Fig. 45 Accessory V-belt routing—Ford/Mercury 2.5L (VIN L) engine with A/C

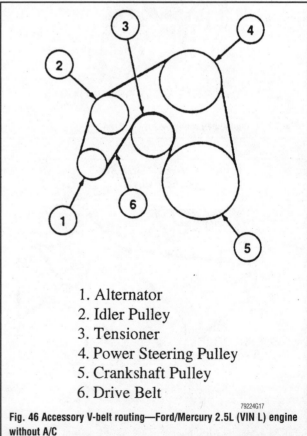

1. Alternator
2. Idler Pulley
3. Tensioner
4. Power Steering Pulley
5. Crankshaft Pulley
6. Drive Belt

Fig. 46 Accessory V-belt routing—Ford/Mercury 2.5L (VIN L) engine without A/C

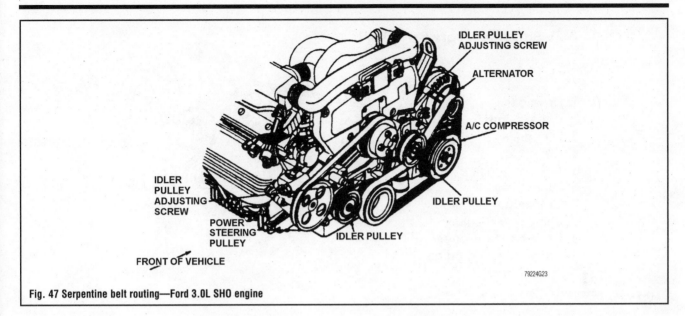

IDLER PULLEY ADJUSTING SCREW

ALTERNATOR

A/C COMPRESSOR

IDLER PULLEY

POWER STEERING PULLEY

IDLER PULLEY

IDLER PULLEY ADJUSTING SCREW

FRONT OF VEHICLE

79224G23

Fig. 47 Serpentine belt routing—Ford 3.0L SHO engine

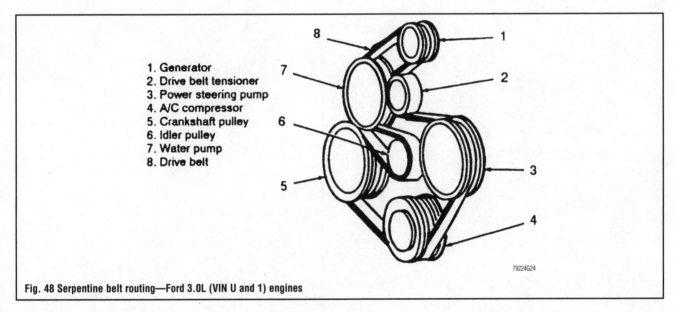

1. Generator
2. Drive belt tensioner
3. Power steering pump
4. A/C compressor
5. Crankshaft pulley
6. Idler pulley
7. Water pump
8. Drive belt

79224G24

Fig. 48 Serpentine belt routing—Ford 3.0L (VIN U and 1) engines

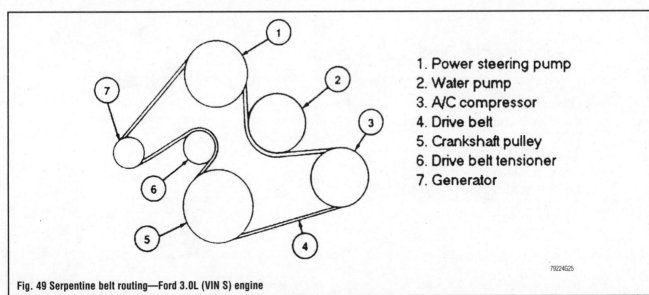

1. Power steering pump
2. Water pump
3. A/C compressor
4. Drive belt
5. Crankshaft pulley
6. Drive belt tensioner
7. Generator

79224G25

Fig. 49 Serpentine belt routing—Ford 3.0L (VIN S) engine

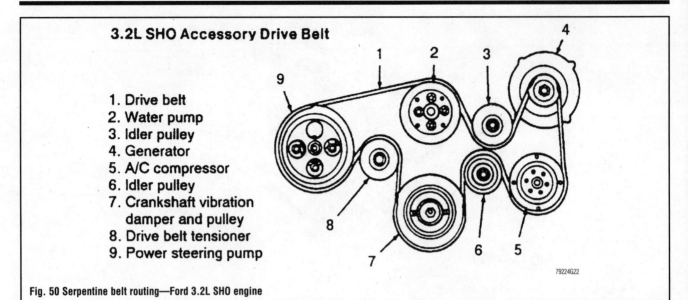

3.2L SHO Accessory Drive Belt

1. Drive belt
2. Water pump
3. Idler pulley
4. Generator
5. A/C compressor
6. Idler pulley
7. Crankshaft vibration damper and pulley
8. Drive belt tensioner
9. Power steering pump

Fig. 50 Serpentine belt routing—Ford 3.2L SHO engine

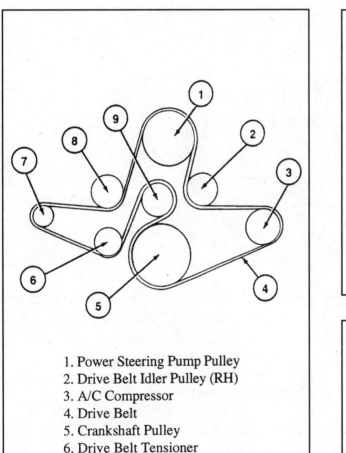

1. Power Steering Pump Pulley
2. Drive Belt Idler Pulley (RH)
3. A/C Compressor
4. Drive Belt
5. Crankshaft Pulley
6. Drive Belt Tensioner
7. Alternator
8. Drive Belt Idler Pulley (LH)
9. Drive Belt Idler Pulley (Center)

Fig. 51 Serpentine belt routing—Ford 3.4L SHO engine

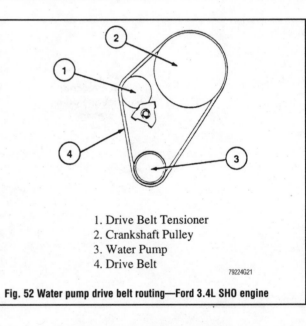

1. Drive Belt Tensioner
2. Crankshaft Pulley
3. Water Pump
4. Drive Belt

Fig. 52 Water pump drive belt routing—Ford 3.4L SHO engine

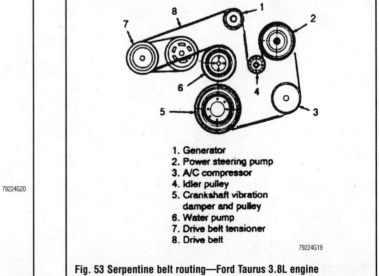

1. Generator
2. Power steering pump
3. A/C compressor
4. Idler pulley
5. Crankshaft vibration damper and pulley
6. Water pump
7. Drive belt tensioner
8. Drive belt

Fig. 53 Serpentine belt routing—Ford Taurus 3.8L engine

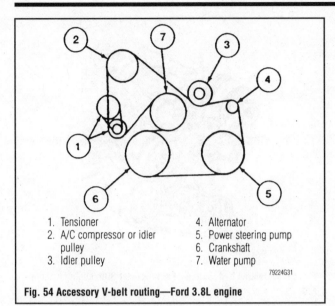

1. Tensioner
2. A/C compressor or idler pulley
3. Idler pulley
4. Alternator
5. Power steering pump
6. Crankshaft
7. Water pump

Fig. 54 Accessory V-belt routing—Ford 3.8L engine

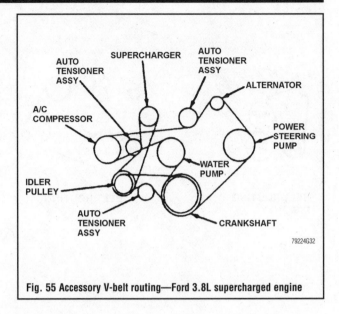

Fig. 55 Accessory V-belt routing—Ford 3.8L supercharged engine

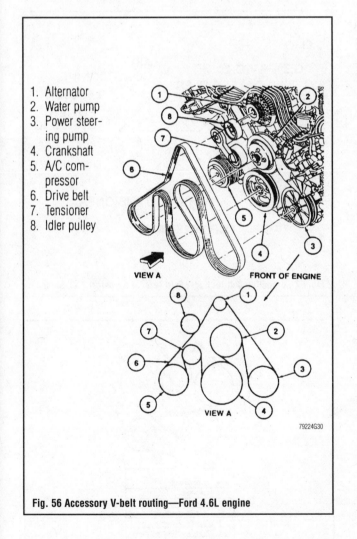

1. Alternator
2. Water pump
3. Power steering pump
4. Crankshaft
5. A/C compressor
6. Drive belt
7. Tensioner
8. Idler pulley

Fig. 56 Accessory V-belt routing—Ford 4.6L engine

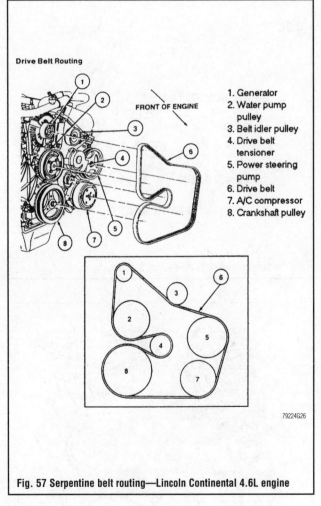

Drive Belt Routing

1. Generator
2. Water pump pulley
3. Belt idler pulley
4. Drive belt tensioner
5. Power steering pump
6. Drive belt
7. A/C compressor
8. Crankshaft pulley

Fig. 57 Serpentine belt routing—Lincoln Continental 4.6L engine

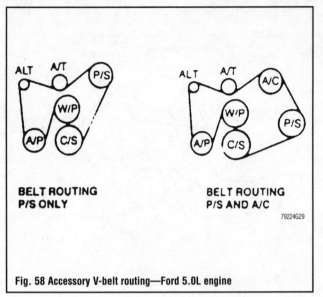

Fig. 58 Accessory V-belt routing—Ford 5.0L engine

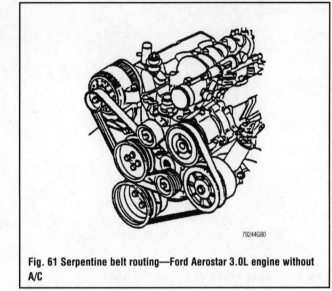

Fig. 61 Serpentine belt routing—Ford Aerostar 3.0L engine without A/C

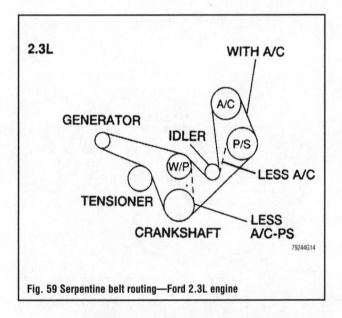

Fig. 59 Serpentine belt routing—Ford 2.3L engine

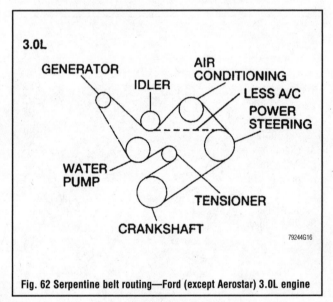

Fig. 62 Serpentine belt routing—Ford (except Aerostar) 3.0L engine

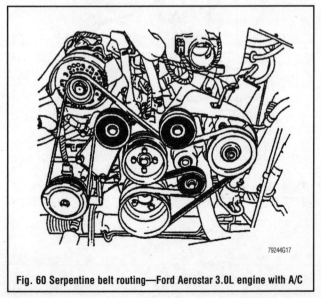

Fig. 60 Serpentine belt routing—Ford Aerostar 3.0L engine with A/C

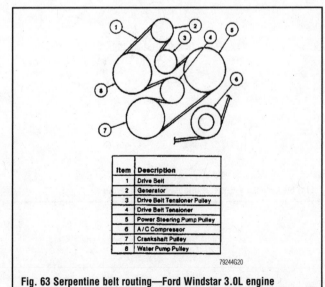

Item	Description
1	Drive Belt
2	Generator
3	Drive Belt Tensioner Pulley
4	Drive Belt Tensioner
5	Power Steering Pump Pulley
6	A/C Compressor
7	Crankshaft Pulley
8	Water Pump Pulley

Fig. 63 Serpentine belt routing—Ford Windstar 3.0L engine

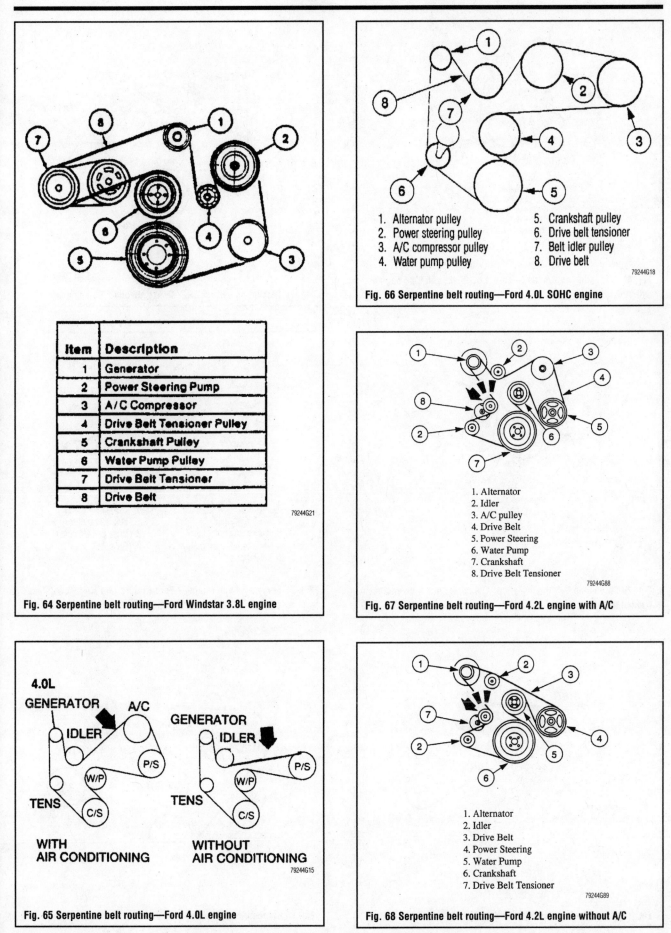

Item	Description
1	Generator
2	Power Steering Pump
3	A/C Compressor
4	Drive Belt Tensioner Pulley
5	Crankshaft Pulley
6	Water Pump Pulley
7	Drive Belt Tensioner
8	Drive Belt

79244G21

Fig. 64 Serpentine belt routing—Ford Windstar 3.8L engine

79244G18

1. Alternator pulley
2. Power steering pulley
3. A/C compressor pulley
4. Water pump pulley
5. Crankshaft pulley
6. Drive belt tensioner
7. Belt idler pulley
8. Drive belt

Fig. 66 Serpentine belt routing—Ford 4.0L SOHC engine

1. Alternator
2. Idler
3. A/C pulley
4. Drive Belt
5. Power Steering
6. Water Pump
7. Crankshaft
8. Drive Belt Tensioner

79244G88

Fig. 67 Serpentine belt routing—Ford 4.2L engine with A/C

4.0L

GENERATOR   A/C
  IDLER
        P/S
    W/P
TENS
      C/S

GENERATOR
  IDLER
        P/S
    W/P
TENS
      C/S

WITH AIR CONDITIONING   **WITHOUT AIR CONDITIONING**

79244G15

Fig. 65 Serpentine belt routing—Ford 4.0L engine

1. Alternator
2. Idler
3. Drive Belt
4. Power Steering
5. Water Pump
6. Crankshaft
7. Drive Belt Tensioner

79244G89

Fig. 68 Serpentine belt routing—Ford 4.2L engine without A/C

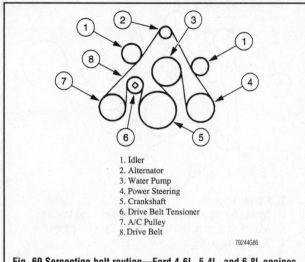

1. Idler
2. Alternator
3. Water Pump
4. Power Steering
5. Crankshaft
6. Drive Belt Tensioner
7. A/C Pulley
8. Drive Belt

79244G86

Fig. 69 Serpentine belt routing—Ford 4.6L, 5.4L, and 6.8L engines with A/C

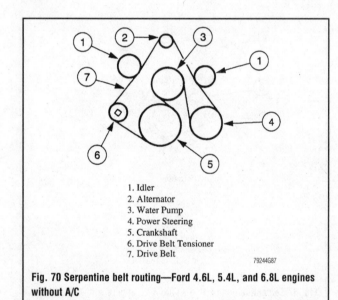

1. Idler
2. Alternator
3. Water Pump
4. Power Steering
5. Crankshaft
6. Drive Belt Tensioner
7. Drive Belt

79244G87

Fig. 70 Serpentine belt routing—Ford 4.6L, 5.4L, and 6.8L engines without A/C

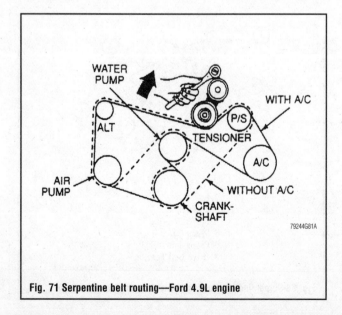

Fig. 71 Serpentine belt routing—Ford 4.9L engine

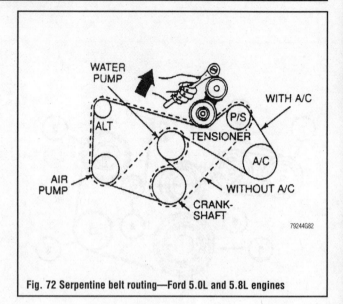

Fig. 72 Serpentine belt routing—Ford 5.0L and 5.8L engines

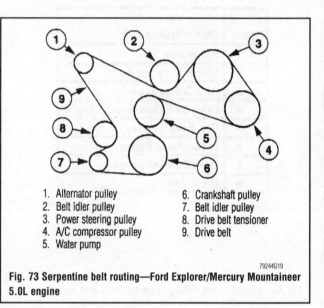

1. Alternator pulley
2. Belt idler pulley
3. Power steering pulley
4. A/C compressor pulley
5. Water pump
6. Crankshaft pulley
7. Belt idler pulley
8. Drive belt tensioner
9. Drive belt

79244G19

Fig. 73 Serpentine belt routing—Ford Explorer/Mercury Mountaineer 5.0L engine

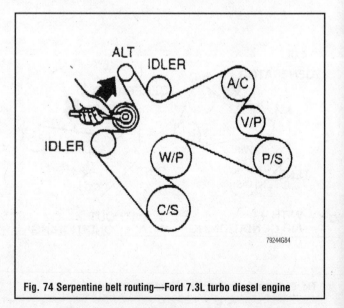

Fig. 74 Serpentine belt routing—Ford 7.3L turbo diesel engine

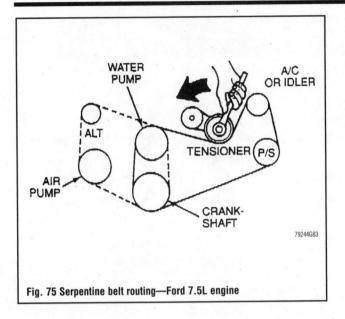

Fig. 75 Serpentine belt routing—Ford 7.5L engine

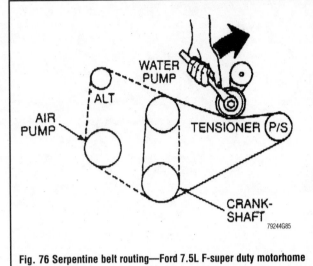

Fig. 76 Serpentine belt routing—Ford 7.5L F-super duty motorhome engine

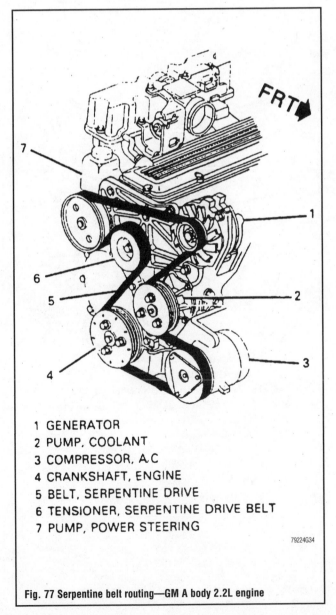

1 GENERATOR
2 PUMP, COOLANT
3 COMPRESSOR, A.C
4 CRANKSHAFT, ENGINE
5 BELT, SERPENTINE DRIVE
6 TENSIONER, SERPENTINE DRIVE BELT
7 PUMP, POWER STEERING

Fig. 77 Serpentine belt routing—GM A body 2.2L engine

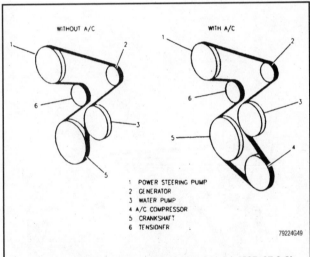

1 POWER STEERING PUMP
2 GENERATOR
3 WATER PUMP
4 A/C COMPRESSOR
5 CRANKSHAFT
6 TENSIONER

Fig. 78 Serpentine belt routing—GM (except A body) 1995–97 2.2L engine

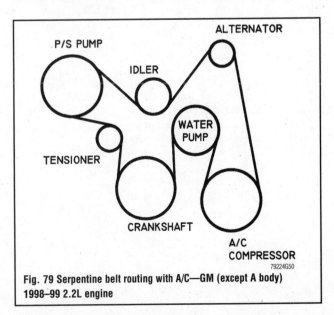

Fig. 79 Serpentine belt routing with A/C—GM (except A body) 1998–99 2.2L engine

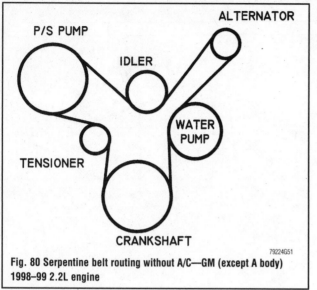

Fig. 80 Serpentine belt routing without A/C—GM (except A body) 1998–99 2.2L engine

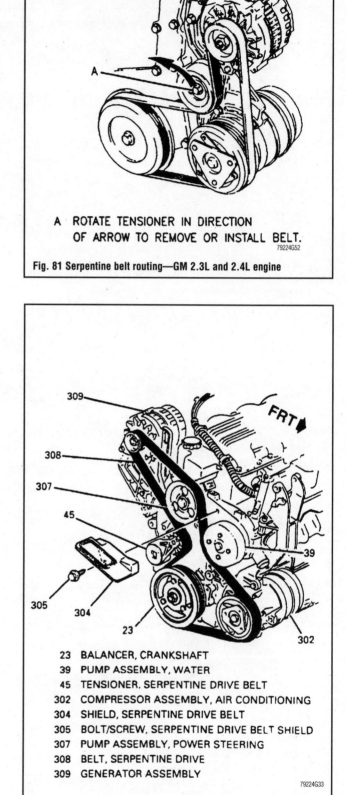

A ROTATE TENSIONER IN DIRECTION OF ARROW TO REMOVE OR INSTALL BELT.

Fig. 81 Serpentine belt routing—GM 2.3L and 2.4L engine

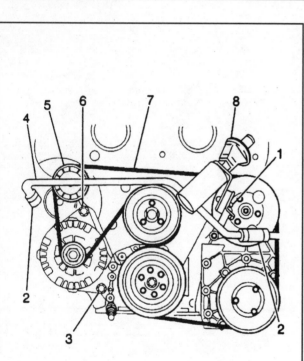

Legend

(1) AIR Injection Crossover Pipe Bushing Nut
(2) AIR Injection Rubber Hose Connection
(3) Lower Generator Bolt
(4) AIR Injection Crossover Pipe
(5) Serpentine Drive Belt Tensioner
(6) AIR Injection Crossover Pipe Support (Generator) Bracket Nut
(7) Serpentine Drive Belt
(8) AIR Injection Diverter Valve

Fig. 82 Serpentine belt routing—GM 3.0L engine

23 BALANCER, CRANKSHAFT
39 PUMP ASSEMBLY, WATER
45 TENSIONER, SERPENTINE DRIVE BELT
302 COMPRESSOR ASSEMBLY, AIR CONDITIONING
304 SHIELD, SERPENTINE DRIVE BELT
305 BOLT/SCREW, SERPENTINE DRIVE BELT SHIELD
307 PUMP ASSEMBLY, POWER STEERING
308 BELT, SERPENTINE DRIVE
309 GENERATOR ASSEMBLY

Fig. 83 Serpentine belt routing—GM A body 3.1L engine

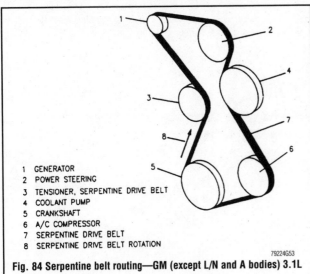

1 GENERATOR
2 POWER STEERING
3 TENSIONER, SERPENTINE DRIVE BELT
4 COOLANT PUMP
5 CRANKSHAFT
6 A/C COMPRESSOR
7 SERPENTINE DRIVE BELT
8 SERPENTINE DRIVE BELT ROTATION

79224G53

Fig. 84 Serpentine belt routing—GM (except L/N and A bodies) 3.1L engine

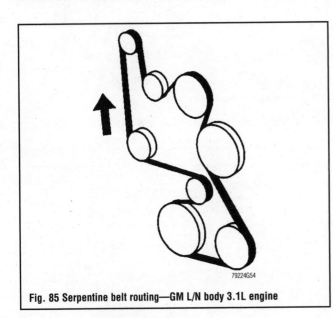

79224G54

Fig. 85 Serpentine belt routing—GM L/N body 3.1L engine

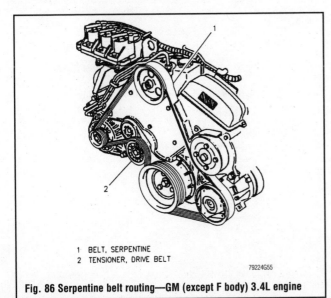

1 BELT, SERPENTINE
2 TENSIONER, DRIVE BELT

79224G55

Fig. 86 Serpentine belt routing—GM (except F body) 3.4L engine

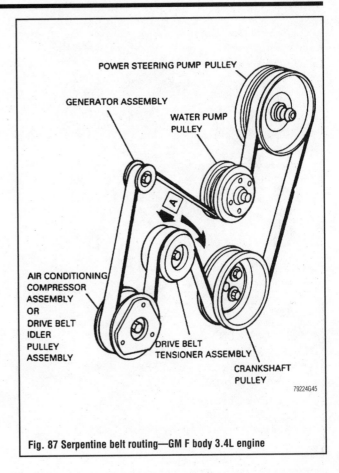

POWER STEERING PUMP PULLEY

GENERATOR ASSEMBLY

WATER PUMP PULLEY

AIR CONDITIONING
COMPRESSOR
ASSEMBLY
OR
DRIVE BELT
IDLER
PULLEY
ASSEMBLY

DRIVE BELT
TENSIONER ASSEMBLY

CRANKSHAFT
PULLEY

79224G45

Fig. 87 Serpentine belt routing—GM F body 3.4L engine

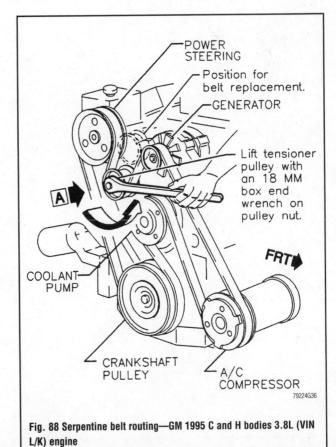

POWER STEERING

Position for belt replacement.

GENERATOR

Lift tensioner pulley with an 18 MM box end wrench on pulley nut.

FRT

COOLANT PUMP

CRANKSHAFT PULLEY

A/C COMPRESSOR

79224G36

Fig. 88 Serpentine belt routing—GM 1995 C and H bodies 3.8L (VIN L/K) engine

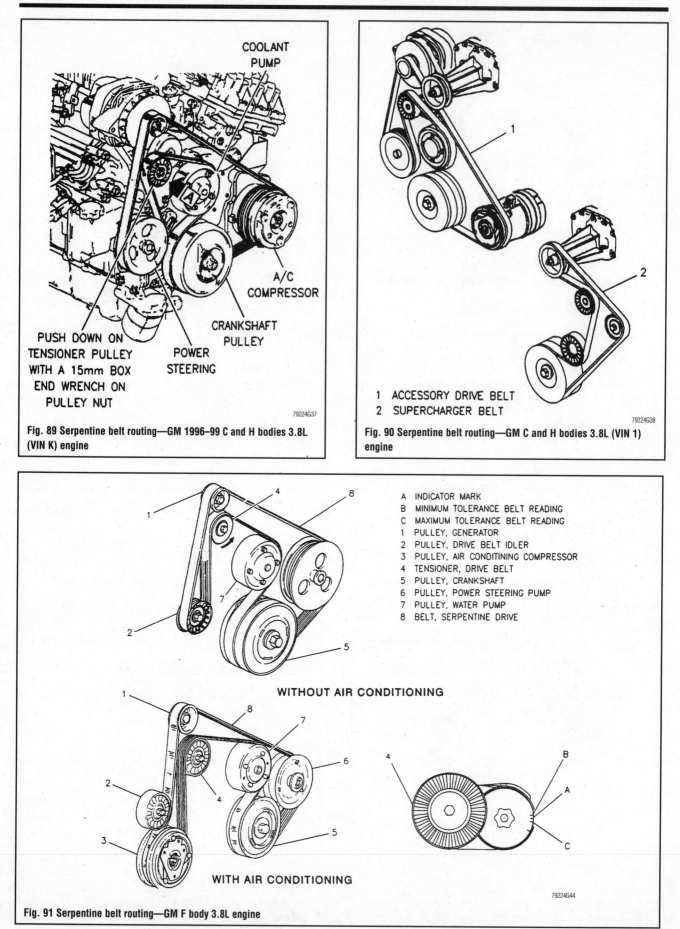

COOLANT PUMP

A/C COMPRESSOR

CRANKSHAFT PULLEY

POWER STEERING

PUSH DOWN ON TENSIONER PULLEY WITH A 15mm BOX END WRENCH ON PULLEY NUT

79224G37

Fig. 89 Serpentine belt routing—GM 1996–99 C and H bodies 3.8L (VIN K) engine

1 ACCESSORY DRIVE BELT
2 SUPERCHARGER BELT

79224G38

Fig. 90 Serpentine belt routing—GM C and H bodies 3.8L (VIN 1) engine

A INDICATOR MARK
B MINIMUM TOLERANCE BELT READING
C MAXIMUM TOLERANCE BELT READING
1 PULLEY, GENERATOR
2 PULLEY, DRIVE BELT IDLER
3 PULLEY, AIR CONDITINING COMPRESSOR
4 TENSIONER, DRIVE BELT
5 PULLEY, CRANKSHAFT
6 PULLEY, POWER STEERING PUMP
7 PULLEY, WATER PUMP
8 BELT, SERPENTINE DRIVE

WITHOUT AIR CONDITIONING

WITH AIR CONDITIONING

79224G44

Fig. 91 Serpentine belt routing—GM F body 3.8L engine

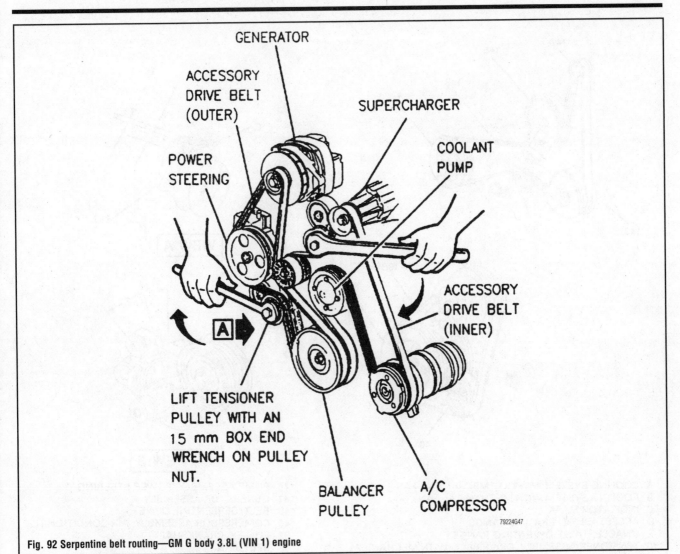

GENERATOR

ACCESSORY
DRIVE BELT
(OUTER)

SUPERCHARGER

COOLANT
PUMP

POWER
STEERING

ACCESSORY
DRIVE BELT
(INNER)

A

LIFT TENSIONER
PULLEY WITH AN
15 mm BOX END
WRENCH ON PULLEY
NUT.

BALANCER
PULLEY

A/C
COMPRESSOR

Fig. 92 Serpentine belt routing—GM G body 3.8L (VIN 1) engine

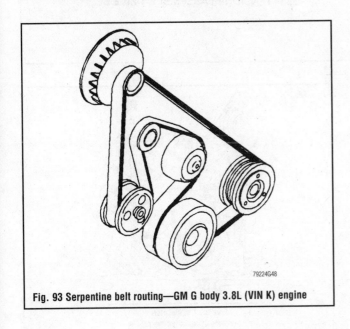

Fig. 93 Serpentine belt routing—GM G body 3.8L (VIN K) engine

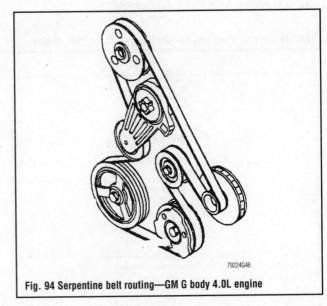

Fig. 94 Serpentine belt routing—GM G body 4.0L engine

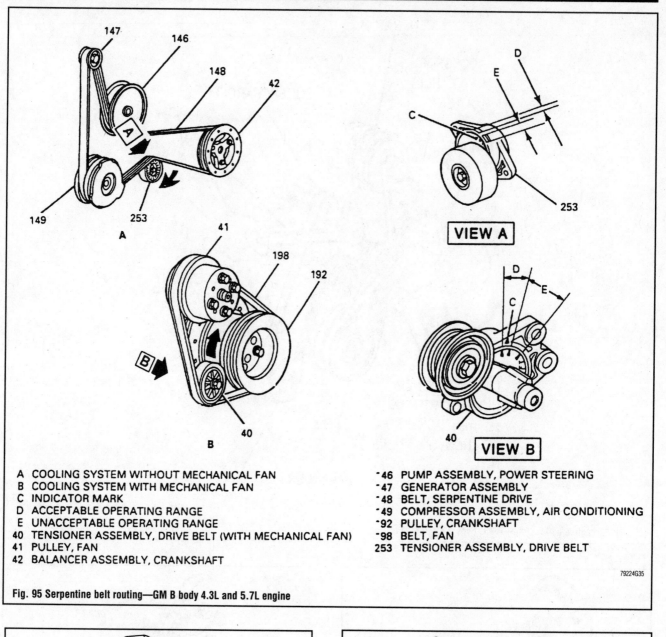

A COOLING SYSTEM WITHOUT MECHANICAL FAN
B COOLING SYSTEM WITH MECHANICAL FAN
C INDICATOR MARK
D ACCEPTABLE OPERATING RANGE
E UNACCEPTABLE OPERATING RANGE
40 TENSIONER ASSEMBLY, DRIVE BELT (WITH MECHANICAL FAN)
41 PULLEY, FAN
42 BALANCER ASSEMBLY, CRANKSHAFT

46 PUMP ASSEMBLY, POWER STEERING
47 GENERATOR ASSEMBLY
48 BELT, SERPENTINE DRIVE
49 COMPRESSOR ASSEMBLY, AIR CONDITIONING
92 PULLEY, CRANKSHAFT
98 BELT, FAN
253 TENSIONER ASSEMBLY, DRIVE BELT

79224G35

Fig. 95 Serpentine belt routing—GM B body 4.3L and 5.7L engine

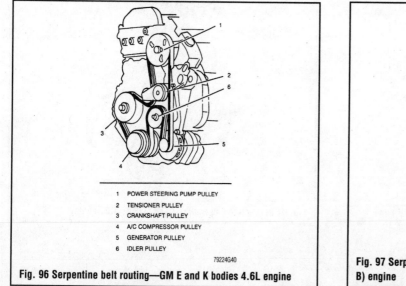

1 POWER STEERING PUMP PULLEY
2 TENSIONER PULLEY
3 CRANKSHAFT PULLEY
4 A/C COMPRESSOR PULLEY
5 GENERATOR PULLEY
6 IDLER PULLEY

79224G40

Fig. 96 Serpentine belt routing—GM E and K bodies 4.6L engine

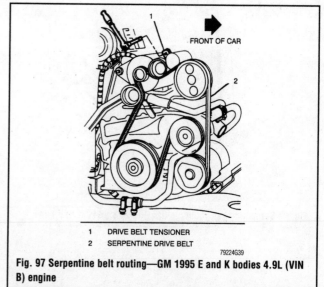

FRONT OF CAR

1 DRIVE BELT TENSIONER
2 SERPENTINE DRIVE BELT

79224G39

Fig. 97 Serpentine belt routing—GM 1995 E and K bodies 4.9L (VIN B) engine

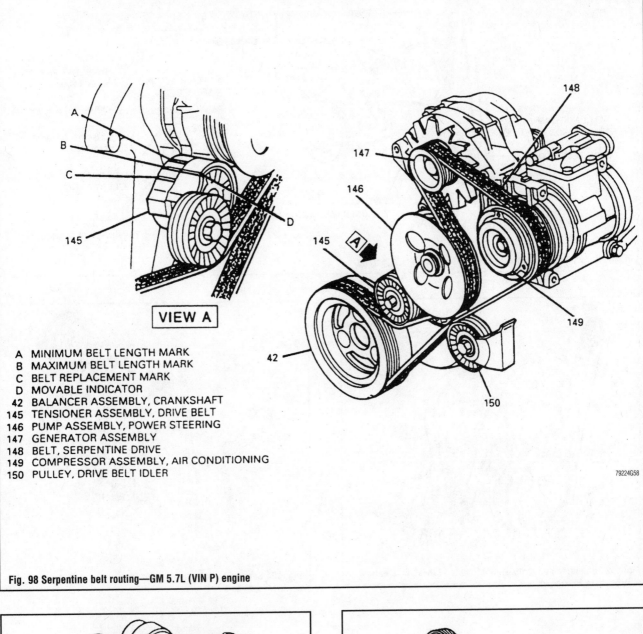

A MINIMUM BELT LENGTH MARK
B MAXIMUM BELT LENGTH MARK
C BELT REPLACEMENT MARK
D MOVABLE INDICATOR
42 BALANCER ASSEMBLY, CRANKSHAFT
145 TENSIONER ASSEMBLY, DRIVE BELT
146 PUMP ASSEMBLY, POWER STEERING
147 GENERATOR ASSEMBLY
148 BELT, SERPENTINE DRIVE
149 COMPRESSOR ASSEMBLY, AIR CONDITIONING
150 PULLEY, DRIVE BELT IDLER

VIEW A

79224G58

Fig. 98 Serpentine belt routing—GM 5.7L (VIN P) engine

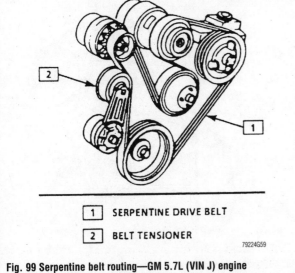

1 SERPENTINE DRIVE BELT

2 BELT TENSIONER

79224G59

Fig. 99 Serpentine belt routing—GM 5.7L (VIN J) engine

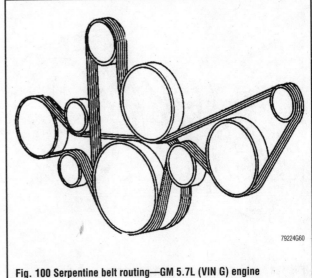

79224G60

Fig. 100 Serpentine belt routing—GM 5.7L (VIN G) engine

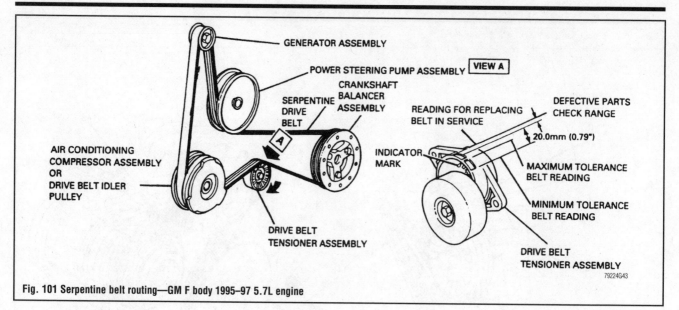

GENERATOR ASSEMBLY

POWER STEERING PUMP ASSEMBLY | VIEW A |

CRANKSHAFT BALANCER ASSEMBLY

SERPENTINE DRIVE BELTS

A

AIR CONDITIONING COMPRESSOR ASSEMBLY OR DRIVE BELT IDLER PULLEY

DRIVE BELT TENSIONER ASSEMBLY

READING FOR REPLACING BELT IN SERVICE

INDICATOR MARK

DEFECTIVE PARTS CHECK RANGE

20.0mm (0.79")

MAXIMUM TOLERANCE BELT READING

MINIMUM TOLERANCE BELT READING

DRIVE BELT TENSIONER ASSEMBLY

79224G43

Fig. 101 Serpentine belt routing—GM F body 1995–97 5.7L engine

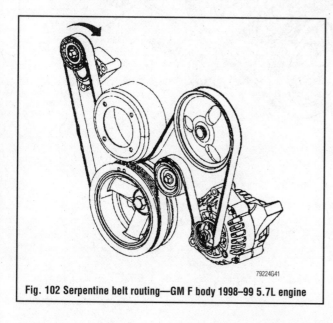

79224G41

Fig. 102 Serpentine belt routing—GM F body 1998–99 5.7L engine

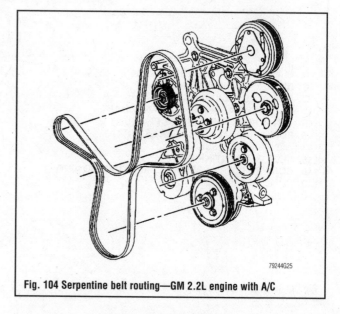

79244G25

Fig. 104 Serpentine belt routing—GM 2.2L engine with A/C

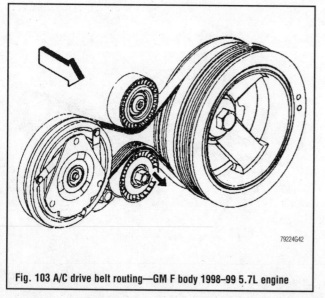

79224G42

Fig. 103 A/C drive belt routing—GM F body 1998–99 5.7L engine

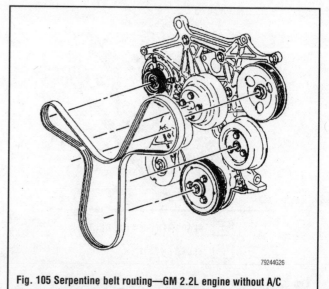

79244G26

Fig. 105 Serpentine belt routing—GM 2.2L engine without A/C

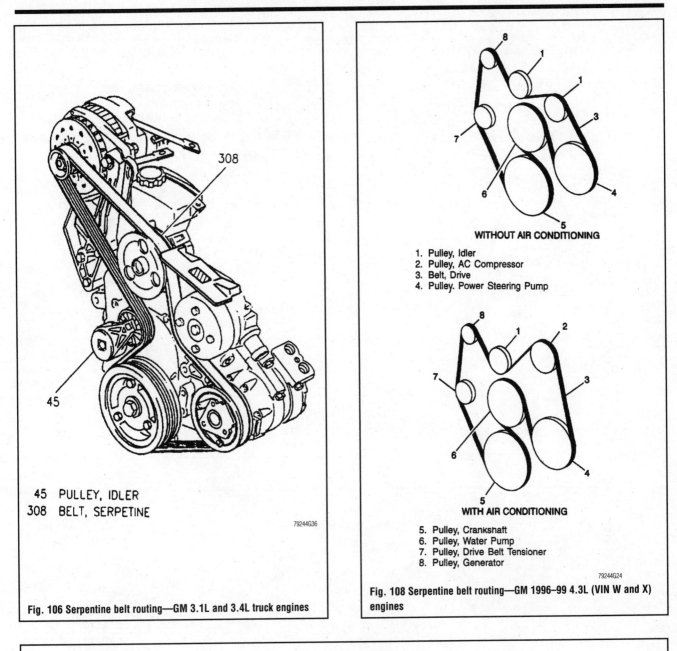

45 PULLEY, IDLER
308 BELT, SERPETINE

79244G36

Fig. 106 Serpentine belt routing—GM 3.1L and 3.4L truck engines

WITHOUT AIR CONDITIONING

1. Pulley, Idler
2. Pulley, AC Compressor
3. Belt, Drive
4. Pulley. Power Steering Pump

WITH AIR CONDITIONING

5. Pulley, Crankshaft
6. Pulley, Water Pump
7. Pulley, Drive Belt Tensioner
8. Pulley, Generator

79244G24

Fig. 108 Serpentine belt routing—GM 1996–99 4.3L (VIN W and X) engines

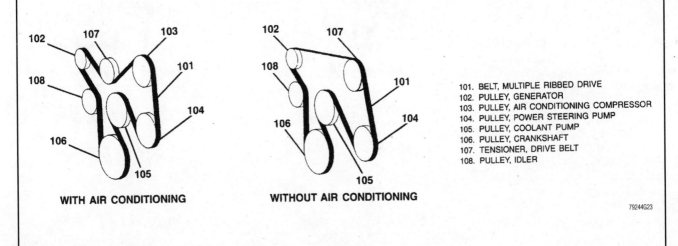

WITH AIR CONDITIONING

WITHOUT AIR CONDITIONING

101. BELT, MULTIPLE RIBBED DRIVE
102. PULLEY, GENERATOR
103. PULLEY, AIR CONDITIONING COMPRESSOR
104. PULLEY, POWER STEERING PUMP
105. PULLEY, COOLANT PUMP
106. PULLEY, CRANKSHAFT
107. TENSIONER, DRIVE BELT
108. PULLEY, IDLER

79244G23

Fig. 107 Serpentine belt routing—GM 1995 4.3L (VIN W and X) truck engines

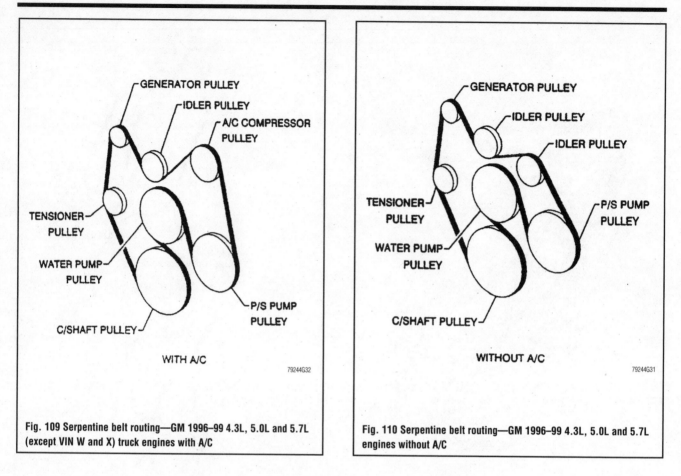

Fig. 109 Serpentine belt routing—GM 1996–99 4.3L, 5.0L and 5.7L (except VIN W and X) truck engines with A/C

Fig. 110 Serpentine belt routing—GM 1996–99 4.3L, 5.0L and 5.7L engines without A/C

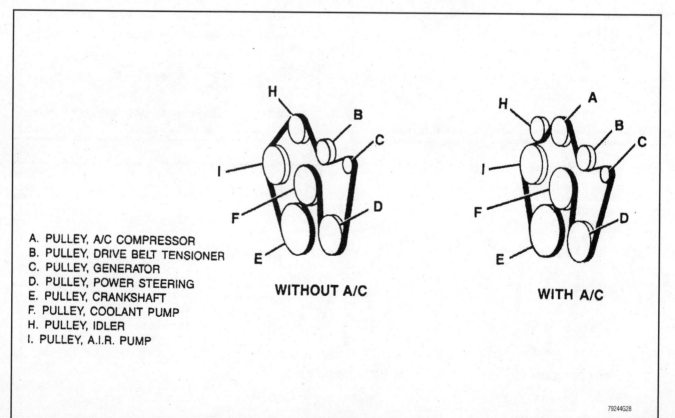

A. PULLEY, A/C COMPRESSOR
B. PULLEY, DRIVE BELT TENSIONER
C. PULLEY, GENERATOR
D. PULLEY, POWER STEERING
E. PULLEY, CRANKSHAFT
F. PULLEY, COOLANT PUMP
H. PULLEY, IDLER
I. PULLEY, A.I.R. PUMP

Fig. 111 Serpentine belt routing—GM 1995 Full-size Truck 4.3L and 5.7L engines with A.I.R.

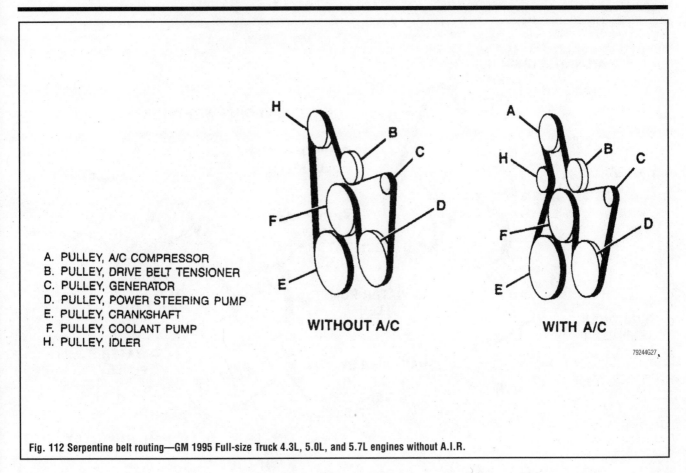

A. PULLEY, A/C COMPRESSOR
B. PULLEY, DRIVE BELT TENSIONER
C. PULLEY, GENERATOR
D. PULLEY, POWER STEERING PUMP
E. PULLEY, CRANKSHAFT
F. PULLEY, COOLANT PUMP
H. PULLEY, IDLER

WITHOUT A/C

WITH A/C

79244G27

Fig. 112 Serpentine belt routing—GM 1995 Full-size Truck 4.3L, 5.0L, and 5.7L engines without A.I.R.

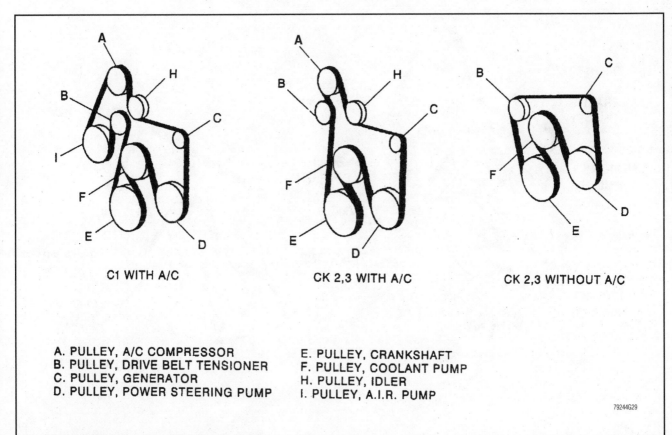

C1 WITH A/C

CK 2,3 WITH A/C

CK 2,3 WITHOUT A/C

A. PULLEY, A/C COMPRESSOR
B. PULLEY, DRIVE BELT TENSIONER
C. PULLEY, GENERATOR
D. PULLEY, POWER STEERING PUMP

E. PULLEY, CRANKSHAFT
F. PULLEY, COOLANT PUMP
H. PULLEY, IDLER
I. PULLEY, A.I.R. PUMP

79244G29

Fig. 113 Serpentine belt routing—GM 1995 7.4L engine

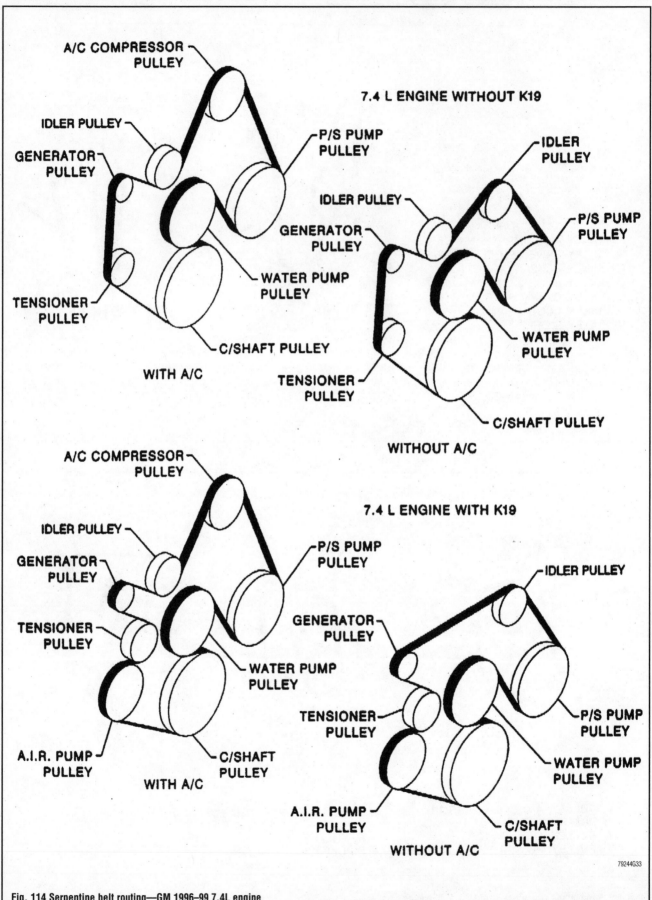

Fig. 114 Serpentine belt routing—GM 1996–99 7.4L engine

79244G33

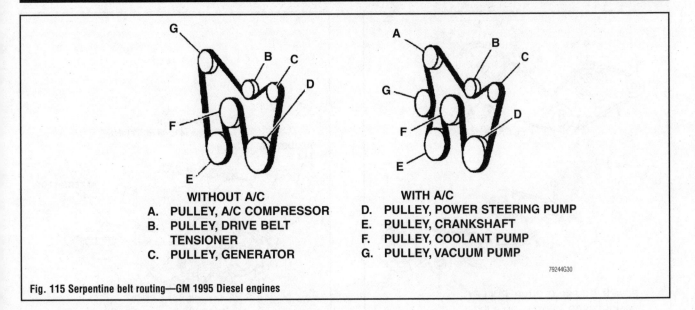

WITHOUT A/C
A. PULLEY, A/C COMPRESSOR
B. PULLEY, DRIVE BELT TENSIONER
C. PULLEY, GENERATOR

WITH A/C
D. PULLEY, POWER STEERING PUMP
E. PULLEY, CRANKSHAFT
F. PULLEY, COOLANT PUMP
G. PULLEY, VACUUM PUMP

79244G30

Fig. 115 Serpentine belt routing—GM 1995 Diesel engines

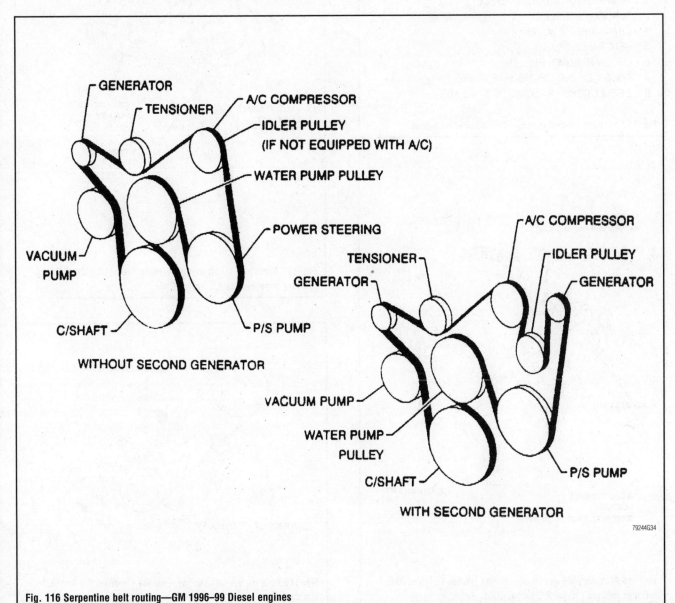

WITHOUT SECOND GENERATOR

WITH SECOND GENERATOR

79244G34

Fig. 116 Serpentine belt routing—GM 1996–99 Diesel engines

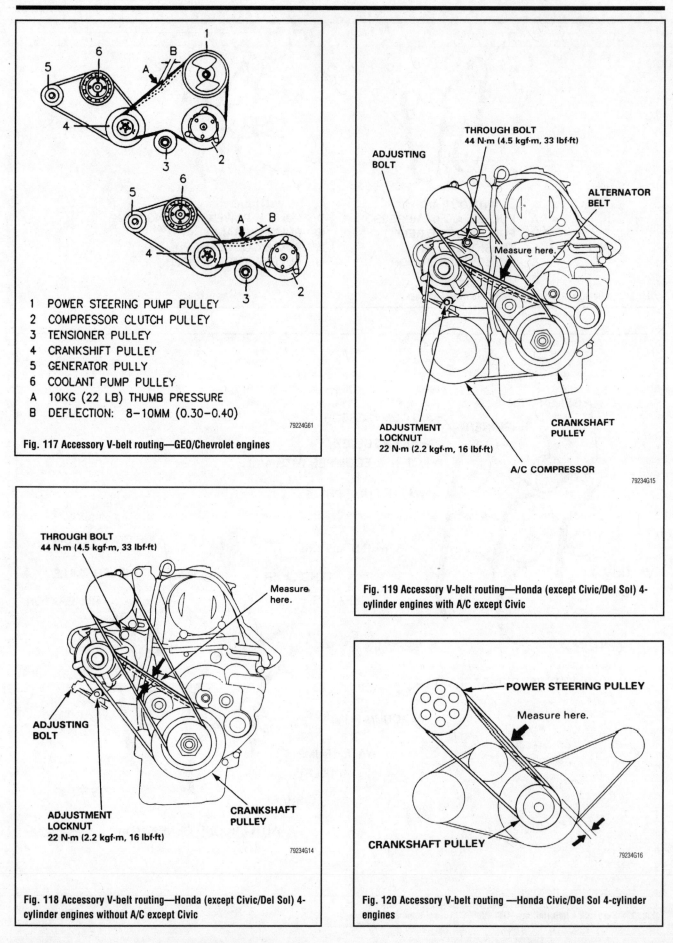

1 POWER STEERING PUMP PULLEY
2 COMPRESSOR CLUTCH PULLEY
3 TENSIONER PULLEY
4 CRANKSHIFT PULLEY
5 GENERATOR PULLY
6 COOLANT PUMP PULLEY
A 10KG (22 LB) THUMB PRESSURE
B DEFLECTION: 8–10MM (0.30–0.40)

79224G61

Fig. 117 Accessory V-belt routing—GEO/Chevrolet engines

THROUGH BOLT
44 N·m (4.5 kgf·m, 33 lbf·ft)

ADJUSTING
BOLT

ALTERNATOR
BELT

Measure here.

ADJUSTMENT
LOCKNUT
22 N·m (2.2 kgf·m, 16 lbf·ft)

CRANKSHAFT
PULLEY

A/C COMPRESSOR

79234G15

Fig. 119 Accessory V-belt routing—Honda (except Civic/Del Sol) 4-cylinder engines with A/C except Civic

THROUGH BOLT
44 N·m (4.5 kgf·m, 33 lbf·ft)

Measure
here.

ADJUSTING
BOLT

ADJUSTMENT
LOCKNUT
22 N·m (2.2 kgf·m, 16 lbf·ft)

CRANKSHAFT
PULLEY

79234G14

Fig. 118 Accessory V-belt routing—Honda (except Civic/Del Sol) 4-cylinder engines without A/C except Civic

POWER STEERING PULLEY

Measure here.

CRANKSHAFT PULLEY

79234G16

Fig. 120 Accessory V-belt routing —Honda Civic/Del Sol 4-cylinder engines

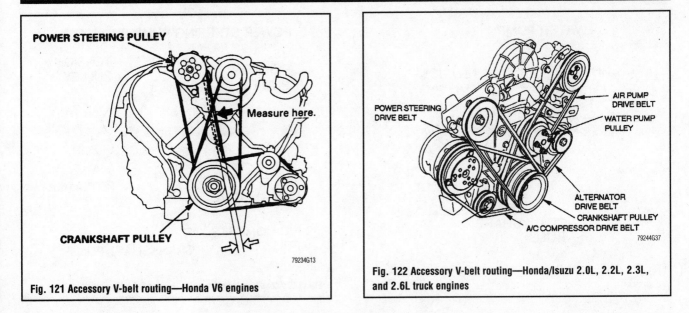

POWER STEERING PULLEY

Measure here.

CRANKSHAFT PULLEY

79234G13

Fig. 121 Accessory V-belt routing—Honda V6 engines

AIR PUMP
DRIVE BELT

POWER STEERING
DRIVE BELT

WATER PUMP
PULLEY

ALTERNATOR
DRIVE BELT

CRANKSHAFT PULLEY

A/C COMPRESSOR DRIVE BELT

79244G37

Fig. 122 Accessory V-belt routing—Honda/Isuzu 2.0L, 2.2L, 2.3L, and 2.6L truck engines

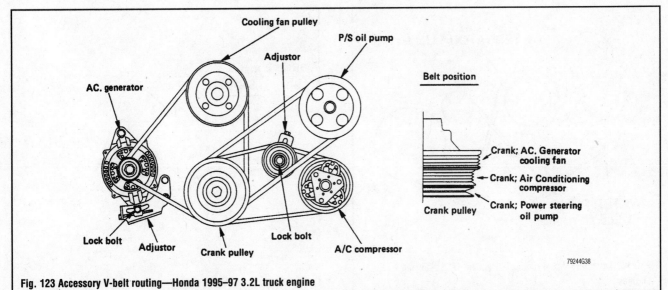

Cooling fan pulley

Adjustor

P/S oil pump

AC. generator

Belt position

Crank; AC. Generator
cooling fan

Crank; Air Conditioning
compressor

Crank pulley

Crank; Power steering
oil pump

Lock bolt

Adjustor

Crank pulley

Lock bolt

A/C compressor

79244G38

Fig. 123 Accessory V-belt routing—Honda 1995–97 3.2L truck engine

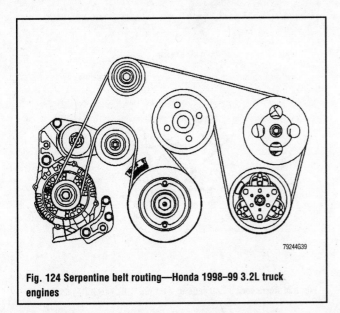

79244G39

Fig. 124 Serpentine belt routing—Honda 1998–99 3.2L truck engines

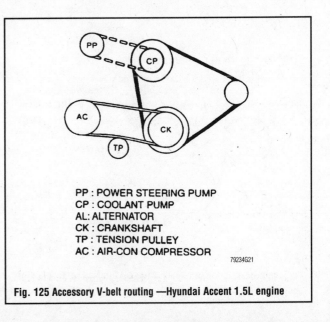

PP : POWER STEERING PUMP
CP : COOLANT PUMP
AL: ALTERNATOR
CK : CRANKSHAFT
TP : TENSION PULLEY
AC : AIR-CON COMPRESSOR

79234G21

Fig. 125 Accessory V-belt routing —Hyundai Accent 1.5L engine

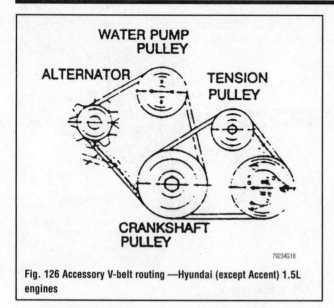

Fig. 126 Accessory V-belt routing —Hyundai (except Accent) 1.5L engines

79234G18

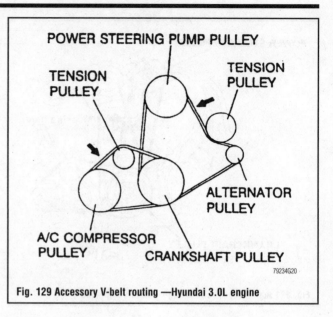

Fig. 129 Accessory V-belt routing —Hyundai 3.0L engine

79234G20

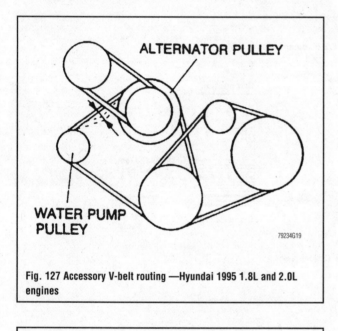

Fig. 127 Accessory V-belt routing —Hyundai 1995 1.8L and 2.0L engines

79234G19

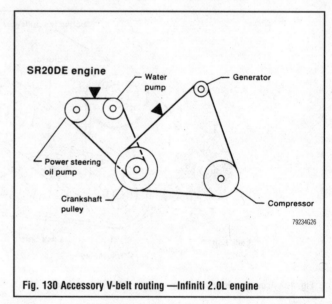

Fig. 130 Accessory V-belt routing —Infiniti 2.0L engine

79234G26

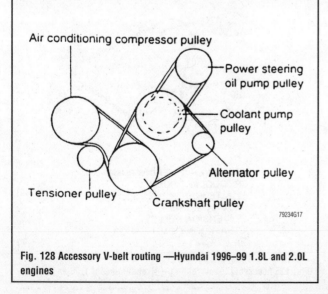

Fig. 128 Accessory V-belt routing —Hyundai 1996-99 1.8L and 2.0L engines

79234G17

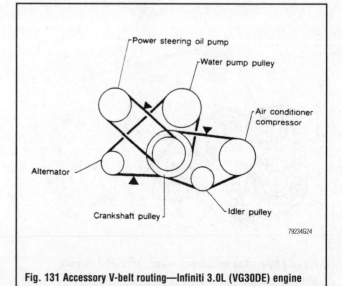

Fig. 131 Accessory V-belt routing—Infiniti 3.0L (VG30DE) engine

79234G24

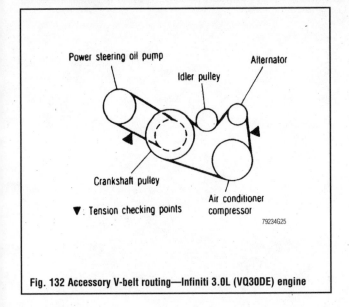

Fig. 132 Accessory V-belt routing—Infiniti 3.0L (VQ30DE) engine

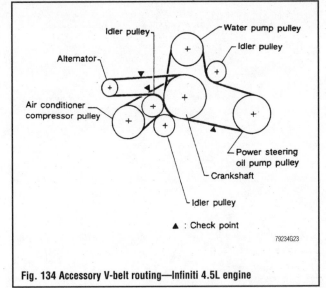

Fig. 134 Accessory V-belt routing—Infiniti 4.5L engine

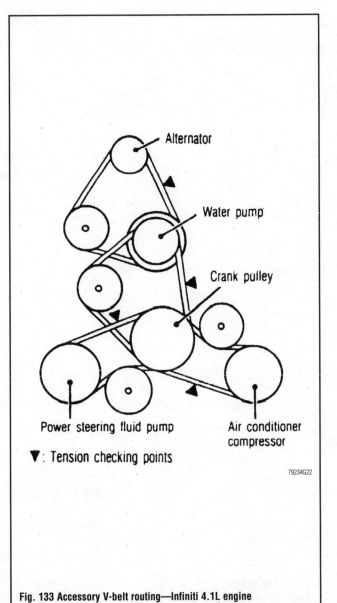

Fig. 133 Accessory V-belt routing—Infiniti 4.1L engine

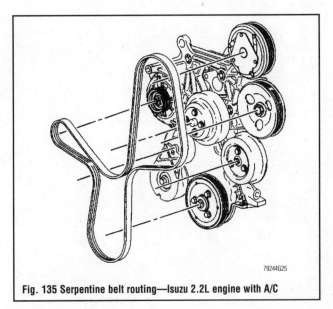

Fig. 135 Serpentine belt routing—Isuzu 2.2L engine with A/C

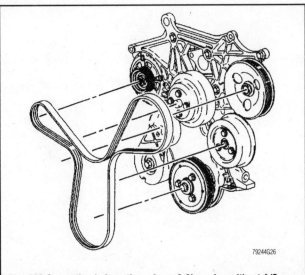

Fig. 136 Serpentine belt routing—Isuzu 2.2L engine without A/C

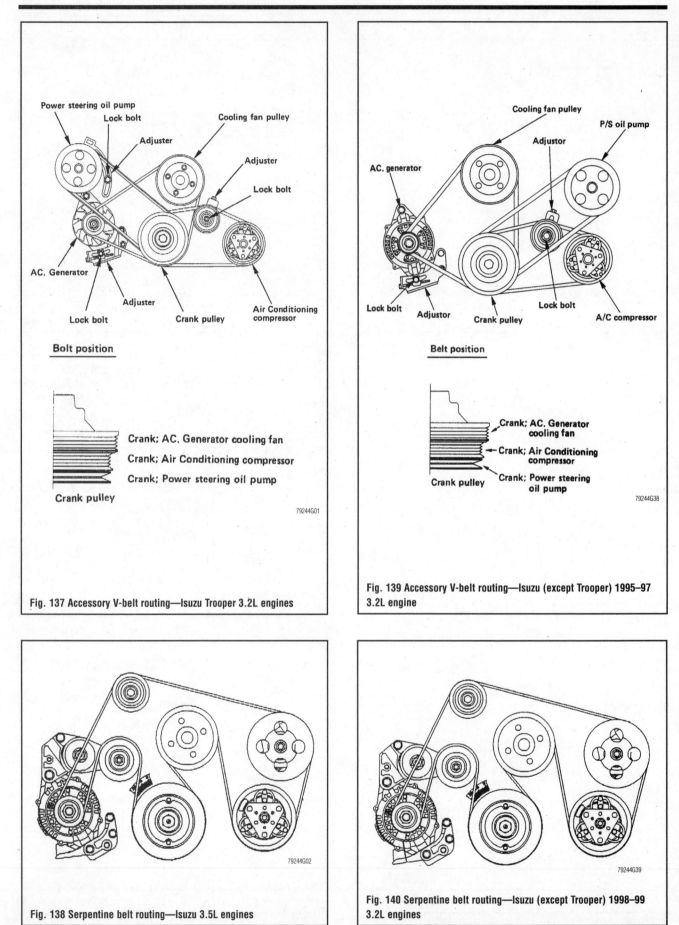

Fig. 137 Accessory V-belt routing—Isuzu Trooper 3.2L engines

Fig. 139 Accessory V-belt routing—Isuzu (except Trooper) 1995–97 3.2L engine

Fig. 138 Serpentine belt routing—Isuzu 3.5L engines

Fig. 140 Serpentine belt routing—Isuzu (except Trooper) 1998–99 3.2L engines

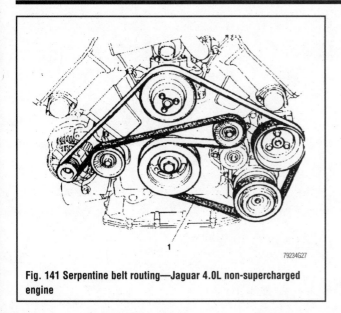

Fig. 141 Serpentine belt routing—Jaguar 4.0L non-supercharged engine

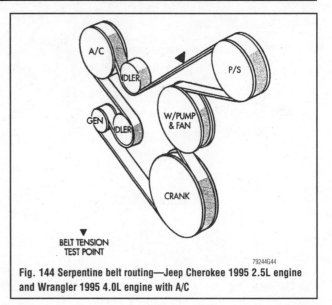

Fig. 144 Serpentine belt routing—Jeep Cherokee 1995 2.5L engine and Wrangler 1995 4.0L engine with A/C

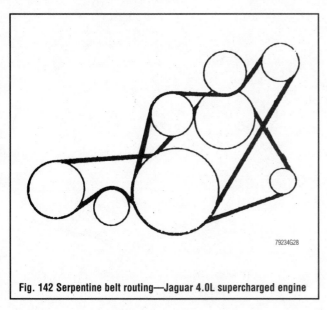

Fig. 142 Serpentine belt routing—Jaguar 4.0L supercharged engine

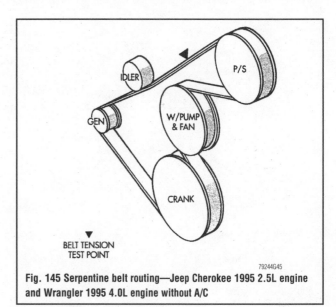

Fig. 145 Serpentine belt routing—Jeep Cherokee 1995 2.5L engine and Wrangler 1995 4.0L engine without A/C

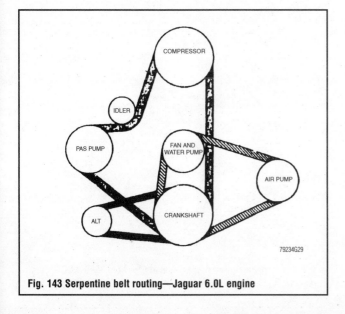

Fig. 143 Serpentine belt routing—Jaguar 6.0L engine

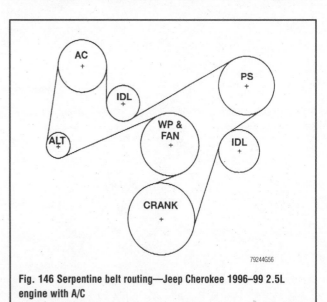

Fig. 146 Serpentine belt routing—Jeep Cherokee 1996–99 2.5L engine with A/C

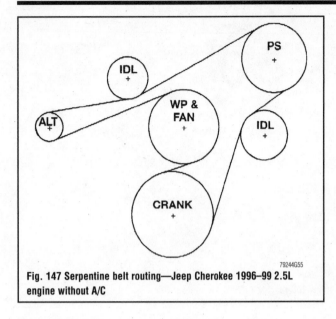

Fig. 147 Serpentine belt routing—Jeep Cherokee 1996–99 2.5L engine without A/C

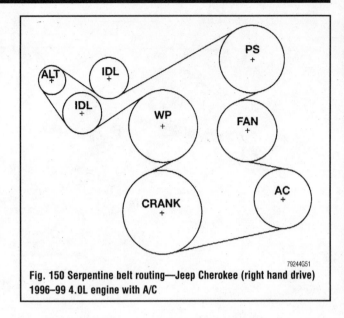

Fig. 150 Serpentine belt routing—Jeep Cherokee (right hand drive) 1996–99 4.0L engine with A/C

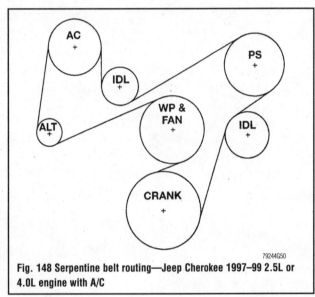

Fig. 148 Serpentine belt routing—Jeep Cherokee 1997–99 2.5L or 4.0L engine with A/C

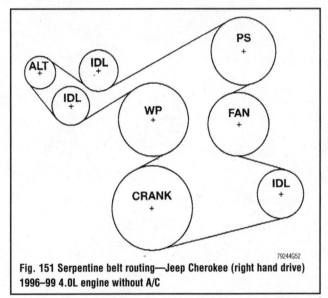

Fig. 151 Serpentine belt routing—Jeep Cherokee (right hand drive) 1996–99 4.0L engine without A/C

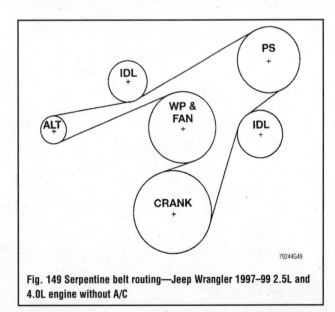

Fig. 149 Serpentine belt routing—Jeep Wrangler 1997–99 2.5L and 4.0L engine without A/C

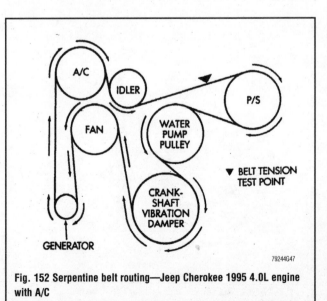

Fig. 152 Serpentine belt routing—Jeep Cherokee 1995 4.0L engine with A/C

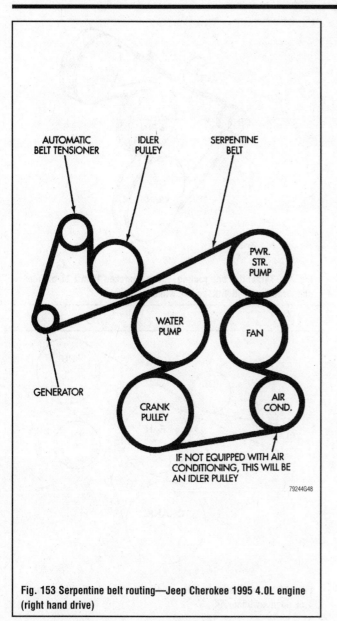

Fig. 153 Serpentine belt routing—Jeep Cherokee 1995 4.0L engine (right hand drive)

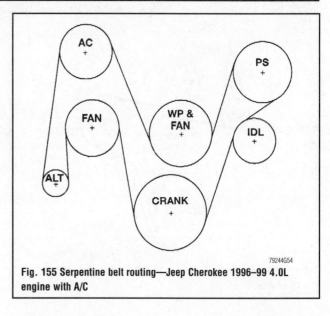

Fig. 155 Serpentine belt routing—Jeep Cherokee 1996–99 4.0L engine with A/C

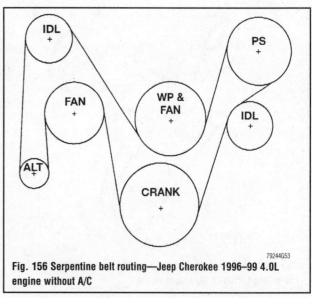

Fig. 156 Serpentine belt routing—Jeep Cherokee 1996–99 4.0L engine without A/C

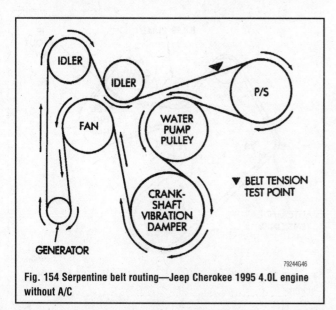

Fig. 154 Serpentine belt routing—Jeep Cherokee 1995 4.0L engine without A/C

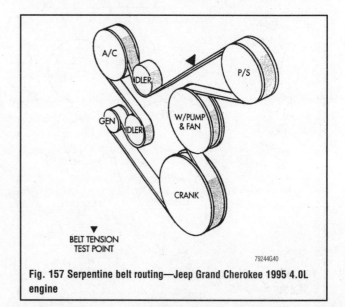

Fig. 157 Serpentine belt routing—Jeep Grand Cherokee 1995 4.0L engine

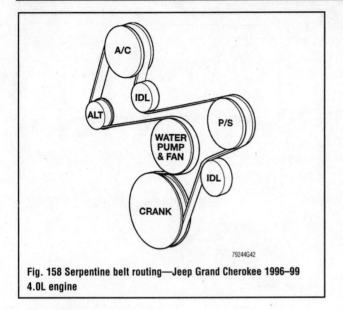

Fig. 158 Serpentine belt routing—Jeep Grand Cherokee 1996–99 4.0L engine

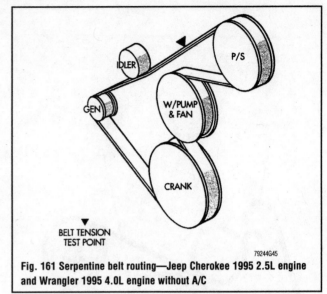

Fig. 161 Serpentine belt routing—Jeep Cherokee 1995 2.5L engine and Wrangler 1995 4.0L engine without A/C

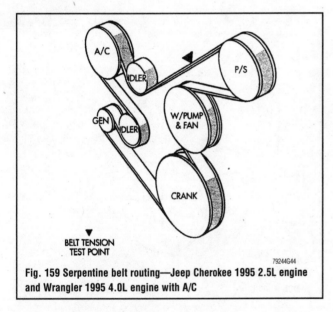

Fig. 159 Serpentine belt routing—Jeep Cherokee 1995 2.5L engine and Wrangler 1995 4.0L engine with A/C

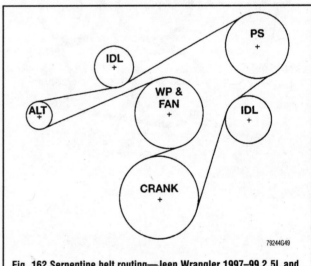

Fig. 162 Serpentine belt routing—Jeep Wrangler 1997–99 2.5L and 4.0L engine without A/C

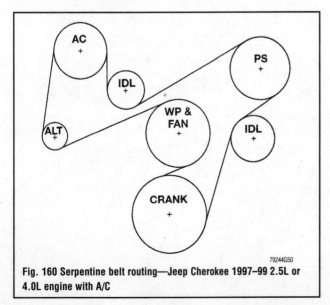

Fig. 160 Serpentine belt routing—Jeep Cherokee 1997–99 2.5L or 4.0L engine with A/C

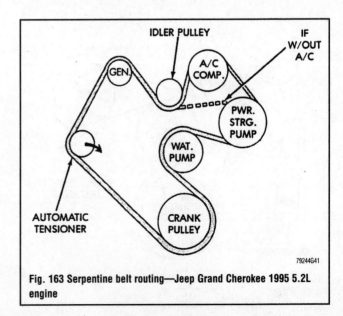

Fig. 163 Serpentine belt routing—Jeep Grand Cherokee 1995 5.2L engine

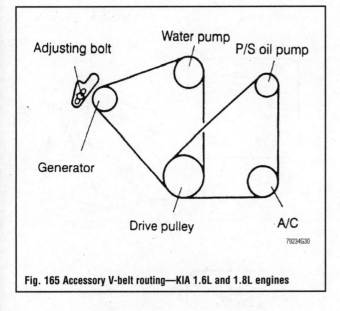

Fig. 164 Serpentine belt routing—Jeep Grand Cherokee 1996–99 5.2L and 5.9L engines

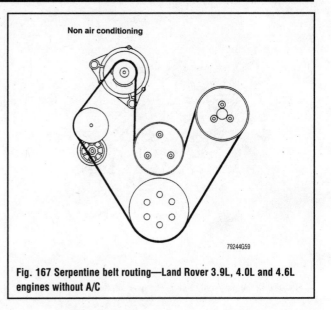

Fig. 167 Serpentine belt routing—Land Rover 3.9L, 4.0L and 4.6L engines without A/C

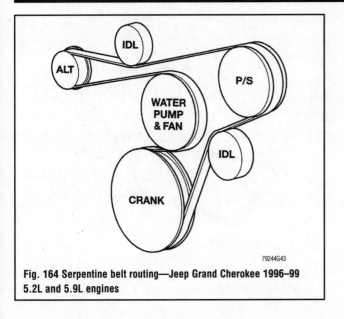

Fig. 165 Accessory V-belt routing—KIA 1.6L and 1.8L engines

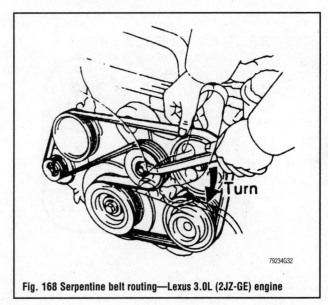

Fig. 168 Serpentine belt routing—Lexus 3.0L (2JZ-GE) engine

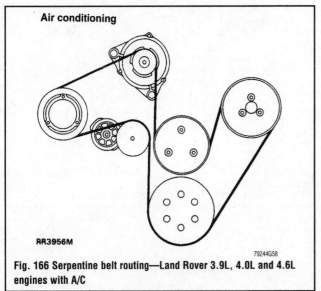

Fig. 166 Serpentine belt routing—Land Rover 3.9L, 4.0L and 4.6L engines with A/C

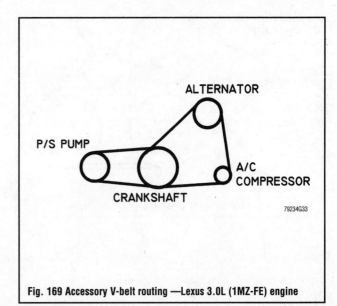

Fig. 169 Accessory V-belt routing —Lexus 3.0L (1MZ-FE) engine

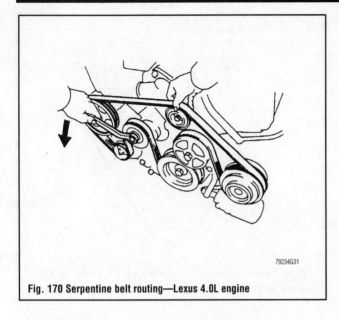

Fig. 170 Serpentine belt routing—Lexus 4.0L engine

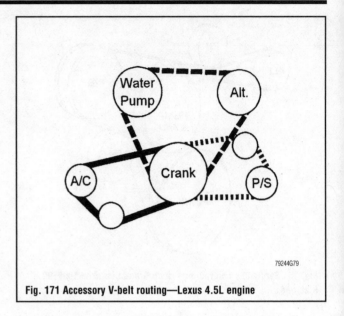

Fig. 171 Accessory V-belt routing—Lexus 4.5L engine

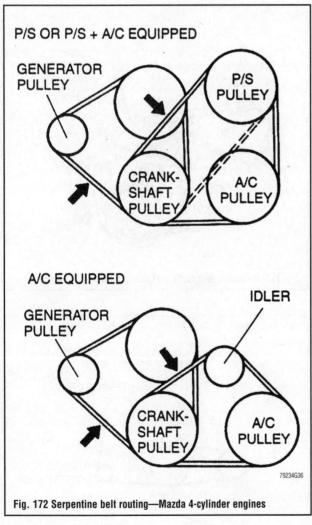

Fig. 172 Serpentine belt routing—Mazda 4-cylinder engines

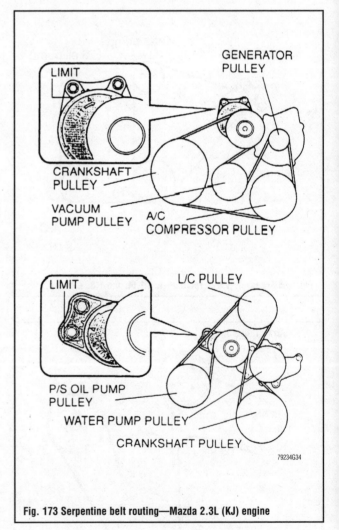

Fig. 173 Serpentine belt routing—Mazda 2.3L (KJ) engine

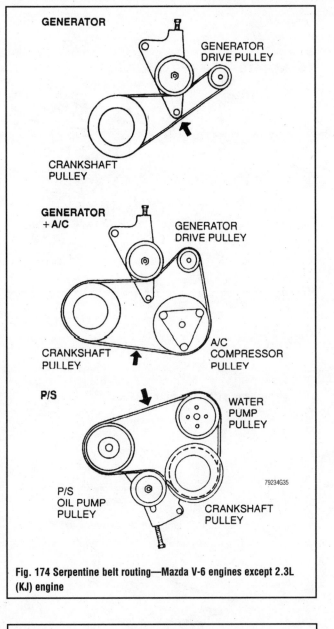

79234G35

Fig. 174 Serpentine belt routing—Mazda V-6 engines except 2.3L (KJ) engine

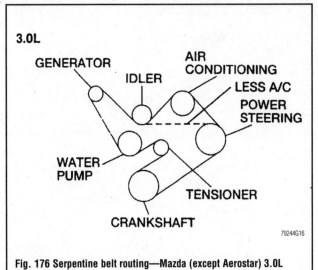

79244G16

Fig. 176 Serpentine belt routing—Mazda (except Aerostar) 3.0L engine

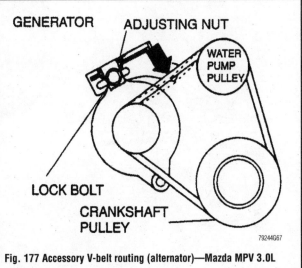

79244G67

Fig. 177 Accessory V-belt routing (alternator)—Mazda MPV 3.0L engine

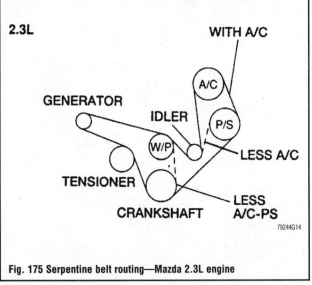

79244G14

Fig. 175 Serpentine belt routing—Mazda 2.3L engine

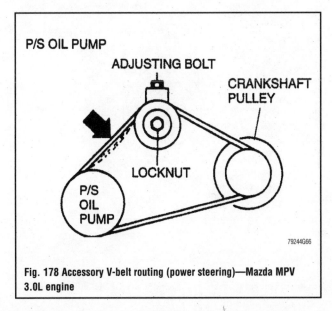

79244G66

Fig. 178 Accessory V-belt routing (power steering)—Mazda MPV 3.0L engine

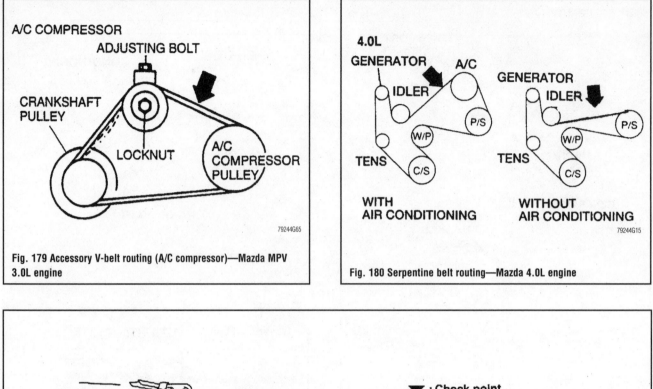

Fig. 179 Accessory V-belt routing (A/C compressor)—Mazda MPV 3.0L engine

Fig. 180 Serpentine belt routing—Mazda 4.0L engine

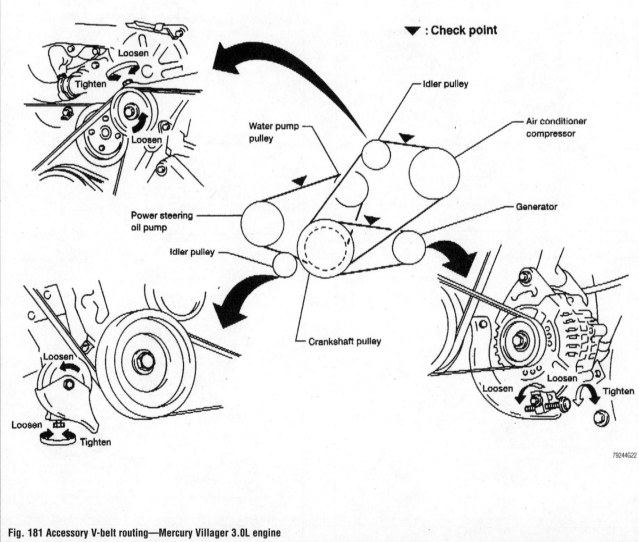

Fig. 181 Accessory V-belt routing—Mercury Villager 3.0L engine

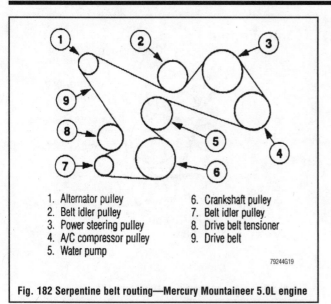

1. Alternator pulley
2. Belt idler pulley
3. Power steering pulley
4. A/C compressor pulley
5. Water pump
6. Crankshaft pulley
7. Belt idler pulley
8. Drive belt tensioner
9. Drive belt

79244G19

Fig. 182 Serpentine belt routing—Mercury Mountaineer 5.0L engine

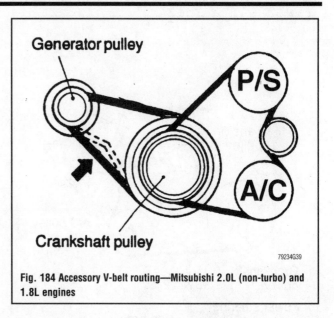

79234G39

Fig. 184 Accessory V-belt routing—Mitsubishi 2.0L (non-turbo) and 1.8L engines

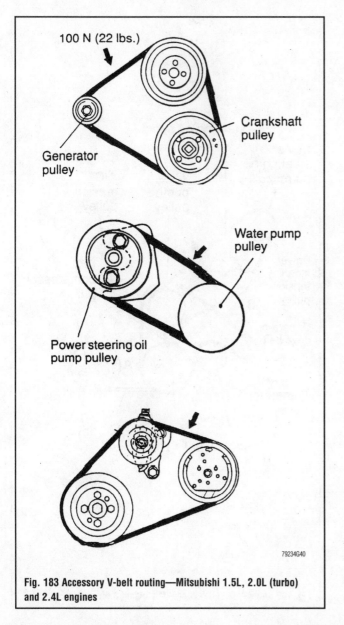

79234G40

Fig. 183 Accessory V-belt routing—Mitsubishi 1.5L, 2.0L (turbo) and 2.4L engines

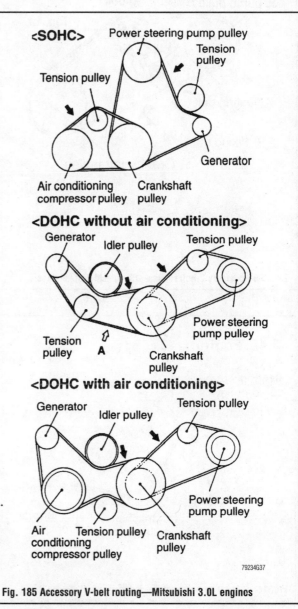

79234G37

Fig. 185 Accessory V-belt routing—Mitsubishi 3.0L engines

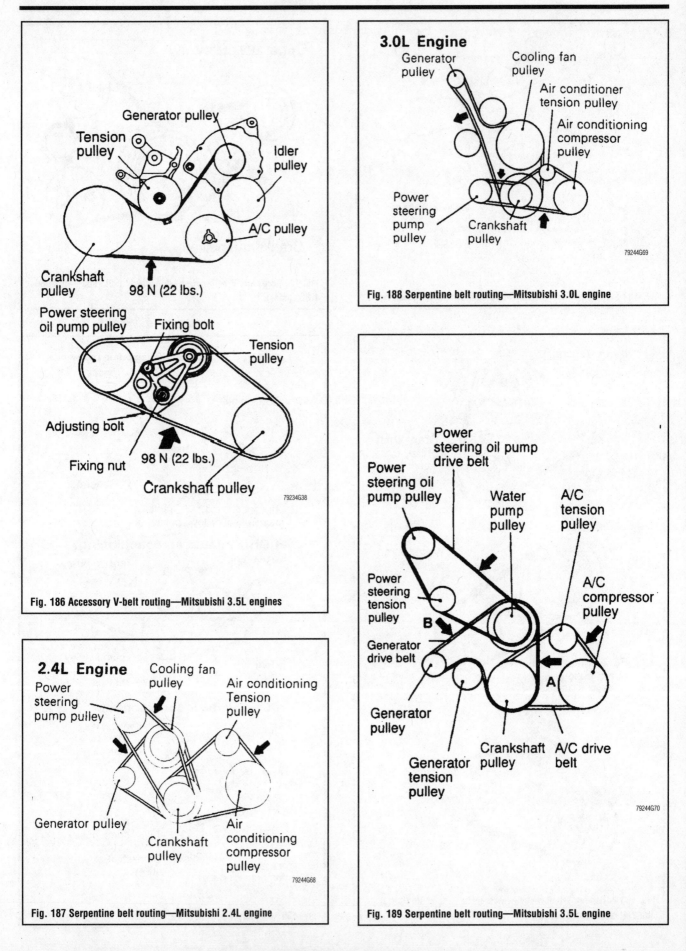

Fig. 186 Accessory V-belt routing—Mitsubishi 3.5L engines

Fig. 187 Serpentine belt routing—Mitsubishi 2.4L engine

Fig. 188 Serpentine belt routing—Mitsubishi 3.0L engine

Fig. 189 Serpentine belt routing—Mitsubishi 3.5L engine

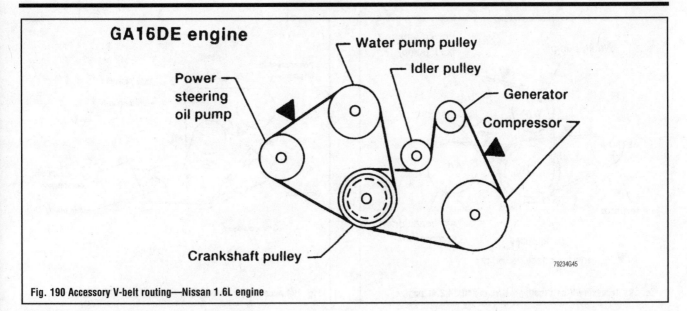

Fig. 190 Accessory V-belt routing—Nissan 1.6L engine

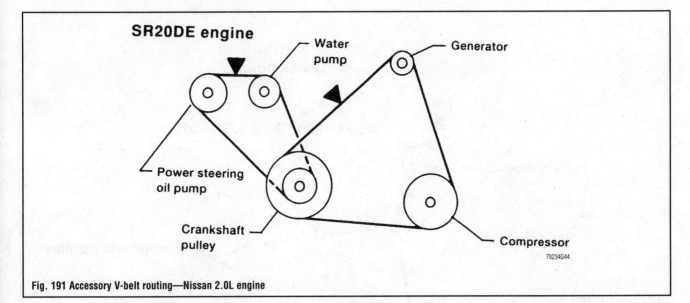

Fig. 191 Accessory V-belt routing—Nissan 2.0L engine

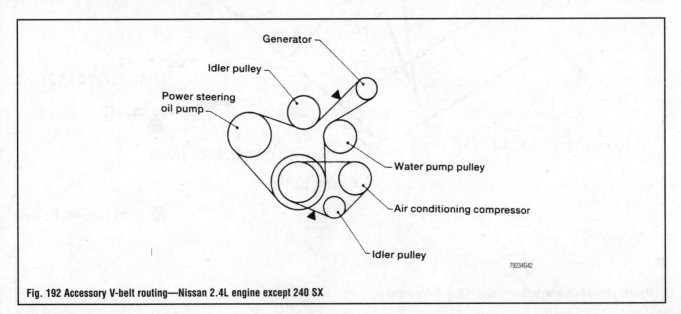

Fig. 192 Accessory V-belt routing—Nissan 2.4L engine except 240 SX

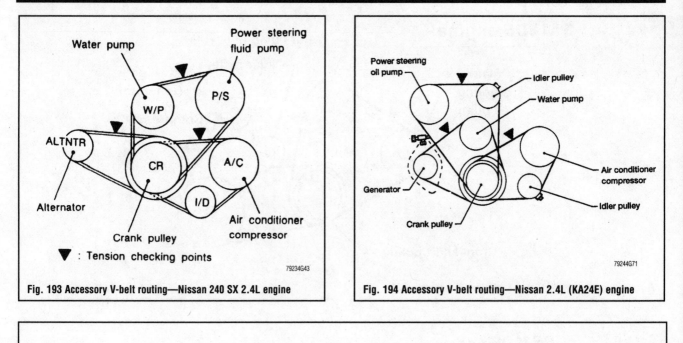

Fig. 193 Accessory V-belt routing—Nissan 240 SX 2.4L engine

Fig. 194 Accessory V-belt routing—Nissan 2.4L (KA24E) engine

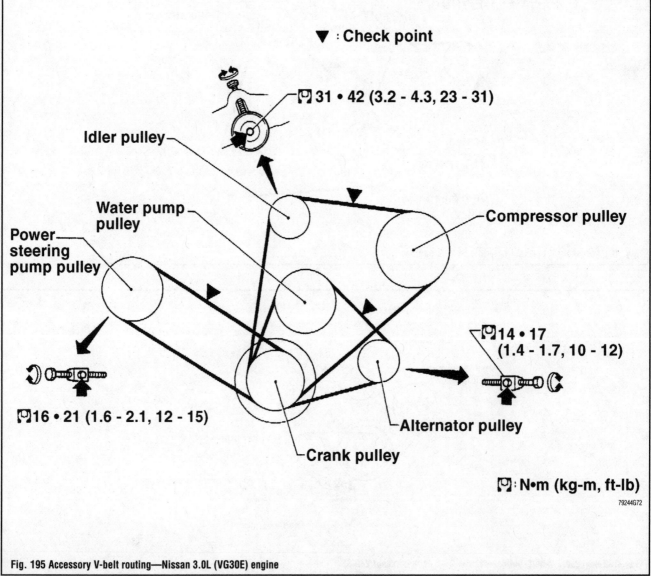

Fig. 195 Accessory V-belt routing—Nissan 3.0L (VG30E) engine

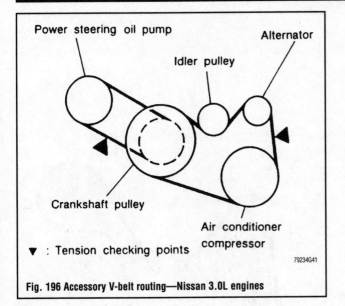

Fig. 196 Accessory V-belt routing—Nissan 3.0L engines

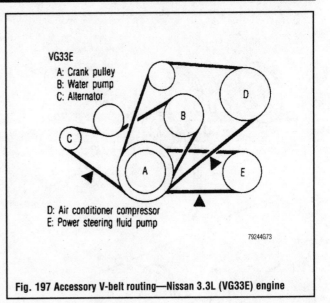

Fig. 197 Accessory V-belt routing—Nissan 3.3L (VG33E) engine

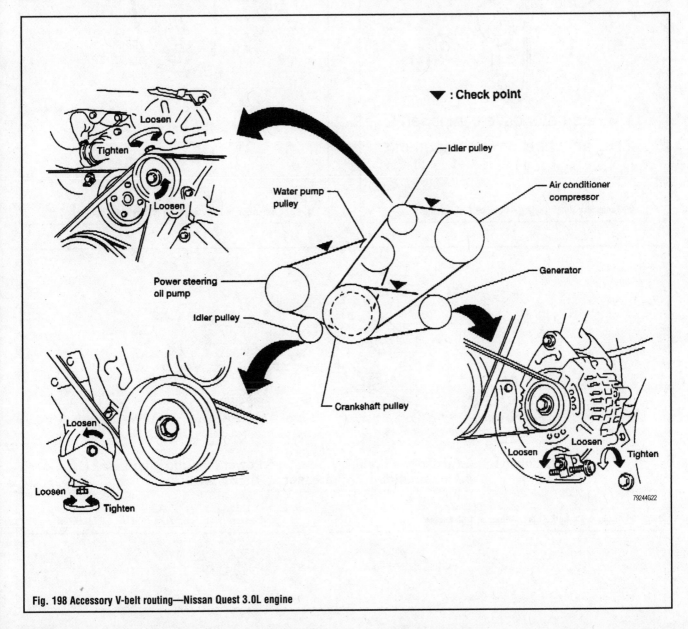

Fig. 198 Accessory V-belt routing—Nissan Quest 3.0L engine

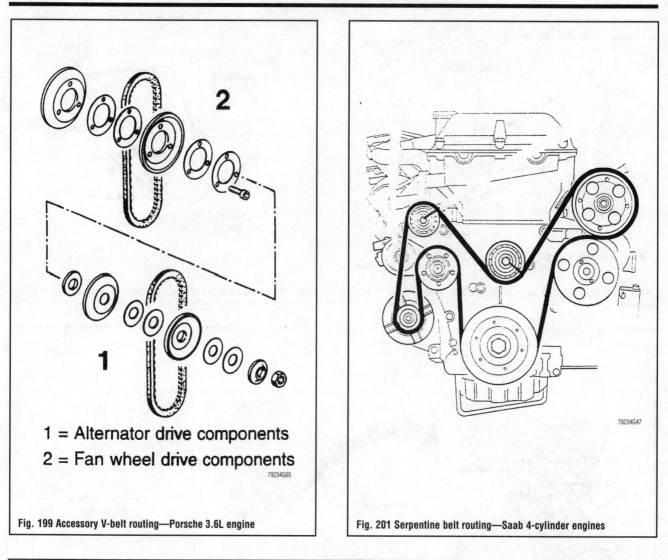

1 = Alternator drive components
2 = Fan wheel drive components

79234G65

Fig. 199 Accessory V-belt routing—Porsche 3.6L engine

Fig. 201 Serpentine belt routing—Saab 4-cylinder engines

79234G47

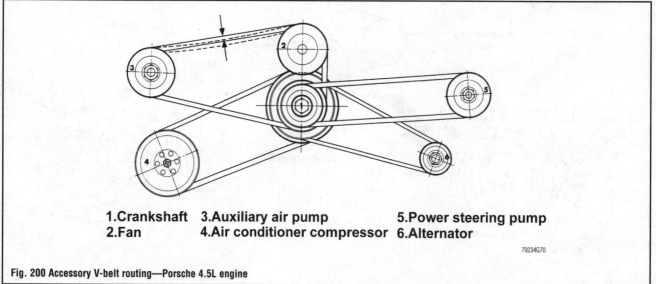

1. Crankshaft   3. Auxiliary air pump   5. Power steering pump
2. Fan   4. Air conditioner compressor   6. Alternator

79234G70

Fig. 200 Accessory V-belt routing—Porsche 4.5L engine

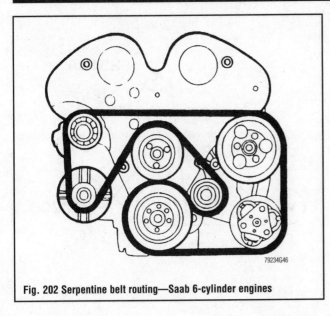

Fig. 202 Serpentine belt routing—Saab 6-cylinder engines

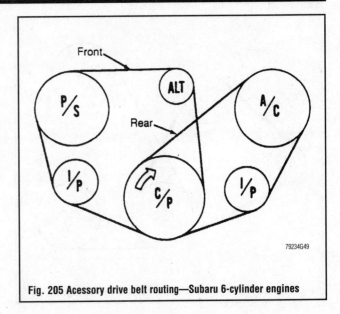

Fig. 205 Acessory drive belt routing—Subaru 6-cylinder engines

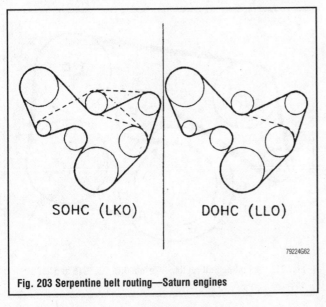

SOHC (LKO)   DOHC (LLO)

Fig. 203 Serpentine belt routing—Saturn engines

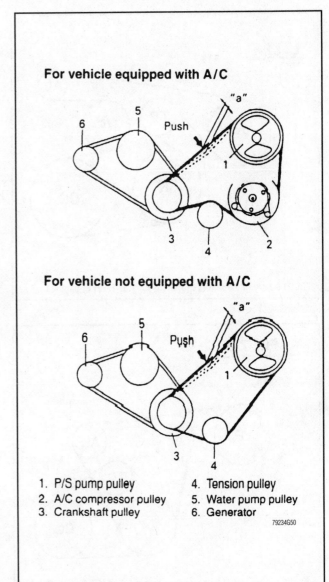

For vehicle equipped with A/C

For vehicle not equipped with A/C

1. P/S pump pulley
2. A/C compressor pulley
3. Crankshaft pulley
4. Tension pulley
5. Water pump pulley
6. Generator

Fig. 206 Accessory V-belt routing—Suzuki engines

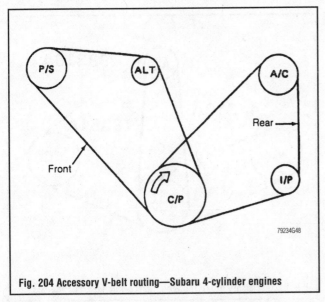

Fig. 204 Accessory V-belt routing—Subaru 4-cylinder engines

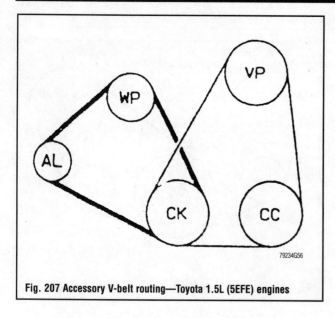

Fig. 207 Accessory V-belt routing—Toyota 1.5L (5EFE) engines

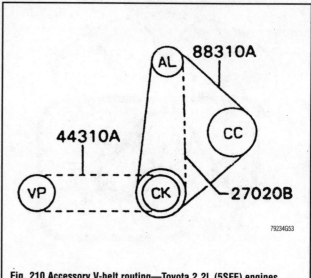

Fig. 210 Accessory V-belt routing—Toyota 2.2L (5SFE) engines

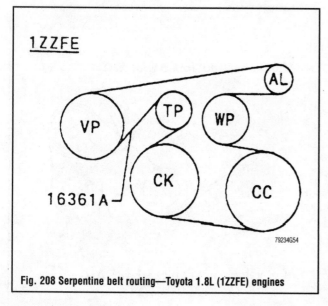

Fig. 208 Serpentine belt routing—Toyota 1.8L (1ZZFE) engines

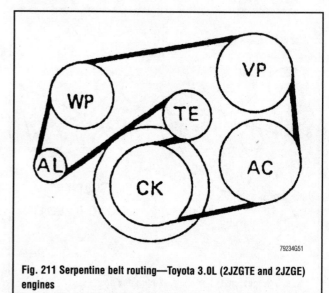

Fig. 211 Serpentine belt routing—Toyota 3.0L (2JZGTE and 2JZGE) engines

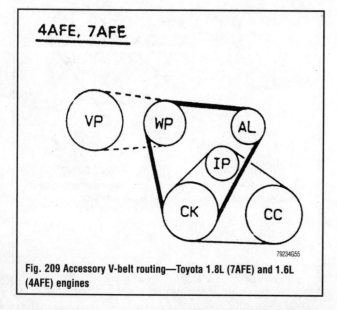

Fig. 209 Accessory V-belt routing—Toyota 1.8L (7AFE) and 1.6L (4AFE) engines

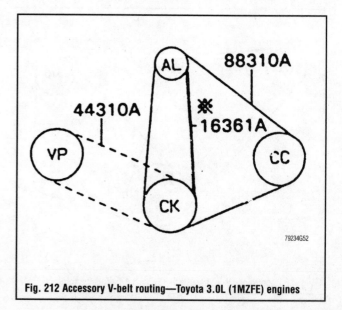

Fig. 212 Accessory V-belt routing—Toyota 3.0L (1MZFE) engines

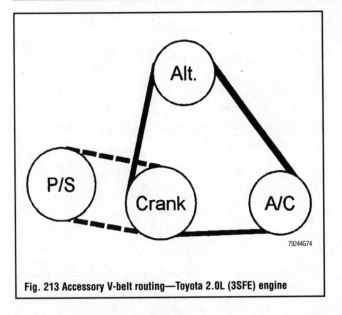

Fig. 213 Accessory V-belt routing—Toyota 2.0L (3SFE) engine

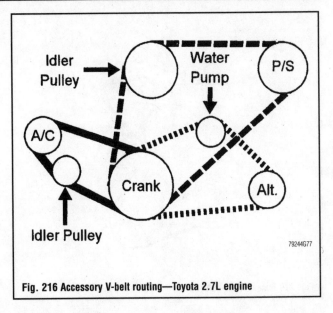

Fig. 216 Accessory V-belt routing—Toyota 2.7L engine

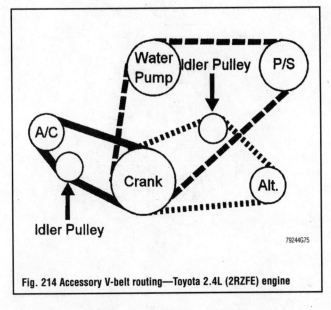

Fig. 214 Accessory V-belt routing—Toyota 2.4L (2RZFE) engine

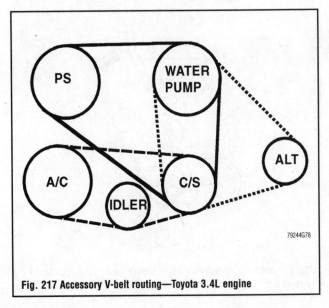

Fig. 217 Accessory V-belt routing—Toyota 3.4L engine

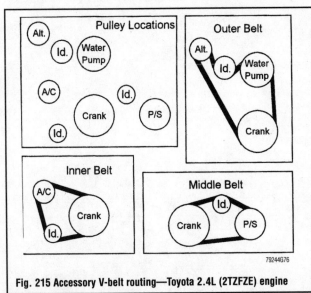

Fig. 215 Accessory V-belt routing—Toyota 2.4L (2TZFZE) engine

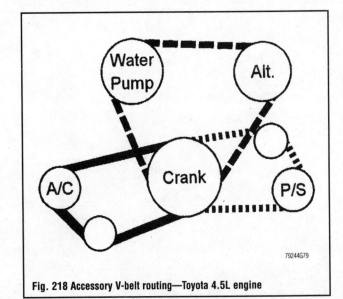

Fig. 218 Accessory V-belt routing—Toyota 4.5L engine

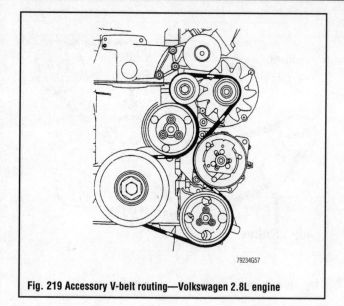

Fig. 219 Accessory V-belt routing—Volkswagen 2.8L engine

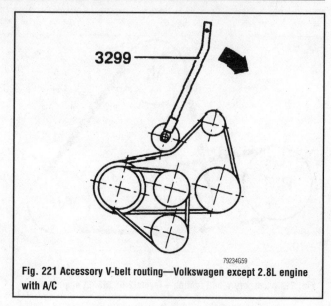

Fig. 221 Accessory V-belt routing—Volkswagen except 2.8L engine with A/C

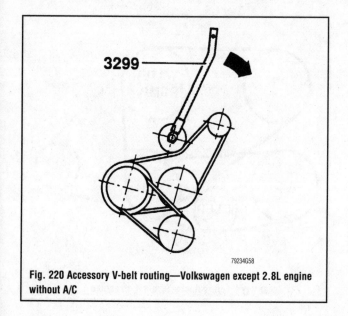

Fig. 220 Accessory V-belt routing—Volkswagen except 2.8L engine without A/C

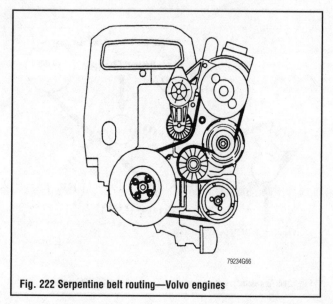

Fig. 222 Serpentine belt routing—Volvo engines

WATER PUMPS

PRECAUTIONS

Before servicing any vehicle, please be sure to read all of the following precautions.

• Never open, service or drain the radiator or cooling system when the engine is hot; serious burns can occur from the steam and hot coolant.

• Observe all applicable safety precautions when working around fuel. Whenever servicing the fuel system, always work in a well ventilated area. Do not allow fuel spray or vapors to come in contact with a spark, open flame, or excessive heat (a hot drop light, for example). Keep a dry chemical fire extinguisher near the work area. Always keep fuel in a container specifically designed for fuel storage; also, always properly seal fuel containers to avoid the possibility of fire or explosion. Refer to the additional fuel system precautions later in this section.

• Fuel injection systems often remain pressurized, even after the engine has been turned **OFF**. The fuel system pressure must be relieved before disconnecting any fuel lines. Failure to do so may result in fire and/or personal injury.

• The EPA warns that prolonged contact with used engine oil may cause a number of skin disorders, including cancer! You should make every effort to minimize your exposure to used engine oil. Protective gloves should be worn when changing oil. Wash your hands and any other exposed skin areas as soon as possible after exposure to used engine oil. Soap and water, or waterless hand cleaner should be used.

• All new vehicles are now equipped with an air bag system, often referred to as a Supplemental Restraint System (SRS) or Supplemental Inflatable Restraint (SIR) system. The system must be disabled before performing service on or around system components, steering column, instrument panel components, wiring and sensors. Failure to follow safety and disabling procedures could result in accidental air bag deployment, possible personal injury and unnecessary system repairs.

• Never operate the engine without the proper amount and type of engine oil; doing so WILL result in severe engine damage.

• Timing belt maintenance is extremely important! Many models utilize an interference-type, non-freewheeling engine. If the timing belt breaks, the valves in the cylinder head may strike the pistons, causing potentially serious (also time-consuming and expensive) engine damage.

• Disconnecting the negative battery cable on some vehicles may interfere with the functions of the on-board computer system(s) and may require the computer to undergo a relearning process once the negative battery cable is reconnected.

DOMESTIC CARS

Chrysler Corporation

CHRYSLER CONCORD, LHS, 300M, NEW YORKER, DODGE INTREPID & EAGLE VISION

The water pump has a die cast aluminum body and a stamped steel impeller. It bolts directly to the chain case cover using an O-ring for sealing. It is driven by the back side of the serpentine belt.

It is normal for a small amount of coolant to drip from the weep hole located on the water pump body (small black spot). If this condition exists, do **NOT** replace the water pump. Only replace the water pump if a heavy deposit or steady flow of brown/green coolant is visible on the water pump body from the weep hole, which would indicate shaft seal failure. Before replacing the water pump, be sure to perform a thorough analysis. Before replacing the water pump, be sure to perform a thorough inspection. A defective pump will not be able to circulate heated coolant through the long heater hose.

2.7L Engine

1. Disconnect the negative battery cable.
2. Drain and recycle the engine coolant.
3. Remove the upper radiator crossmember.
4. Remove the fan module.
5. Remove the accessory drive belts.

➡**The water pump is driven by the primary timing chain.**

6. Remove the crankshaft damper, timing chain cover, timing chain, and all guides.
7. Remove the water pump mounting bolts.
8. Remove the water pump, then clean the mounting surface.
To install:
9. Install the water pump and gasket, then tighten the mounting bolts to 105 inch lbs. (12 Nm).
10. Install the all guides, timing chain , and timing chain cover.
11. Install the crankshaft damper and tighten the center bolt to 125 ft. lbs. (170 Nm).
12. Install the accessory drive belts.
13. Install the fan module and the upper radiator crossmember.
14. Fill the cooling system.
15. Connect the negative battery cable.

3.2L and 3.5L Engines

1. Disconnect the negative battery cable.

✳✳ CAUTION

Do not open the radiator drain or the coolant pressure bottle cap with the system hot and under pressure or serious burns from coolant may occur.

2. Place a drain pan under the radiator. Open the radiator drain located at the lower right side of the radiator. Do **NOT** use pliers to open the plastic drain.
3. Remove the coolant pressure bottle cap and open the thermostat bleed valve.
4. Remove the timing belt using the recommended procedure. It is good practice to turn the crankshaft until the engine is at TDC No. 1 cylinder, compression stroke (firing position). This aligns all the timing marks and serves as a reference point for all the work that follows.
5. Remove the water pump mounting bolts and pump. Discard the O-ring seal.
6. Clean the gasket sealing surfaces. Do not scratch the aluminum surfaces.
To install:
7. Install a new O-ring and wet with clean coolant. Be sure to keep the new O-ring free of any oil or grease.
8. Install the water pump and O-ring to the engine block.
9. Install the retaining bolts and tighten to 105 inch lbs. (12 Nm).
10. Rotate the pump and check for freedom of movement.
11. Install the timing belt using the recommended procedure. Verify that all valve timing marks align. This is most important. An engine out-of-time will be seriously damaged when first started.
12. Be sure that the radiator drain is closed. Open the thermostat bleed valve. Install a ¼ in. (6mm) clear hose about 48 in. (1.2m) long to the end of the bleed valve and the other end into a clean container. The intent is to keep coolant off of the drive belt(s).
13. Slowly refill the coolant pressure bottle until a steady stream of coolant flows out of the thermostat bleed valve. Gently squeeze the upper radiator hose until all of the air is removed from the system.
14. Close the bleed valve and continue to fill up the coolant pressure bottle to the proper level. Install the cap back on the bottle and remove the hose from the bleed valve.
15. Reconnect the negative battery cable. Start the engine and allow to run until normal operating temperature is reached.
16. Check the cooling system for leaks and correct coolant level. Be sure that the thermostat bleed valve is closed once the cooling system has been bled of any trapped air.

3.3L Engines

1. Disconnect the negative battery cable.

✳✳ CAUTION

Do not remove the radiator drain with the system hot and under pressure or serious burns from coolant may occur.

2. Place a drain pan under the radiator. Open the radiator drain located at the lower right side of the radiator. Do **NOT** use pliers to open the plastic drain.
3. Remove the coolant pressure bottle cap and open the thermostat bleed valve.
4. Remove the serpentine belt. If necessary, remove the right front lower fender shield.
5. Remove the water pump pulley bolts and pulley.
6. Remove the water pump mounting bolts and pump. Discard the O-ring seal.
7. Clean the gasket sealing surfaces. Do not scratch the aluminum surfaces.
To install:
8. Install a new O-ring into the O-ring groove and the water pump to the timing chain case. Be sure to keep the O-ring free of any oil or grease.
9. Install the retaining bolts and tighten to 105 inch lbs. (12 Nm).
10. Rotate the pump and check for freedom of movement.
11. Install the pump pulley and tighten the bolts to 250 inch lbs. (30 Nm).
12. Install the serpentine belt and right lower fender shield.
13. Be sure that that the radiator drain is closed. Open the thermostat bleed valve. Install a ¼ in. (6mm) diameter clear hose about 48 in. (1.2m) long, to the end of the bleed valve and the other end into a clean container. The intent is to keep coolant off the drive belt(s).
14. Slowly refill the coolant pressure bottle until a steady stream of coolant flows out of the thermostat bleed valve. Gently squeeze the upper radiator hose until all of the air is removed from the system.
15. Close the bleed valve and continue to fill up the coolant pressure bottle to the proper level. Install the cap back on the bottle and remove the hose from the bleed valve.
16. Reconnect the negative battery cable. Start the engine and allow to run until normal operating temperature is reached.
17. Check the cooling system for leaks and correct coolant level. Be sure that the thermostat bleed valve is closed once the cooling system has been bled of any trapped air.

CHRYSLER SEBRING COUPE & DODGE AVENGER

The water pump is driven by the timing belt from the crankshaft. It is good practice to turn the engine crankshaft by hand (clockwise) to set the engine to TDC No. 1 cylinder compression stroke (firing position) before starting work. This should align all timing marks and serve as a reference point for later work.

2.0L Engine

1. Disconnect the negative battery cable.

➡**This procedure requires removing the engine timing belt and the auto tensioner. The factory specifies that the timing marks should always be aligned before removing the timing belt. Set the engine at TDC on No. 1 compression stroke. This should align all timing marks on the crankshaft sprocket and both camshaft sprockets.**

2. Raise and safely support the vehicle to a level that allows access from above and underneath.

3. Remove the right inner splash shield.

4. Remove the accessory drive belts.

5. Place a drain pan under the radiator drain plug. Drain and properly contain the cooling system.

6. Support the engine using a floor jack and block of wood, then remove the right motor mount.

7. Remove the timing belt, tensioner and camshaft sprockets.

✳✳ WARNING

With the timing belt removed, DO NOT rotate the camshaft or crankshaft or damage to the engine could occur.

8. Remove the rear timing belt cover to access the water pump.

9. Remove the water pump attaching bolts.

10. Remove the water pump.

To install:

11. Thoroughly clean all sealing surfaces. Replace the water pump if there are any cracks, signs of coolant leakage from the shaft seal, loose or rough turning bearings, damaged impeller or sprocket or sprocket flange loose or damaged.

12. Install a new rubber O-ring into the water pump.

➡**Be sure the O-ring is properly seated in the water pump groove before tightening the screws. An improperly located O-ring may cause damage to the O-ring and cause a coolant leak.**

13. Install the water pump and tighten the bolts to 105 inch lbs. (12 Nm).

14. Using a cooling system pressure tester, pressurize the cooling system to 15 psi and check for leaks. If okay, release the pressure and continue the engine assembly process.

15. Rotate the water pump by hand to check for freedom of movement.

16. Install the rear timing belt cover.

17. Install the camshaft sprocket(s), timing belt and tensioner. DO NOT allow the camshafts to turn while the sprockets bolts are being tightened to maintain timing mark alignment.

✳✳ WARNING

Do not attempt to compress the tensioner plunger with the tensioner assembly installed in the engine. This will cause damage to the tensioner and other related components. The tensioner MUST be compressed in a vise.

18. Install the timing belt covers.

19. Install the right engine mount bracket and engine mount.

20. Remove the floor jack and wood block from underneath the engine.

21. Install the crankshaft damper.

22. Install the right inner splash shield.

23. Lower the vehicle.

24. Install and tension the accessory drive belts.

25. Refill the cooling system using the correct quantity and type of coolant. Bleed the cooling system.

26. Start the engine and check for proper operation.

27. Check and top off cooling system, if necessary.

2.5L Engine

1. Disconnect the negative battery cable.

2. Place a large drain pan under the radiator drain plug. Drain and properly contain the engine coolant.

➡**This procedure requires removing the engine timing belt and the auto tensioner. To help assure proper alignment at assembly, it may be helpful to set the engine at TDC on No. 1 compression stroke. This should align all timing marks on the crankshaft sprocket and both camshaft sprockets.**

3. Remove the accessory drive belts and crankshaft damper.

4. Remove the right engine mount. This requires safely supporting the engine with a floor jack and wood block so the mount can be removed.

5. Remove the timing belt covers.

6. Remove the timing belt and tensioner.

7. Remove the water pump mounting bolts.

8. Separate the water pump from the water inlet pipe and remove the pump.

To install:

9. Thoroughly clean all sealing surfaces. Inspect the pump for damage or cracks, signs of coolant leakage at the vent and excessive looseness or rough turning bearing. Any problems require a new pump.

10. Install a new O-ring on the water inlet pipe. Wet the O-ring with water to make installation easier. DO NOT use oil or grease on the O-ring.

11. Install a new gasket on the water pump and fit the pump inlet opening over the water pipe. Press the assembly together to force the pipe into the water pump.

12. Install the water pump-to-engine bolts and tighten to 20 ft. lbs. (27 Nm).

13. Install the timing belt and timing belt tensioner. Set the timing belt tension.

14. Install the timing belt covers. Install the right engine mount. Remove the floor jack and engine block from underneath the engine.

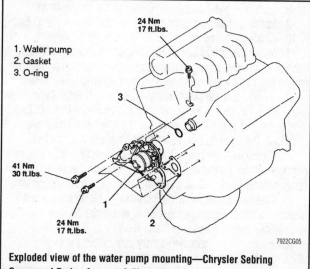

1. Water pump
2. Gasket
3. O-ring

24 Nm
17 ft.lbs.

41 Nm
30 ft.lbs.

24 Nm
17 ft.lbs.

7922CG05

Exploded view of the water pump mounting—Chrysler Sebring Coupe and Dodge Avenger 2.5L engine

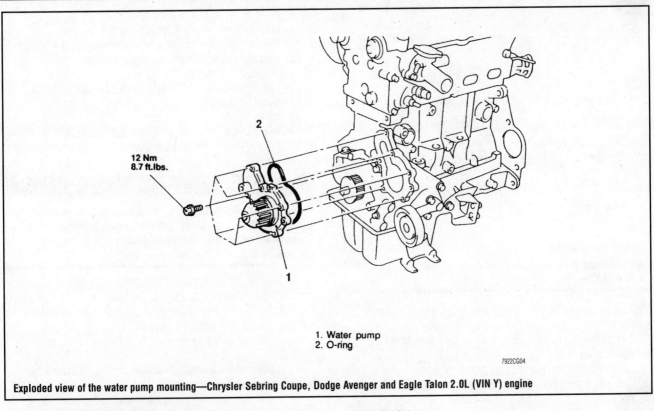

12 Nm
8.7 ft.lbs.

2

1

1. Water pump
2. O-ring

7922CG04

Exploded view of the water pump mounting—Chrysler Sebring Coupe, Dodge Avenger and Eagle Talon 2.0L (VIN Y) engine

15. Install the crankshaft damper.
16. Install the accessory drive belts and set to proper tension.
17. Connect the negative battery cable.
18. Fill and bleed the engine cooling system.
19. Start the engine and verify proper operation.

EAGLE TALON

1. Drain the engine coolant into a suitable container.
2. Remove the timing belt, as described in the timing belt unit repair section in the beginning of this manual.

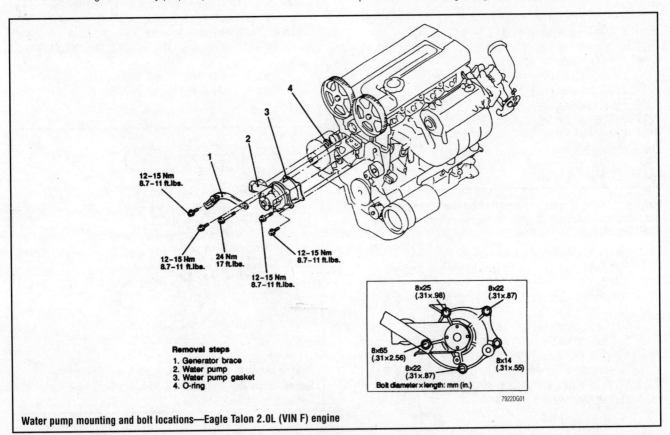

4

3

2

1

12–15 Nm
8.7–11 ft.lbs.

12–15 Nm
8.7–11 ft.lbs.

24 Nm
17 ft.lbs.

12–15 Nm
8.7–11 ft.lbs.

12–15 Nm
8.7–11 ft.lbs.

8x25
(.31x.98)

8x22
(.31x.87)

8x65
(.31x2.56)

8x22
(.31x.87)

8x14
(.31x.55)

Bolt diameter x length: mm (in.)

Removal steps
1. Generator brace
2. Water pump
3. Water pump gasket
4. O-ring

7922DG01

Water pump mounting and bolt locations—Eagle Talon 2.0L (VIN F) engine

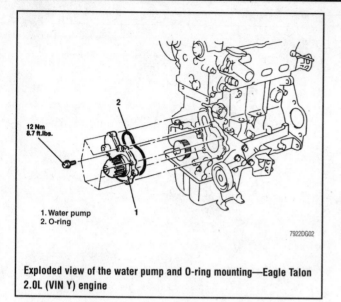

Exploded view of the water pump and O-ring mounting—Eagle Talon 2.0L (VIN Y) engine

3. If necessary, remove the alternator brace from the water pump.

4. Unfasten the retainers, then remove any brackets for access to the rear cover.

5. If necessary, remove the timing belt rear cover.

6. Remove the water pump mounting bolts.

7. Remove the water pump, gasket and O-ring. Discard the gasket and O-ring and replace with new ones during installation.

To install:

8. Install a new O-ring on the water inlet pipe. Coat the O-ring with water or coolant. Do not allow oil or other grease to contact the O-ring.

9. Use a new gasket and install the water pump on the engine block. Tighten the mounting bolts to 8.7–11 ft. lbs. (12–15 Nm). Install the alternator brace on the water pump. Tighten the brace pivot bolt to 17 ft. lbs. (24 Nm).

10. If removed, install the timing belt rear cover.

11. Install the timing belt.

12. Install the remaining components.

13. Refill the engine with coolant.

14. Connect the negative battery cable, start the engine and check for leaks.

DODGE/PLYMOUTH NEON

➡After all components are installed on the engine, a DRB scan tool is necessary to perform the camshaft and crankshaft timing relearn procedure.

1. Disconnect the negative battery cable.

2. Raise and safely support the vehicle. Remove the right inner splash shield.

3. Remove the accessory drive belts and power steering pump.

4. Drain the cooling system into a suitable container.

5. Securely support the engine from the bottom, then remove the right engine mount.

6. Remove the power steering pump bracket bolts, then set the pump and bracket assembly aside, but the power steering lines do not need to be disconnected.

7. Remove the right engine mount bracket.

8. Remove the timing belt tensioner and timing belt.

9. Remove the camshaft sprocket(s) and inner timing belt cover.

10. Unfasten the water pump-to-engine attaching screws, then remove the water pump from the engine.

11. Remove and discard the water pump O-ring, and thoroughly clean the mating surfaces.

To install:

12. Install a new O-ring gasket in the water pump O-ring groove. Hold the O-ring in place with a few small dabs of suitable silicone sealant.

✳✳ WARNING

Before proceeding, be sure the O-ring gasket is properly seated in the water pump groove before tightening the screws. A improperly installed O-ring could cause a coolant leak.

13. Position the water pump to the block and install the retainers. Tighten the retainers to 9 ft. lbs. (12 Nm). Use a pressure tester to pressurize the cooling system to 15 psi and check the water pump shaft seal and O-ring for leaks.

14. Rotate the pump by hand to check for freedom of movement.

15. Install the inner timing belt cover, timing belt and tensioner.

16. Install the right engine mount bracket and engine mount.

17. Refill the cooling system with the proper type and amount of coolant.

18. Install the power steering pump and accessory drive belts.

19. Connect the negative battery cable.

20. Use a DRB or equivalent scan tool to perform the camshaft and crankshaft timing relearn procedure, as follows:

a. Connect the scan tool to the Data Link Connector (located under the instrument panel, near the steering column).

b. Turn the ignition switch **ON**, and access the "miscellaneous" screen.

c. Select the "re-learn cam/crank" option, then follow the instructions on the scan tool screen.

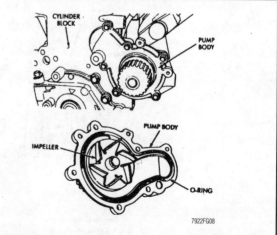

Properly install the O-ring to insure a tight seal—Chrysler Cirrus, Sebring Convertible, Dodge Stratus, Plymouth Breeze and Dodge/Plymouth Neon engines

CHRYSLER CIRRUS, SEBRING CONVERTIBLE, DODGE STRATUS & PLYMOUTH BREEZE

2.0L and 2.4L Engines

This engine uses a die-cast aluminum body water pump with a stamped steel impeller. The water pump bolts directly to the block. The cylinder block to water pump sealing is provided by a large rubber O-ring. The water pump is driven by the timing belt which must be removed to service the water pump.

1. Disconnect the negative battery cable from the left shock tower. The ground cable is equipped with a insulator grommet which should be placed on the stud to prevent the negative battery cable from accidentally grounding.

➡This procedure requires removing the engine timing belt and the auto tensioner. The factory specifies that the timing marks should always be aligned before removing the timing belt. Set the engine at TDC on No. 1 compression stroke. This should align all timing marks on the crankshaft sprocket and both camshaft sprockets.

2. Raise and safely support the vehicle.
3. Remove the right inner splash shield.
4. Remove the accessory drive belts.

> ✳✳ CAUTION
>
> Never open, service or drain the radiator or cooling system when hot; serious burns can occur from the steam and hot coolant.

5. Place a drain pan under the radiator drain plug. Drain and properly contain the cooling system.
6. Support the engine and remove the right motor mount.
7. Remove the power steering pump mounting bracket bolts and place the pump/bracket assembly off to one side. Do not disconnect the power steering fluid lines.
8. Remove the right engine mount bracket.
9. Remove the timing belt front covers.
10. Loosen the timing belt tensioner screws and remove the belt tensioner and timing belt.

> ✳✳ WARNING
>
> With the timing belt removed, DO NOT rotate the camshaft or crankshaft or damage to the engine could occur.

11. Remove the camshaft sprockets. With the timing belt removed, remove both camshaft sprocket bolts. Do not allow the camshafts to turn when the camshaft sprockets are being removed.
12. Remove the rear timing belt cover to access the water pump.
13. Remove the water pump attaching bolts.
14. Remove the water pump.

To install:

15. Thoroughly clean all sealing surfaces. Replace the water pump if there are any cracks, signs of coolant leakage from the shaft seal, loose or rough tuning bearing, damaged impeller or sprocket or sprocket flange loose or damaged.
16. Install a new rubber O-ring into the water pump.

> ✳✳ WARNING
>
> Be sure the O-ring is properly seated in the water pump groove before tightening the screws. An improperly located O-ring may cause damage to the O-ring and cause a coolant leak.

17. Install the water pump and tighten the bolts to 105 inch lbs. (12 Nm).
18. Pressurize the cooling system to 15 psi (103.4 kPa) and check for leaks. If okay, release the pressure and continue the engine assembly process.
19. Reinstall the rear timing belt cover.
20. Reinstall the camshaft sprockets and tighten the attaching bolts to 75 ft. lbs. (101 Nm). DO NOT allow the camshafts to turn while the sprockets bolts are being tighten to maintain timing mark alignment.

> ✳✳ WARNING
>
> Do not attempt to compress the tensioner plunger with the tensioner assembly installed in the engine. This will cause damage to the tensioner and other related components. The tensioner MUST be compressed in a vise.

21. Reinstall the timing belt tensioner and timing belt. Be sure to properly tension the timing belt.
22. Reinstall the front upper and lower timing belt covers.
23. Reinstall the right engine mount bracket and engine mount.
24. Reinstall the crankshaft damper and tighten the center bolt to 105 ft. lbs. (142 Nm).
25. Reinstall the right inner splash shield.
26. Lower the vehicle.
27. Reinstall the power steering pump bracket and power steering pump. Tighten the bracket mounting bolts to 40 ft. lbs. (54 Nm).
28. Reinstall the drive belts. Properly tension the drive belts.
29. Refill the cooling system using a mixture of 50/50 water and ethylene glycol antifreeze. Bleed the cooling system.
30. Start the engine and check for proper operation.
31. Check and top off cooling system, if necessary.

2.5L Engine

The water pump bolts directly to the engine block using a gasket for pump-to-block sealing. The pump is serviced as a unit. The 2.5L engine uses metal piping beyond the lower radiator hose to route coolant to the suction side of the water pump, located in the "V" of the cylinder banks. These pipes also have connections for thermostat bypass and heater return coolant hoses. The pipes use O-rings for sealing.

The water pump is driven by the timing belt which must be removed to service the water pump. Timing belt covers must be removed to access the timing belt.

1. Disconnect the negative battery cable from the left shock tower. The ground cable is equipped with a insulator grommet which should be placed on the stud to prevent the negative battery cable from accidentally grounding.

> ✳✳ CAUTION
>
> Never open, service or drain the radiator or cooling system when hot; serious burns can occur from the steam and hot coolant.

2. Place a large drain pan under the radiator drain plug. Drain and properly contain the engine coolant.

➡**This procedure requires removing the engine timing belt and the auto tensioner. To help assure proper alignment at assembly, it may be helpful to set the engine at TDC on No. 1 compression stroke. This should align all timing marks on the crankshaft sprocket and both camshaft sprockets.**

3. Remove the accessory drive belts and crankshaft damper.
4. Remove the right engine mount. This requires safely supporting the engine so the mount can be removed.
5. Remove the timing belt covers in this order: upper left cover, upper right cover, the lower cover.
6. Remove the timing belt and tensioner.
7. Remove the water pump mounting bolts.
8. Separate the water pump from the water inlet pipe and remove the pump.

To install:

9. Thoroughly clean all sealing surfaces. Inspect the pump for damage or cracks, signs of coolant leakage at the vent and excessive looseness or rough turning bearing. Any problems require a new pump.
10. Install a new O-ring on the water inlet pipe. Wet the O-ring with water to make installation easier. DO NOT use oil or grease on the O-ring.
11. Install a new gasket on the water pump and fit the pump inlet opening over the water pipe. Press the assembly together to force the pipe into the water pump.
12. Reinstall the water pump to engine bolts and tighten to 20 ft. lbs. (27 Nm).
13. Reinstall the timing belt and timing belt tensioner. Set the timing belt tension.
14. Reinstall the timing belt covers. Reinstall the right engine mount.
15. Reinstall the crankshaft damper.
16. Install the accessory drive belts and set to proper tension.
17. Reconnect the negative battery cable.
18. Refill and bleed the engine cooling system.
19. Start the engine and verify proper operation.

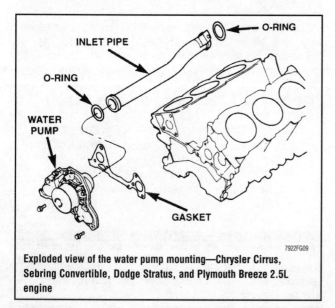

Exploded view of the water pump mounting—Chrysler Cirrus, Sebring Convertible, Dodge Stratus, and Plymouth Breeze 2.5L engine

Ford Motor Company

FORD ASPIRE

❊❊ CAUTION

Some models covered by this manual may be equipped with a Supplemental Restraint System (SRS), which uses an air bag. Whenever working near any of the SRS components, such as the impact sensors, the air bag module, steering column and instrument panel, properly disable the SRS.

1. Disconnect the negative battery cable.
2. Remove the timing belt.

❊❊ CAUTION

Never open, service or drain the radiator or cooling system when hot; serious burns can occur from the steam and hot coolant.

3. Drain the cooling system into a suitable container.
4. Remove the two bolts attaching the inlet tube to the water pump housing. Remove the inlet tube and gasket.

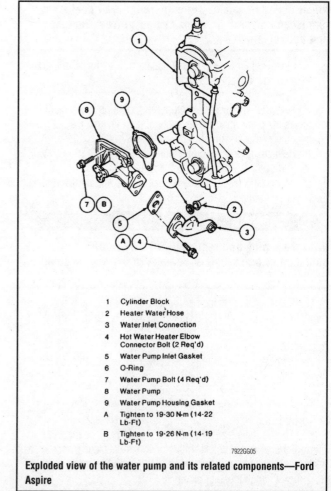

1	Cylinder Block
2	Heater Water Hose
3	Water Inlet Connection
4	Hot Water Heater Elbow Connector Bolt (2 Req'd)
5	Water Pump Inlet Gasket
6	O-Ring
7	Water Pump Bolt (4 Req'd)
8	Water Pump
9	Water Pump Housing Gasket
A	Tighten to 19-30 N·m (14-22 Lb-Ft)
B	Tighten to 19-26 N·m (14-19 Lb-Ft)

Exploded view of the water pump and its related components—Ford Aspire

5. Remove the four water pump-to-cylinder block retaining bolts and remove the water pump.

6. Remove all existing gasket material from the cylinder block and inlet tube gasket surfaces.

To install:

7. Coat both sides of the new water pump and inlet tube gaskets with a suitable water resistant sealer. Apply the gaskets to the engine and inlet tube surfaces. Make certain the gasket holes are aligned with the bolt holes.

8. Position the water pump against the gasket. Be sure the holes in the water pump are aligned with the gasket holes and that the pump does not shift the position of the gasket.

9. Install the four water pump retaining bolts and tighten to 14–19 ft. lbs. (19–26 Nm).

10. Position the inlet tube and gasket against the water pump housing and install the attaching bolts. Tighten the bolts to 14–22 ft. lbs. (19–30 Nm).

11. Install the timing belt.

12. Fill the cooling system.

13. Connect the negative battery cable.

14. Start the engine and allow it to reach normal operating temperature. Check for coolant leaks and proper operation.

FORD PROBE

2.0L Engine

1. Disconnect the negative battery cable.

2. Drain the cooling system into a suitable container.

3. Remove the accessory drive belts.

4. Disconnect the Power Steering Pressure (PSP) switch electrical connector.

5. Remove the power steering pump drive belt idler pulley shield bolts and remove the shield.

6. Remove the power steering pump through-bolt and lockbolt, and position it out of the way.

7. Remove the cylinder head cover.

8. Raise and safely support the vehicle.

9. Remove the water pump pulley using pulley tool T92C-6312-AH or equivalent, to hold the pulley while the bolts are removed.

10. Remove the splash shields.

11. Remove the timing belt.

12. Remove the power steering pump lower bracket from the water pump.

13. Remove the 5 water pump mounting bolts and remove the water pump.

To install:

14. Clean all gasket mating surfaces.

15. Install a new gasket on the water pump and install the water pump on the engine. Install the mounting bolts and tighten to 14–19 ft. lbs. (19–25 Nm).

16. Install the power steering pump lower bracket from the water pump.

17. Install the water pump pulley and bolts. Hold the pulley with the tool and tighten the bolts to 71–88 inch lbs. (8–10 Nm).

18. Install the timing belt.

19. Install the splash shields and tighten the bolts to 71–88 inch lbs. (8–10 Nm).

20. Lower the vehicle and install the cylinder head cover. Tighten the bolts in 2–3 steps to 52–69 inch lbs. (6–7 Nm) in the proper sequence.

21. Place the power steering pump in position. Install the through-bolt and tighten to 32–45 ft. lbs. (43–61 Nm). Install the lockbolt and tighten to 23–34 ft. lbs. (31–46 Nm).

22. Connect the PSP switch electrical connector.

23. Install the accessory drive belts and adjust the tension.

24. Install the steering idler pulley shield and tighten the bolts to 61–86 ft. lbs. (7–9 Nm).

25. Connect the negative battery cable.

26. Fill and bleed the cooling system.

27. Run the engine and bring to normal operating temperature. Check for leaks.

2.5L Engine

1. Disconnect the negative battery cable.

2. Drain the cooling system into a suitable container.

3. Remove the timing belt covers and the timing belt.

4. Use pulley removal tool T92C-6312-AH or equivalent, to hold the water pump pulley and remove the bolts. Remove the water pump pulley.

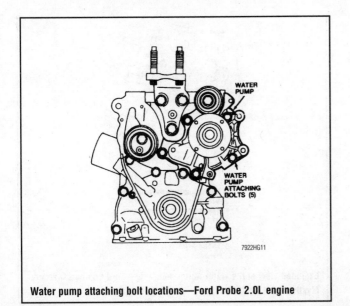

Water pump attaching bolt locations—Ford Probe 2.0L engine

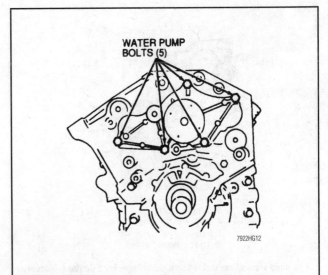

Water pump attaching bolt locations—Ford Probe 2.5L engine

5. Position a drain pan under the water pump.

6. Remove the 3 front engine support insulator mounting bracket bolts.

7. Remove the 5 water pump mounting bolts and remove the water pump.

To install:

8. Clean the mating surfaces of the water pump and the engine block.

9. Install a new O-ring onto the water pump.

10. Install the water pump and tighten the bolts 14–18 ft. lbs. (19–25 Nm).

11. Install the 3 engine support mounting bracket bolts.

12. Install the water pump pulley with the bolts. Hold the pulley with the tool and tighten the bolts to 71–88 inch lbs. (8–10 Nm).

13. Install the timing belt and timing belt covers.

14. Connect the negative battery cable.

15. Fill and bleed the cooling system.

16. Run the engine and bring to normal operating temperature. Check for leaks.

FORD CONTOUR, MERCURY MYSTIQUE & 1999 MERCURY COUGAR

> ※※ **CAUTION**
>
> **Some models covered by this manual may be equipped with a Supplemental Restraint System (SRS), which uses an air bag. Whenever working near any of the SRS components, such as the impact sensors, the air bag module, steering column and instrument panel, properly disable the SRS.**

2.0L Engine

1. Disconnect the negative battery cable.

> ※※ **CAUTION**
>
> **Never open, service or drain the radiator or cooling system when hot; serious burns can occur from the steam and hot coolant.**

2. Drain the engine cooling system.

3. Raise and safely support the vehicle.

4. Remove the lower radiator hose from the water pump.

5. Lower the vehicle.

6. Remove the accessory drive belt.

7. Remove the timing belt covers and the timing belt using the recommended procedure.

8. Remove the four water pump retaining bolts.

9. Remove the water pump.

To install:

10. Thoroughly clean all sealing surfaces.

11. Install a new water pump gasket and the water pump onto the cylinder block.

12. Tighten the retaining bolts to 12–15 ft. lbs. (16–20 Nm).

13. Reinstall the timing belt and the timing belt covers using the recommended procedure.

14. Reinstall the accessory drive belt.

15. Raise and safely support the vehicle.

16. Reinstall the lower radiator hose.

17. Lower the vehicle.

18. Fill the engine cooling system.

19. Reconnect the negative battery cable.

20. Start the engine and top off the coolant as necessary. Check for leaks.

2.5L Engine

➡Before continuing with this procedure, be sure three new water pump retaining bolts (W701544) are available. Due to their torque-to-yield design, the bolts stretch and cannot be reused.

1. Disconnect the negative battery cable.

> ※※ **CAUTION**
>
> **Never open, service or drain the radiator or cooling system when hot; serious burns can occur from the steam and hot coolant.**

2. Drain the engine cooling system.

3. Remove the water pump pulley shield.

4. Remove the water pump drive belt.

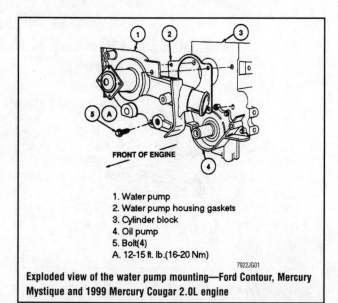

1. Water pump
2. Water pump housing gaskets
3. Cylinder block
4. Oil pump
5. Bolt(4)
A. 12-15 ft. lb.(16-20 Nm)

7922JG01

Exploded view of the water pump mounting—Ford Contour, Mercury Mystique and 1999 Mercury Cougar 2.0L engine

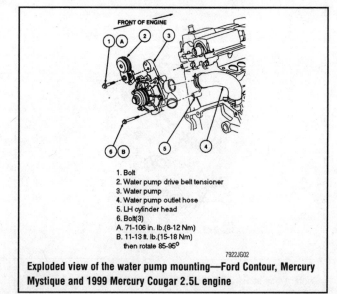

1. Bolt
2. Water pump drive belt tensioner
3. Water pump
4. Water pump outlet hose
5. LH cylinder head
6. Bolt(3)
A. 71-106 in. lb.(8-12 Nm)
B. 11-13 ft. lb.(15-18 Nm)
   then rotate 85-95°

7922JG02

Exploded view of the water pump mounting—Ford Contour, Mercury Mystique and 1999 Mercury Cougar 2.5L engine

5. Remove the water pump inlet and outlet hoses from the water pump.

6. Remove the three water pump to left cylinder head retaining bolts.

7. Remove the water pump and water pump housing from the vehicle.

8. Remove the water pump to water pump housing retaining bolts and separate the water pump from the water pump housing.

To install:

9. Thoroughly clean all sealing surfaces.

10. Install the water pump to the water pump housing using a new gasket and install the retaining bolts. Tighten the retaining bolts to 16–18 ft. lbs. (22–25 Nm).

11. Position the water pump and water pump housing and install three new torque-to-yield retaining bolts into the left cylinder head.

12. Tighten the new retaining bolts to 11–13 ft. lbs. (15–18 Nm), then rotate the retaining bolts 85–95 degrees.

13. Reinstall the water pump inlet and outlet hoses to the water pump.

14. Reinstall the water pump drive belt.

15. Reinstall the water pump shield.

16. Fill the engine cooling system.

17. Reconnect the negative battery cable.

18. Start the engine and top off the coolant as necessary. Check for leaks.

FORD TAURUS & MERCURY SABLE

❊❊ CAUTION

Do not remove the radiator cap or open the cooling system until the engine has cooled. Removing the radiator cap or opening the cooling system prior to the engine cooling could cause severe burns from scalding engine coolant.

1995 3.0L (DOHC) and 3.2L SHO Engines

1. Disconnect the battery cables, negative cable first and remove the battery and the battery tray.

2. Drain the cooling system and remove the accessory drive belts.

3. Remove the left-hand drive belt idler pulley.

4. Disconnect the electrical connector from the ignition module and ground strap.

5. Loosen the clamps on the upper intake connector tube, then remove the retaining bolts and the connector tube.

6. Remove the upper outer timing belt cover.

7. Raise and safely support the vehicle. Remove the right wheel and tire assembly. Remove the splash shield.

8. Remove the crankshaft pulley using a suitable puller.

9. Remove the lower outer timing belt cover. Disconnect the crankshaft position sensor wiring harness and move it aside.

10. Remove the center timing belt cover. Remove the right-hand drive belt tensioner idler pulley.

11. Remove the water pump attaching bolts and remove the water pump.

To install:

12. Lightly oil all bolt threads before installation. Clean gasket surfaces on pump and engine block.

13. Position a new gasket on the water pump and use a gasket sealer to hold the gasket in place. Install water pump and retaining bolts. Tighten to 12–17 ft. lbs. (16–23 Nm).

14. To complete the installation, reverse the removal procedures.

15. Tighten the crankshaft pulley retaining bolt to 112–127 ft. lbs. (152–172 Nm). Tighten the upper intake connector tube retaining bolts to 11–17 ft. lbs. (15–23 Nm).

16. Reconnect the battery cables.

17. Be sure draincock is closed and refill cooling system. Run the engine and check for leaks.

3.0L (OHV) Engine

1. Disconnect the negative battery cable. Allow the engine to cool. Remove the radiator cap and drain the cooling system.

2. Loosen four retaining bolts securing the water pump pulley to the water pump hub.

3. Remove the accessory drive belts. Remove the idler pulley or automatic tensioner, as required.

4. Disconnect and remove the heater hose from the water pump.

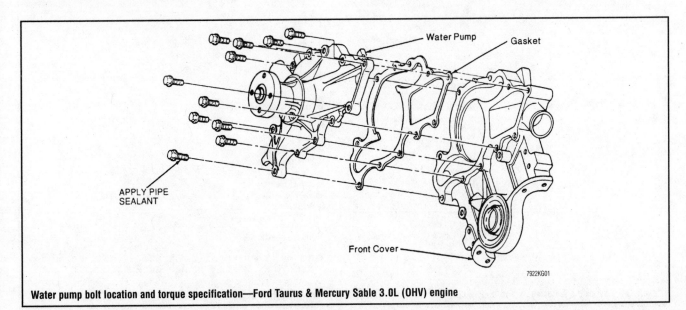

APPLY PIPE SEALANT

Water Pump

Gasket

Front Cover

7922KG01

Water pump bolt location and torque specification—Ford Taurus & Mercury Sable 3.0L (OHV) engine

5. Remove the engine control sensor wiring from the locating stud bolt, if equipped.

6. Remove the water pump-to-engine retaining bolts and lift the water pump and pulley up and out of the vehicle.

To install:

7. Clean the gasket surfaces on the water pump and engine front cover. Install a new gasket on the water pump using gasket adhesive.

8. Place the water pump in position on the engine with the pulley and four retaining bolts loosely installed on the hub.

9. Apply pipe sealant and install the bolts in the water pump housing. Tighten the bolts designated by reference No. 1 to 15–22 ft. lbs. (20–30 Nm) and the bolts designated by reference No. 2 to 72–96 inch lbs. (8–12 Nm).

➡**The bolts are of different lengths and must be installed in the correct locations.**

10. Install the remaining components in the reverse order of removal. Tighten the water pump pulley bolts to 15–22 ft. lbs. (20–30 Nm).

11. Fill the cooling system. Connect the negative battery cable.

12. Start the engine and allow it to reach normal operating temperature. Check for leaks and proper operation.

1996–99 3.0L (DOHC) Engine

1. Disconnect the negative battery cable. Drain the engine cooling system.

2. Remove the water pump drive belt. Remove the radiator and heater hoses from the water pump.

3. Remove the four nuts securing the water pump to the engine and remove the water pump.

To install:

4. Clean the water pump to engine gasket sealing surfaces.

5. Install the water pump using a new gasket and install the four retaining nuts. Tighten the retaining nuts to 15–22 ft. lbs. (20–30 Nm).

6. Install the remaining components in the reverse order of removal.

7. Fill the engine cooling system, then connect the negative battery cable.

8. Start the engine and allow it to reach normal operating temperature, then check for coolant leaks and proper engine operation.

3.4L Engine

1. Disconnect the negative battery cable. Drain the engine cooling system.

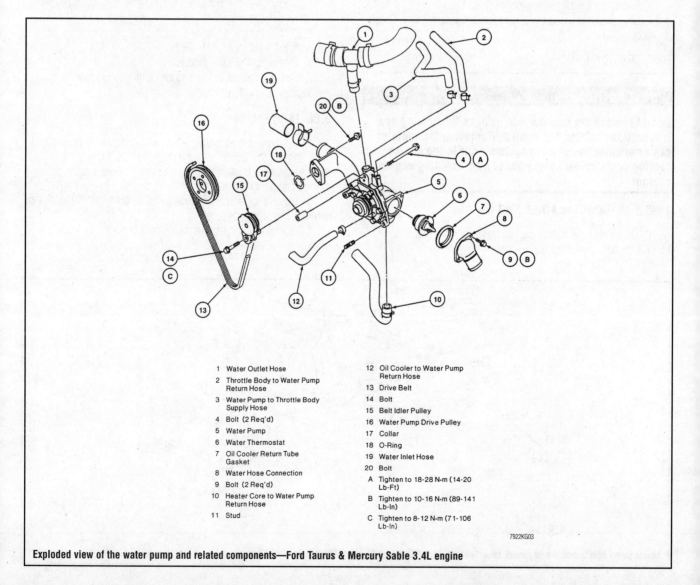

1 Water Outlet Hose	12 Oil Cooler to Water Pump Return Hose
2 Throttle Body to Water Pump Return Hose	13 Drive Belt
3 Water Pump to Throttle Body Supply Hose	14 Bolt
4 Bolt (2 Req'd)	15 Belt Idler Pulley
5 Water Pump	16 Water Pump Drive Pulley
6 Water Thermostat	17 Collar
7 Oil Cooler Return Tube Gasket	18 O-Ring
8 Water Hose Connection	19 Water Inlet Hose
9 Bolt (2 Req'd)	20 Bolt
10 Heater Core to Water Pump Return Hose	A Tighten to 18-28 N·m (14-20 Lb-Ft)
11 Stud	B Tighten to 10-16 N·m (89-141 Lb-In)
	C Tighten to 8-12 N·m (71-106 Lb-In)

7922KG03

Exploded view of the water pump and related components—Ford Taurus & Mercury Sable 3.4L engine

2. Remove the two flange bolts and four cap nuts and remove the engine appearance cover.

3. Remove the battery and battery tray.

4. Remove the water pump drive belt.

5. Disconnect the throttle body return and supply hoses, heater core return hose and the oil cooler return hose from the water pump.

6. Remove the two water hose connection (thermostat housing) retaining bolts and remove the thermostat.

7. Remove the bolts and collar retaining the water pump housing to the left side cylinder head.

8. Disconnect the water outlet and inlet hoses from the water pump and remove the water pump from the engine.

9. Remove the belt idler pulley retaining nut and remove the idler pulley from the water pump.

To install:

10. Clean the water pump to engine and housing gasket sealing surfaces.

11. Install the belt idler pulley and retaining nut and tighten the retaining bolt to 89–141 inch lbs. (10–16 Nm).

12. Apply a silicone lubricant to the water pump inlet and outlet hoses, install them on the water pump housing and clamp securely.

13. Apply a silicone lubricant to the O-ring and install it between the water pump housing and left side cylinder head.

14. Install the bolts retaining the water pump housing to the left side cylinder head. Tighten the two retaining bolts at the water inlet to 71–106 inch lbs. (8–12 Nm). Tighten the two remaining retaining bolts to 14–20 ft. lbs. (18–28 Nm).

15. Install the water thermostat, oil cooler return tube gasket and water hose connection to the water pump housing and tighten the two retaining bolts to 71–106 inch lbs. (8–12 Nm).

16. Connect the throttle body return and supply hoses, heater core return hose and the oil cooler return hose to the water pump.

17. Install the water pump drive belt.

18. Install the engine appearance cover and tighten the two flange bolts and four cap nuts to 71–106 inch lbs. (8–12 Nm).

19. Fill the engine cooling system, then connect the negative battery cable.

20. Start the engine and allow it to reach normal operating temperature, then check for coolant leaks and proper engine operation.

3.8L (OHV) engines

1. Disconnect the negative battery cable. Allow the engine to cool before proceeding.

2. Remove the radiator cap and drain the cooling system by opening the radiator draincock.

3. Support the engine using engine support bar D88L-6000-A or equivalent. Remove the lower nut on both right engine mounts. Raise the engine.

4. Loosen the accessory drive belt idler. Remove the drive belt and water pump pulley. Remove the air suspension pump, if equipped.

5. Remove the power steering pump mounting bracket attaching bolts. Leaving hoses connected, place pump/bracket assembly aside in a position to prevent fluid from leaking out.

6. If equipped with air conditioning, remove the compressor front support bracket. Leave the compressor in place.

7. Disconnect coolant bypass and heater hoses at the water pump.

8. Remove the water pump-to-engine block attaching bolts and remove the pump from the vehicle. Discard the gasket and replace with new.

To install:

9. Lightly oil all bolt and stud threads before installation, except those that require sealant. Thoroughly clean the water pump and front cover gasket contact surfaces.

10. Apply a coating of contact adhesive to both surfaces of the new gasket. Position a new gasket on water pump sealing surface.

11. Position water pump on the front cover and install attaching bolts. Tighten to 15–22 ft. lbs. (20–30 Nm).

12. Install the remaining components in the reverse order of removal.

13. Remove the engine support bar. Fill cooling system to the proper level. Start engine and check for coolant leaks.

LINCOLN CONTINENTAL

⁎⁎ CAUTION

Never open, service or drain the radiator or cooling system when hot; serious burns can occur from the steam and hot coolant.

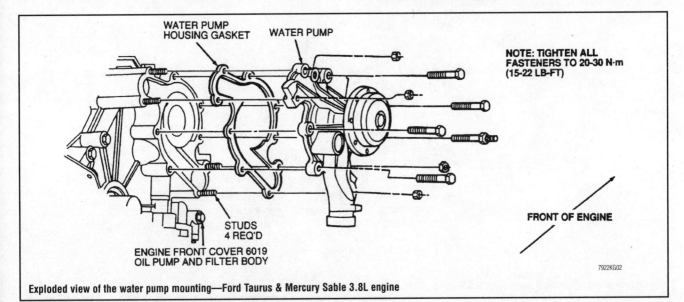

Exploded view of the water pump mounting—Ford Taurus & Mercury Sable 3.8L engine

WATER PUMP HOUSING GASKET   WATER PUMP

NOTE: TIGHTEN ALL FASTENERS TO 20-30 N·m (15-22 LB-FT)

FRONT OF ENGINE

STUDS 4 REQ'D

ENGINE FRONT COVER 6019 OIL PUMP AND FILTER BODY

7922KG02

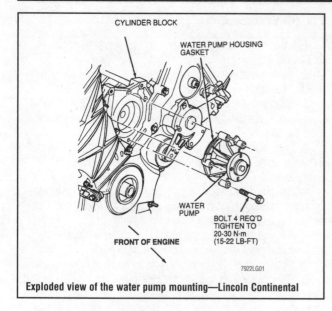

Exploded view of the water pump mounting—Lincoln Continental

1. Disconnect the negative battery cable. Drain the engine cooling system.

2. Remove the coolant recovery reservoir assembly.

3. Loosen the four bolts retaining the water pump pulley to the water pump.

4. Release the drive belt tensioner and remove the drive belt.

5. Remove the four bolts retaining the water pump pulley to the water pump and remove the pulley.

6. Remove the four bolts retaining the water pump to the cylinder block and remove the water pump.

To install:

7. Replace the water pump O-ring and clean the sealing surface of the cylinder block and the water pump.

8. Lubricate the water pump O-ring with fresh coolant and install the water pump into position. Be sure the water pump is fully seated. Install the four water pump retaining bolts and tighten to 15–22 ft. lbs. (20–30 Nm).

9. Install the water pump pulley on the water pump with the four retaining bolts. Tighten to 15–22 ft. lbs. (20–30 Nm).

10. Install the remaining components in the reverse order of removal.

11. Fill the cooling system to the proper level. Connect the negative battery cable, then start the engine and check for coolant leaks.

FORD ESCORT, ESCORT ZX2 & MERCURY TRACER

> **⁂ CAUTION**
>
> **Never open, service or drain the radiator or cooling system when hot; serious burns can occur from the steam and hot coolant. Also, when draining engine coolant, keep in mind that cats and dogs are attracted to ethylene glycol antifreeze and could drink any that is left in an uncovered container or in puddles on the ground. This will prove fatal in sufficient quantities. Always drain coolant into a sealable container. Coolant should be reused unless it is contaminated or is several years old.**

1.8L Engine

1. Disconnect the negative battery cable.
2. Drain the cooling system.
3. Remove the timing belt.
4. Raise and safely support the vehicle.
5. Remove the engine oil dipstick tube bracket bolt(s) from the water pump.

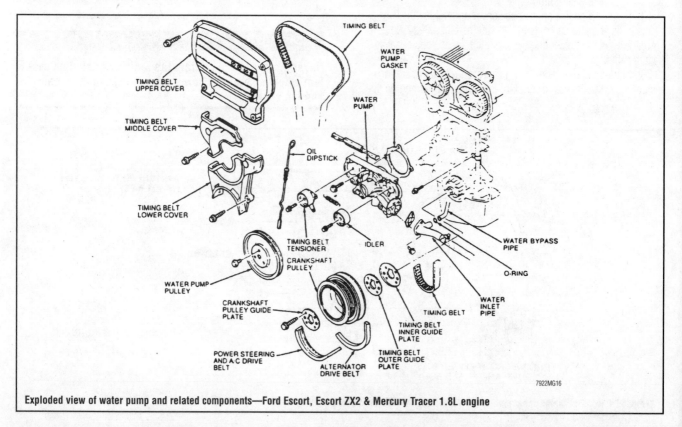

Exploded view of water pump and related components—Ford Escort, Escort ZX2 & Mercury Tracer 1.8L engine

6. Remove the two bolts and the gasket from the water inlet pipe.

7. Remove all but the uppermost water pump mounting bolt.

8. Lower the vehicle.

9. Remove the remaining bolts and the water pump assembly.

10. If the water pump is being reused, remove all gasket material from the water pump.

11. Remove all gasket material from the engine block.

To install:

12. Install a new gasket onto the water pump.

13. Place the water pump into its mounting position, then install the uppermost bolt.

14. Raise and safely support the vehicle.

15. Install the remaining water pump mounting bolts and tighten all bolts to 14–19 ft. lbs. (19–25 Nm).

16. Install a new gasket onto the water inlet pipe.

17. Install the two bolts from the water inlet pipe to the water pump and tighten to 14–19 ft. lbs. (19–25 Nm).

18. Install the bolt to the engine oil dipstick tube bracket.

19. Lower the vehicle.

20. Install the timing belt.

21. Fill the cooling system.

22. Connect the negative battery cable.

23. Start the engine and allow it to reach operating temperature. Check for coolant leaks.

24. Check the coolant level and add coolant, as necessary.

1.9L Engine

1. Disconnect the negative battery cable.

2. Drain the cooling system.

3. Remove the timing belt cover and the timing belt.

4. Raise and safely support the vehicle.

5. Remove the lower radiator hose.

6. Remove the heater hose from the water pump.

7. Lower the vehicle.

8. Support the engine with a suitable floor jack.

9. Remove the right engine mount attaching bolts and roll the engine mount aside.

10. Remove the water pump attaching bolts.

11. Using the floor jack, raise the engine enough to provide clearance for removing the water pump.

12. Remove the water pump and the gasket from the engine through the top of the engine compartment.

To install:

13. Be sure the mating surfaces of the cylinder block and water pump are clean and free of gasket material.

14. If the water pump is to be replaced, transfer the timing belt tensioner components to the new water pump.

15. With the engine supported and raised with a suitable floor jack, place the water pump and the gasket on the cylinder block and install the four attaching bolts. Tighten the bolts to 15–22 ft. lbs. (20–30 Nm).

16. Install the timing belt and cover.

17. Roll the engine mount into position and install the mount bolts. Remove the floor jack.

18. Raise and safely support the vehicle.

19. Install the lower radiator hose and connect the heater hose to the pump.

20. Lower the vehicle.

21. Connect the negative battery cable.

22. Refill the cooling system.

23. Start the engine, allow it to reach normal operating temperature and check for coolant leaks.

24. Check the coolant level and add as necessary.

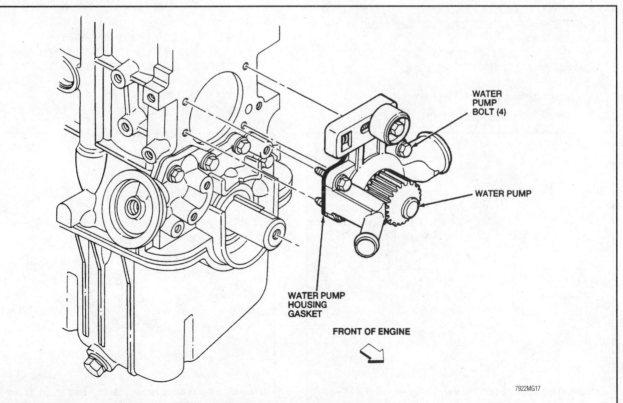

WATER PUMP BOLT (4)

WATER PUMP

WATER PUMP HOUSING GASKET

FRONT OF ENGINE

7922MG17

Exploded view of water pump mounting—Ford Escort, Escort ZX2 & Mercury Tracer 1.9L engine

2.0L Engine

SOHC SPLIT PORT INJECTION (SPI)

1. Disconnect the negative battery cable.
2. Raise the vehicle and support it with safety stands.

> **⁂ CAUTION**
>
> **Never open, service or drain the radiator or cooling system when hot; serious burns can occur from the steam and hot coolant. Also, when draining engine coolant, keep in mind that cats and dogs are attracted to ethylene glycol antifreeze and could drink any that is left in an uncovered container or in puddles on the ground. This will prove fatal in sufficient quantities. Always drain coolant into a sealable container. Coolant should be reused unless it is contaminated or is several years old.**

3. Drain and recycle the engine coolant.
4. Remove the timing belt.
5. Unfasten the timing belt tensioner bolt and remove the tensioner.
6. Disconnect the lower radiator hose from the water pump.
7. Lower the vehicle and disconnect the heater hose from the water pump.
8. Unfasten the three bolts and one stud from the water pump.
9. Remove the water pump.

To install:

> **⁂ CAUTION**
>
> **Do not use any abrasive grinding discs to remove gasket material. Use a plastic manual gasket scraper to remove the gasket residue. Be careful not to scratch or gouge the aluminum sealing surfaces when cleaning them.**

10. Clean the gasket surfaces thoroughly until all traces of the old gasket residue are removed. Inspect the gasket mating surfaces, both must be clean and flat.
11. Install a new gasket and the water pump.
12. Install the water pump bolts and stud to 15–22 ft. lbs. (20–30 Nm).
13. Connect the heater hose to the pump.
14. Raise the vehicle and support it with safety stands.

15. Connect the lower radiator hose to the pump.
16. Install the timing belt tensioner and tighten the bolt to 15–22 ft. lbs. (20–30 Nm).
17. Install the timing belt and lower the vehicle.
18. Fill the cooling system with the proper amount and mixture of coolant.
19. Start the engine and check for leaks.

DOHC ZETEC

1. Disconnect the negative battery cable.

> **⁂ CAUTION**
>
> **Never open, service or drain the radiator or cooling system when hot; serious burns can occur from the steam and hot coolant. Also, when draining engine coolant, keep in mind that cats and dogs are attracted to ethylene glycol antifreeze and could drink any that is left in an uncovered container or in puddles on the ground. This will prove fatal in sufficient quantities. Always drain coolant into a sealable container. Coolant should be reused unless it is contaminated or is several years old.**

2. Drain and recycle the engine coolant.
3. Raise the vehicle and support it with safety stands.
4. Unfasten the splash shield bolts and remove the shield.
5. Loosen the water pump pulley retaining bolts, then remove the drive belt.
6. Remove the A/C compressor bolts and move the compressor aside.
7. Unfasten the water pump bolts and remove the pump from the middle timing belt cover.

To install:

8. Install the water pump and tighten the bolts to 17 ft. lbs. (24 Nm).
9. Place the A/C compressor in position and tighten the mounting bolts.
10. Install the drive belt and tighten the water pump pulley bolts.
11. Install the splash shield and tighten the bolts.
12. Fill the cooling system with the proper amount and mixture of coolant.
13. Connect the negative battery cable.
14. Start the engine and check for leaks.

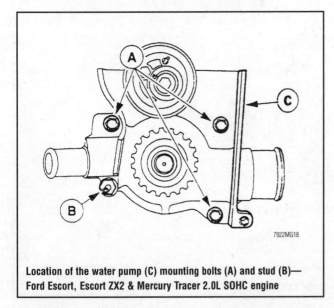

Location of the water pump (C) mounting bolts (A) and stud (B)— Ford Escort, Escort ZX2 & Mercury Tracer 2.0L SOHC engine

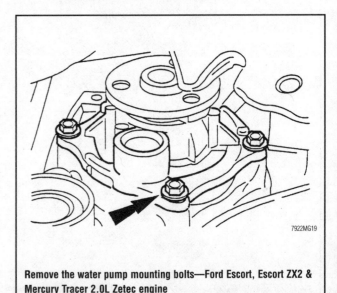

Remove the water pump mounting bolts—Ford Escort, Escort ZX2 & Mercury Tracer 2.0L Zetec engine

FORD MUSTANG

3.8L Engine

> ❊❊ **CAUTION**
>
> Never open, service or drain the radiator or cooling system when hot; serious burns can occur from the steam and hot coolant.

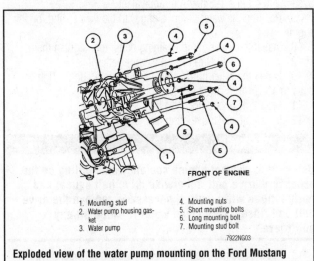

1. Mounting stud
2. Water pump housing gasket
3. Water pump
4. Mounting nuts
5. Short mounting bolts
6. Long mounting bolt
7. Mounting stud bolt

.7922NG03

Exploded view of the water pump mounting on the Ford Mustang 3.8L engine—during removal, note the original positions of the different length bolts for reassembly

1. Drain the engine cooling system.
2. Remove the electric cooling fan from the radiator.
3. Slightly loosen the water pump pulley bolts.
4. Remove the accessory drive belt.
5. Remove the water pump pulley retaining bolts, then separate the pulley from the water pump flange.
6. Remove the ignition coil bracket mounting nuts and bolts, then position the ignition coil (with all wires still attached) aside.
7. Remove the power steering pump pulley, then remove the water pump-to-power steering pump brace.
8. Remove the heater water outlet tube retaining bolts, then separate the outlet tube from the water pump.
9. Detach the lower radiator hose from the water pump.

> ❊❊ **WARNING**
>
> If it is necessary to utilize a prytool to separate the water pump from the engine, use caution not to damage the water pump and engine mating surfaces.

10. Loosen the water pump bolts, then pull the water pump off of the engine. Remove and discard the old water pump gasket.

To install:

➥ **Lightly oil all bolt and stud threads prior to assembly, except those specifying sealant, with clean engine oil.**

11. Clean the gasket mating surfaces of the water pump and engine of all old gasket material and dirt.
12. Apply adhesive, such as Ford Gasket and Trim Adhesive F3AZ-19B508-B or equivalent, to the water pump gasket, then position the water pump and the new gasket on the engine.

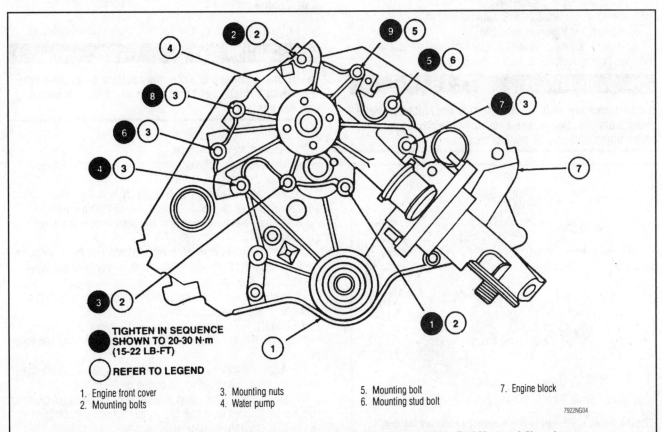

● **TIGHTEN IN SEQUENCE SHOWN TO 20-30 N·m (15-22 LB-FT)**

○ **REFER TO LEGEND**

1. Engine front cover
2. Mounting bolts
3. Mounting nuts
4. Water pump
5. Mounting bolt
6. Mounting stud bolt
7. Engine block

7922NG04

When tightening the mounting fasteners, be sure to adhere to the sequence shown for proper sealing—Ford Mustang 3.8L engine

➡The threads of the No. 1 water pump retaining bolt must be coated with a Teflon® sealant, such as Ford Pipe Sealant with Teflon® D8AZ-19554-A or equivalent, prior to installation.

13. Install the water pump retaining bolts, studs and nuts, then tighten, in the sequence shown in the accompanying illustration, the retaining bolts and studs to 15–22 ft. lbs. (20–30 Nm) and the nuts to 71–106 inch lbs. (8–12 Nm).

14. Inspect the O-ring seal on the heater water outlet tube for damage, and replace it if necessary. Then, install the outlet tube to the water pump. Tighten the retaining bolts to 71–106 inch lbs. (8–12 Nm).

15. Attach the lower radiator hose to the water pump. Ensure that the radiator hose clamps are tightened properly.

16. Install the water pump-to-power steering pump brace, then install the power steering pump pulley.

17. Install the water pump pulley and retaining bolts. Tighten the bolts to 15–21 ft. lbs. (21–29 Nm).

18. Install the accessory drive belt.

19. Install the electric cooling fan.

20. Fill the cooling system with the proper amount and type of coolant.

21. Start the engine and check for coolant leaks.

4.6L Engines

❊❊ CAUTION

Never open, service or drain the radiator or cooling system when hot; serious burns can occur from the steam and hot coolant.

1. Drain the engine cooling system.
2. Remove the electric cooling fan from the radiator.
3. Remove the accessory drive belt.
4. Remove the water pump pulley by loosening the retaining bolts and pulling it off of the water pump flange.

❊❊ WARNING

If it is necessary to utilize a prytool to separate the water pump from the engine, use caution not to damage the water pump and engine mating surfaces.

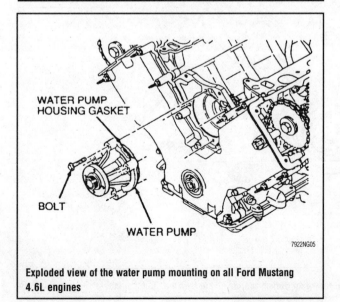

WATER PUMP HOUSING GASKET

BOLT

WATER PUMP

7922NG05

Exploded view of the water pump mounting on all Ford Mustang 4.6L engines

5. Loosen the water pump bolts, then separate the water pump from the engine. Remove and discard the old water pump O-ring.

To install:

➡Lightly oil all bolt and stud threads prior to assembly, except those specifying sealant, with clean engine oil.

6. Clean the mating surfaces of the water pump and engine of all corrosion and dirt.

7. Install a new O-ring in the groove on the water pump, then lubricate the O-ring with coolant, such as Ford Premium Cooling System Fluid E2Fz-19549-AA or equivalent.

8. Position the water pump, along with the new O-ring, on the engine.

9. Install the water pump retaining bolts, then tighten them 15–22 ft. lbs. (20–30 Nm) in a crisscross pattern.

10. Install the water pump pulley and retaining bolts. Tighten the bolts to 15–21 ft. lbs. (21–29 Nm).

11. Install the accessory drive belt.

12. Install the electric cooling fan.

❊❊ WARNING

Use care to prevent engine coolant from spilling on the accessory drive belt, otherwise drive belt squeal and early fatigue will result. If necessary, remove the drive belt and rinse it with clean water to clean off any antifreeze.

13. Fill the cooling system with the proper amount and type of coolant.

14. Start the engine and check for coolant leaks.

5.0L Engine

1. Disconnect the negative battery cable for safety purposes.

❊❊ CAUTION

Never open, service or drain the radiator or cooling system when hot; serious burns can occur from the steam and hot coolant.

2. Drain the cooling system.
3. Remove the air inlet tube.
4. Remove the fan shroud attaching bolts and position the shroud over the fan.
5. Remove the fan and clutch assembly from the water pump shaft, then remove the clutch, fan and shroud from the vehicle.
6. Remove the accessory drive belt, then remove the water pump pulley.
7. Remove all accessory brackets that attach to the water pump.
8. Disconnect the lower radiator hose, heater hose and water pump bypass hose from the water pump.
9. Remove the water pump attaching bolts, then remove the water pump and discard the gasket.

To install:

10. Clean all old gasket material from the timing cover and water pump.

11. Apply a suitable waterproof sealing compound to both sides of a new gasket, then position the gasket on the timing cover.

12. Position the water pump carefully over the gasket (making sure not to dislodge it). Install and tighten the pump mounting bolts to 12–18 ft. lbs. (16–24 Nm).

13. Connect the hoses and accessory brackets to the water pump.

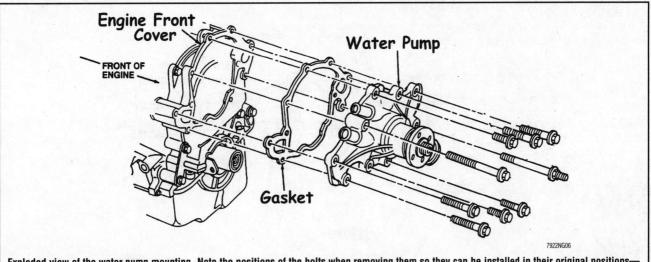

Exploded view of the water pump mounting. Note the positions of the bolts when removing them so they can be installed in their original positions—Ford Mustang 5.0L engines

14. Install the pulley on the water pump shaft.
15. Install the shroud along with the clutch and fan assembly.
16. Route and install the accessory drive belt.
17. Connect the negative battery cable and refill the cooling system.
18. Run the engine and check for leaks.

LINCOLN MARK VIII

✳✳ CAUTION

Never open, service or drain the radiator or cooling system when hot; serious burns can occur from the steam and hot coolant.

1. Disconnect the negative battery cable and drain the cooling system.
2. Remove the four bolts retaining the pulley to the water pump shaft.
3. Release belt tensioner and remove the drive belts. Remove the four bolts retaining water pump pulley and remove the pulley.
4. Remove the four water pump retaining bolts and remove the water pump.
   To install:
5. Install a new water pump O-ring seal.
6. Installation is the reverse of the removal procedure. Clean all mating surfaces prior to installation. Tighten the water pump bolts and pulley retaining bolts to 15–22 ft. lbs. (20–30 Nm).
7. Fill and bleed the cooling system. Operate the engine until normal operating temperatures have been reached and check for leaks.

FORD THUNDERBIRD & 1995–97 MERCURY COUGAR

3.8L Engines

✳✳ CAUTION

Never open, service or drain the radiator or cooling system when hot; serious burns can occur from the steam and hot coolant. Also, when draining engine coolant,
keep in mind that cats and dogs are attracted to ethylene glycol antifreeze and could drink any that is left in an uncovered container or in puddles on the ground. This will prove fatal in sufficient quantities. Always drain coolant into a sealable container. Coolant should be reused unless it is contaminated or is several years old.

1. Disconnect the negative battery cable and drain the cooling system.
2. On all except supercharged engine, remove the fan/clutch assembly and shroud.
3. Rotate the main accessory drive belt tensioner. Remove the main drive belt and water pump pulley.
4. Remove the power steering pump pulley and remove the water pump to power steering pump brace.

➡**On supercharged engines, it may be necessary to remove the charge air cooler to gain access to the power steering pump pulley.**

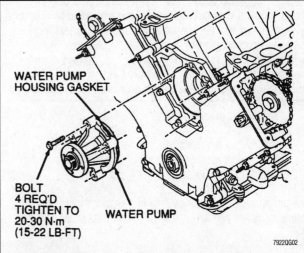

Exploded view of the water pump mounting—Lincoln Mark VIII, Ford Thunderbird and 1995–97 Mercury Cougar 4.6L engine

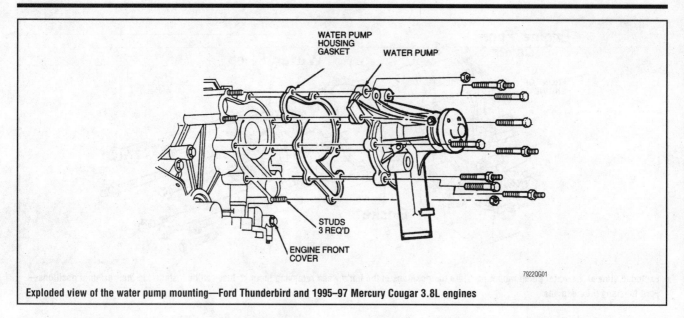

WATER PUMP HOUSING GASKET

WATER PUMP

STUDS 3 REQ'D

ENGINE FRONT COVER

7922QG01

Exploded view of the water pump mounting—Ford Thunderbird and 1995–97 Mercury Cougar 3.8L engines

5. On all except supercharged engine, disconnect the coolant bypass hose(s) and the heater hose at the water pump. On supercharged engine, disconnect the oil cooler tube and bypass hose and remove the upper crankshaft sensor cover.

6. Disconnect the lower radiator hose. Remove the water pump retaining bolts and the pump. If a prybar is used to assist removal, be careful not to damage the mating surfaces.

7. Installation is the reverse of the removal procedure. Clean all gasket mating surfaces prior to installation. Tighten the water pump retaining bolts to 15–22 ft. lbs. (20–30 Nm). Fill and bleed the cooling system. Operate the engine until normal operating temperatures have been reached and check for leaks.

➡ **The threads of the No. 1 water pump retaining bolt must be coated with pipe sealant before installing.**

4.6L Engine

❉❉ **CAUTION**

Never open, service or drain the radiator or cooling system when hot; serious burns can occur from the steam and hot coolant. Also, when draining engine coolant, keep in mind that cats and dogs are attracted to ethylene glycol antifreeze and could drink any that is left in an uncovered container or in puddles on the ground. This will prove fatal in sufficient quantities. Always drain coolant into a sealable container. Coolant should be reused unless it is contaminated or is several years old.

1. Disconnect the negative battery cable and drain the cooling system.

2. Release the drive belt tensioner and remove the belt.

3. Remove the water pump pulley bolts and pulley.

4. Remove the pump-to-block bolts and pump.

5. Installation is the reverse of the removal procedure. Clean all gasket mating surfaces prior to installation. Tighten the water pump bolts and pulley retaining bolts to 15–22 ft. lbs. (20–30 Nm).

6. Fill and bleed the cooling system. Operate the engine until normal operating temperatures have been reached and check for leaks.

FORD CROWN VICTORIA, LINCOLN TOWN CAR & MERCURY GRAND MARQUIS

1. Disconnect the negative battery cable.

❉❉ **CAUTION**

Never open, service or drain the radiator or cooling system when hot; serious burns can occur from the steam and hot coolant.

2. Drain the cooling system, remove the cooling fan and the shroud.

3. Release the belt tensioner and remove the accessory drive belt.

4. Remove the four bolts retaining the water pump pulley to the water pump and remove the pulley.

5. Remove the four bolts retaining the water pump to the engine assembly and remove the water pump.

To install:

6. Installation is the reverse of the removal procedure. Be sure to clean the sealing surfaces of the water pump and block and use a

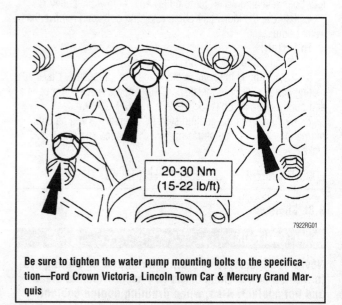

20-30 Nm (15-22 lb/ft)

7922RG01

Be sure to tighten the water pump mounting bolts to the specification—Ford Crown Victoria, Lincoln Town Car & Mercury Grand Marquis

new O-ring. Lubricate the O-ring with clean anti-freeze prior to installation.

7. Tighten the water pump-to-engine bolts and the pulley-to-water pump bolts to 15–22 ft. lbs. (20–30 Nm). Fill and bleed the cooling system. Operate the engine until normal operating temperatures have been reached and check for leaks.

General Motors Corporation

A BODY—BUICK CENTURY, OLDSMOBILE CUTLASS CIERA & CUTLASS CRUISER

2.2L Engine

1. Disconnect the negative battery cable.

✲✲ CAUTION

Never open, service or drain the radiator or cooling system when hot; serious burns can occur from the steam and hot coolant.

2. Drain the cooling system into a suitable container.
3. Loosen, but do not remove, the water pump pulley bolts.
4. Remove the serpentine belt.
5. Remove the alternator mounting bolts and set the alternator aside.
6. Remove the water pump pulley bolts and remove the water pump pulley.
7. Remove the four water pump mounting bolts and remove the water pump.
   To install:
8. Clean all the gasket surfaces completely.
9. Apply a thin bead of sealer around the outer edge of the water pump gasket seating area and place he gasket on the pump.
10. Install the water pump on the engine and tighten the four mounting bolts to 18 ft. lbs. (25 Nm).
11. Install the water pump pulley and tighten the mounting bolts finger-tight.
12. Install the alternator in the mounting bracket.
13. Install the serpentine belt.

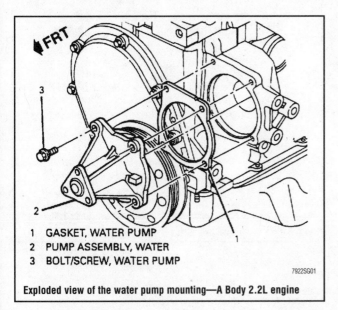

1 GASKET, WATER PUMP
2 PUMP ASSEMBLY, WATER
3 BOLT/SCREW, WATER PUMP
7922SG01

Exploded view of the water pump mounting—A Body 2.2L engine

14. Tighten the water pump pulley mounting bolts to 22 ft. lbs. (30 Nm).
15. Connect the negative battery cable.
16. Refill and bleed the cooling system.

3.1L Engine

1. Disconnect the negative battery cable.

✲✲ CAUTION

Never open, service or drain the radiator or cooling system when hot; serious burns can occur from the steam and hot coolant.

2. Drain the cooling system into a suitable container.
3. Loosen, but do not remove, the water pump pulley bolts.
4. Remove the serpentine belt.
5. Remove the water pump pulley bolts and remove the water pump pulley.
6. Remove the five water pump mounting bolts and remove the water pump.
   To install:
7. Clean all the gasket surfaces completely.
8. Apply a thin bead of sealer around the outside edge of the water pump along the gasket sealing area and install the gasket onto the water pump.
9. Install the water pump on the engine and tighten the water pump mounting bolts to 89 inch lbs. (10 Nm).
10. Install the water pump pulley and tighten the pulley bolts finger-tight.
11. Install the serpentine belt.
12. Tighten the water pump pulley bolts to 18 ft. lbs. (25 Nm).
13. Connect the negative battery cable.
14. Refill and bleed the cooling system.

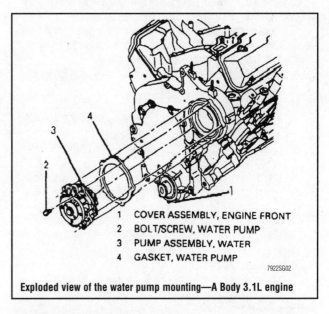

1 COVER ASSEMBLY, ENGINE FRONT
2 BOLT/SCREW, WATER PUMP
3 PUMP ASSEMBLY, WATER
4 GASKET, WATER PUMP
7922SG02

Exploded view of the water pump mounting—A Body 3.1L engine

B BODY—BUICK ROADMASTER, CADILLAC FLEETWOOD, CHEVROLET CAPRICE & IMPALA SS

1. Disconnect the negative battery cable.
2. Disengage the wiring harness from the cooling fan assembly, then remove the assembly from the vehicle.

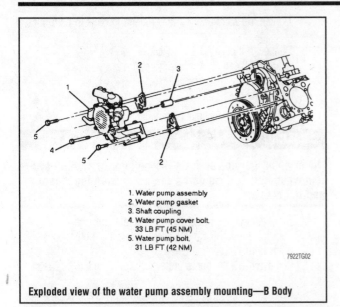

1. Water pump assembly
2. Water pump gasket
3. Shaft coupling
4. Water pump cover bolt.
   33 LB FT (45 NM)
5. Water pump bolt.
   31 LB FT (42 NM)

7922TG02

Exploded view of the water pump assembly mounting—B Body

✳✳ CAUTION

Never open, service or drain the radiator or cooling system when hot; serious burns can occur from the steam and hot coolant.

3. Drain the engine cooling system, removing the block drain plug and the knock sensor to assure proper draining. Reinstall the drain plug and knock sensor, as soon as the system is empty.

4. Disconnect the upper and lower radiator hoses from the water pump assembly.

5. Remove the heater hose assemblies from the water pump and from the throttle body.

6. Disengage the coolant sensor wiring harness, then reposition the ignition coil and bracket assembly.

7. Remove the shorter water pump retaining bolt from the center of each pump mating flange, then remove the longer pump bolts from either side of the center bolts.

8. Carefully remove the water pump assembly and gaskets along with the pump shaft coupling.

To install:

9. Thoroughly clean the gasket mating surfaces of any remaining gasket material.

10. Install the water pump shaft coupling along with the water pump and gaskets.

11. Install the longer pump bolts and tighten 33 ft. lbs. (45 Nm), then install the shorter bolts and tighten to 31 ft. lbs. (42 Nm).

12. Reposition the ignition coil and bracket assembly.

13. Engage the coolant sensor electrical connector.

14. Install the heater and radiator hoses to the throttle body and water pump, as applicable.

15. Install the air cleaner and intake duct assemblies.

16. Install the engine cooling fan assembly and engage the wiring harness connector.

17. Connect the negative battery cable and properly fill the engine cooling system.

C & H BODIES—BUICK LE SABRE, PARK AVE., OLDSMOBILE EIGHTY EIGHT, EIGHTY EIGHT LS, LSS, REGENCY & PONTIAC BONNEVILLE

3.8L (VIN L and K) Engines

1. Disconnect the negative battery cable.

✳✳ CAUTION

Never open, service or drain the radiator or cooling system when hot; serious burns can occur from the steam and hot coolant.

2. Drain the cooling system into an approved container.

3. Remove the accessory drive belt following the procedure in the belt removal section.

4. Disconnect the coolant hoses from the water pump.

5. Remove the water pump pulley bolts. (The long bolt can be removed by lining the head of the bolt up with the hole in the frame rail). Remove the pulley.

6. The following step is necessary only on 1995 engines:

   a. Support the engine using engine support fixture tool J 28467-A or equivalent.

   b. Remove the front engine mount.

7. Remove the water pump mounting bolts and remove the water pump.

To install:

8. Clean all the sealing surfaces.

9. Apply a thin bead of sealer around the outside edge of the water pump and install the gasket on the pump.

10. Install the water pump on the engine. Tighten the water pump-to-engine block bolts to 22 ft. lbs. (30 Nm) and the water pump-to-front cover bolts to 11 ft. lbs. (15 Nm) plus an additional 80 degree turn.

11. Install the front engine mount, if removed.

12. Install the water pump pulley. Tighten the water pump pulley bolts to 115 inch lbs. (13 Nm).

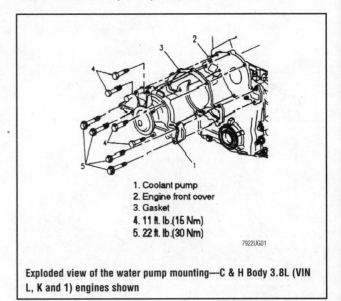

1. Coolant pump
2. Engine front cover
3. Gasket
4. 11 ft. lb.(15 Nm)
5. 22 ft. lb.(30 Nm)

7922UG01

Exploded view of the water pump mounting—C & H Body 3.8L (VIN L, K and 1) engines shown

13. Connect the coolant hoses to the water pump and secure clamps.

14. Install the drive belt following the procedure in the belt section.

15. Refill and bleed the cooling system following proper procedures.

16. Run the engine and check for leaks.

17. Recheck the coolant level when the engine has cooled.

3.8L (VIN 1) Engine

1995 MODELS

1. Disconnect the negative battery cable.

> ⁂ **CAUTION**
>
> **Never open, service or drain the radiator or cooling system when hot; serious burns can occur from the steam and hot coolant.**

2. Drain the cooling system into an approved container.
3. Remove the accessory drive and supercharger belts.
4. Remove the alternator and brace.
5. Disconnect the hoses and pipes from the water pump.
6. Remove the pulley bracket assembly.
7. Raise and safely support the vehicle.
8. Remove the power steering pump and lines.
9. Lower the vehicle.
10. Support the engine using engine support fixture J 28467-A or equivalent, and remove the front engine mount.
11. Remove the water pump pulley.
12. Remove the water pump mounting bolts and remove the water pump from the vehicle.

To install:

13. Clean all sealing surfaces.
14. Apply a thin bead of sealant to the gasket mating surface of the water pump and install a new gasket to the pump.
15. Install the water pump on the engine. Tighten the pump-to-block bolts to 22 ft. lbs. (30 Nm) and the pump-to-front cover bolts to 11 ft. lbs. (15 Nm) plus an additional 80 degree turn.
16. Install the water pump pulley and tighten the bolts to 115 inch lbs. (13 Nm).
17. Install the front engine mount following the proper procedure.
18. Raise and safely support the vehicle.
19. Install the power steering pump and lines following the proper procedure.
20. Lower the vehicle.
21. Install the pulley bracket assembly.
22. Connect the hoses and pipes from the water pump.
23. Install the alternator and brace.
24. Install the supercharger and accessory drive and belts.
25. Connect the negative battery cable.
26. Fill and bleed the cooling system following the proper procedure.
27. Run the engine and check for leaks.
28. Recheck the coolant level when the engine has cooled.

1996–99 MODELS

1. Disconnect the negative battery cable.
2. Drain the cooling system into an approved container.
3. Remove the A/C compressor splash shield.
4. Remove the supercharger and accessory drive belts.

5. Remove the coil pack and position out of the way.
6. Remove the supercharger belt tensioner.
7. Support the engine using engine support fixture J 28467-A or equivalent, and remove the front engine mount.
8. Remove the power steering pump.
9. Remove the engine mount bracket and the idler pulley.
10. Remove the water pump pulley.
11. Remove the water pump mounting bolts and remove the water pump.

To install:

12. Clean all sealing surfaces.
13. Apply a thin bead of sealer around the outside edge of the water pump and install a new gasket on the pump.
14. Install the water pump on the engine and tighten the short bolts to 11 ft. lbs. (15 Nm). Tighten the long bolts to 22 ft. lbs. (30 Nm).
15. Install the water pump pulley and tighten bolts to 115 inch lbs. (13 Nm).
16. Install the engine mount bracket and the idler pulley.
17. Install the power steering pump.
18. Install the front engine mount.
19. Install the supercharger belt tensioner.
20. Install the ignition coil pack assembly.
21. Install the supercharger and accessory drive belts.
22. Install the A/C compressor splash shield.
23. Connect the negative battery cable.
24. Fill and bleed the cooling system following the proper procedure.
25. Run the engine and check for leaks.
26. Recheck the coolant level when the engine has cooled.

E & K BODIES—CADILLAC DEVILLE, CONCOURS, ELDORADO & SEVILLE

➡ **On 4.6L engines, there was a design change to the water pump inlet housing. If it is a black plastic housing, the housing must be replaced. There are no seals available for the plastic housings. The new housings are made of aluminum. When ordering parts, be sure to ask for a water pump, water pump seal and an inlet housing seal and housing if needed.**

1. Disconnect the negative battery cable.
2. Drain the coolant into a suitable container.
3. Remove the accessory drive belt. Remove the water pump pulley.
4. On 4.6L engines, remove the following:
   a. Remove the air cleaner.
   b. Remove the lower radiator hose and remove the thermostat bypass hose from the coolant inlet housing.
5. On 4.9L engines, remove the coolant overflow tank.
6. Remove the water pump mounting bolts.
7. On 4.6L engines, remove the water pump from the vehicle by rotating clockwise with tool J-38816-A or equivalent. Remove the O-ring and clean out the groove.

To install:

8. Clean all gasket mating surfaces.
9. On 4.6L engines, install the O-ring seal into the groove, then install the water pump, turning it counterclockwise until it stops, using tool J-38816-A or equivalent.
10. Install the water pump housing and tighten the bolts to 88 inch lbs. (10 Nm).
11. On 4.9L engines, install the new water pump gasket on the front cover studs with the raised sealing surface facing outward.

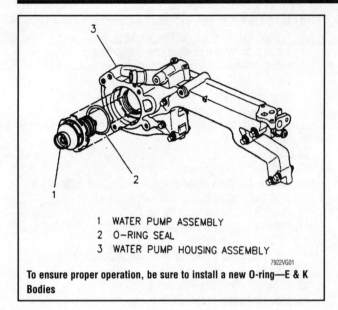

1 WATER PUMP ASSEMBLY
2 O—RING SEAL
3 WATER PUMP HOUSING ASSEMBLY

7922VG01

To ensure proper operation, be sure to install a new O-ring—E & K Bodies

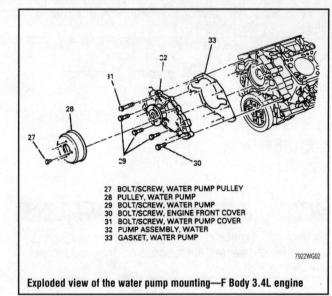

27 BOLT/SCREW, WATER PUMP PULLEY
28 PULLEY, WATER PUMP
29 BOLT/SCREW, WATER PUMP
30 BOLT/SCREW, ENGINE FRONT COVER
31 BOLT/SCREW, WATER PUMP COVER
32 PUMP ASSEMBLY, WATER
33 GASKET, WATER PUMP

7922WG02

Exploded view of the water pump mounting—F Body 3.4L engine

12. Install the water pump on the engine and tighten the mounting bolts.
13. Install the water pump pulley. Install the pulley bolts finger-tight.
14. Install the accessory drive belt.
15. Tighten the water pump pulley bolts to 115 inch lbs. (13 Nm).
16. On 4.6L engines, install the air cleaner.
17. On 4.9L engines, install and connect the coolant reservoir.
18. Connect the negative battery cable.
19. Refill and bleed the cooling system.
20. Run the engine and check for leaks.

F BODY—CHEVROLET CAMARO, Z28, PONTIAC FIREBIRD & TRANS AM

3.4L Engine

1. Disconnect the negative battery cable.

✳✳ CAUTION

Never open, service or drain the radiator or cooling system when hot; serious burns can occur from the steam and hot coolant.

2. Drain the cooling system into a suitable container.
3. Remove the air intake duct.
4. Remove the top coil of the ignition coil pack.
5. Loosen the tensioner pulley bolt.
6. Disconnect the heater hose from the water pump.
7. Loosen the water pump pulley bolts.
8. Unfasten the power steering bracket bolts.
9. Remove the serpentine drive belt.
10. Remove the water pump pulley.
11. Remove the power steering pump bracket and swing the bracket and pump aside. Do NOT disconnect the power steering lines from the pump.
12. Unfasten the water pump and engine front cover bolts, then remove the water pump assembly from the vehicle.
13. Remove and discard the gasket and thoroughly clean the gasket mating surfaces.

To install:

14. Position a new water pump gasket, then install the pump. Secure using all of the water pump retaining bolts, except the bottom three, and tighten to 33 ft. lbs. (45 Nm).
15. Install the engine front cover bolts and tighten to 97 inch lbs. (11 Nm).
16. Install the three bottom water pump retaining screws and tighten to 89 inch lbs. (10 Nm).
17. Place the power steering pump into position and install the pump bracket.
18. Install the water pump pulley.
19. Install the serpentine drive belt.
20. Install the power steering pump bracket retaining bolts and tighten to 30 ft. lbs. (40 Nm).
21. Connect the heater hose to the water pump.
22. Install the top coil to the ignition coil pack.
23. Attach the electrical connector to the alternator.
24. Install the air inlet duct and air cleaner assembly.
25. Connect the negative battery cable.
26. Fill the cooling system. Start the engine and check for leaks.

3.8L Engine

1. Disconnect the negative battery cable.

✳✳ CAUTION

Never open, service or drain the radiator or cooling system when hot; serious burns can occur from the steam and hot coolant.

2. Drain the cooling system into a suitable container.
3. Remove the serpentine belt.
4. Disconnect the radiator inlet hose from the water pump.
5. Unfasten the water pump pulley bolts and remove the pulley.
6. Remove the water pump pulley retaining bolts, then remove the pump from the vehicle.
7. Remove and discard the gasket and thoroughly clean the gasket mating surfaces.

To install:

8. Position a new water pump gasket, then install the pump. Secure using the water pump retaining bolts and tighten to 11 ft. lbs. (15 Nm), plus an additional 80 degree rotation.

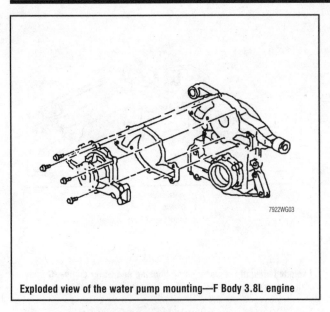

Exploded view of the water pump mounting—F Body 3.8L engine

9. Install the water pump pulley and retaining bolts. Tighten to 115 inch lbs. (13 Nm).
10. Install the serpentine drive belt.
11. Connect the radiator inlet hose to the water pump.
12. Connect the negative battery cable.
13. Fill the cooling system. Start the engine and check for leaks.

5.7L Engines

1. Unplug the electrical connector from the cooling fan.
2. Remove both cooling fan assemblies.

✳✳ CAUTION

Never open, service or drain the radiator or cooling system when hot; serious burns can occur from the steam and hot coolant.

3. Place a suitable drain pan under the vehicle, then remove the block coolant drain plug and the knock sensor. Drain the radiator.
4. Install the drain plug and knock sensor.
5. Remove the air intake duct and air cleaner.
6. Disconnect the upper and lower radiator hoses from the water pump.
7. Remove the heater hoses from the water pump and throttle body.
8. Unplug the electrical connector from the coolant sensor.
9. Unfasten the retainers and reposition the ignition coil and bracket.
10. If equipped, remove the air pump and bracket.
11. Unfasten the water pump attaching bolts, then remove the water pump and discard the gasket. Remove the shaft coupling and water pump driveshaft seals, if necessary.
12. Thoroughly clean the gasket mating surfaces

To install:

13. If removed, install the shaft coupling and water pump driveshaft using J 39089 or equivalent water pump driveshaft O-ring installer.
14. Install the water pump with a new gasket. Tighten the bolts to 30–32 ft. lbs. (41–43 Nm).
15. If equipped, install the air pump and bracket.
16. Reposition and secure the ignition coil and bracket.
17. Attach the coolant sensor connector.
18. Connect the heater hoses to the water pump and throttle body.
19. Connect the upper and lower radiator hoses to the water pump.
20. Install the air cleaner and air inlet duct.
21. Install the electric cooling fan(s), then attach the electrical connector.
22. Refill the cooling system.
23. Connect the negative battery cable, then start the engine and check for leaks.

G BODY—BUICK RIVIERA & OLDSMOBILE AURORA

3.8L (VIN 1 and K) Engines

1. Disconnect the negative battery cable.

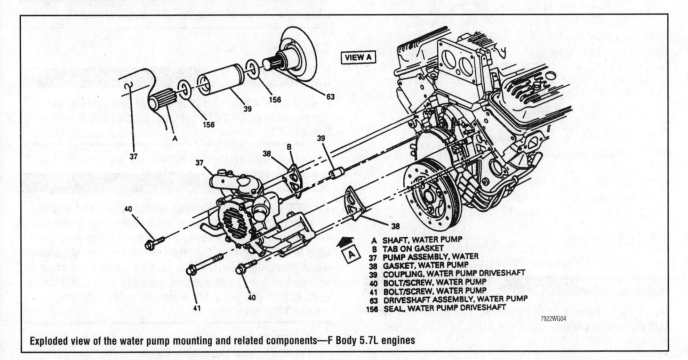

VIEW A

A	SHAFT, WATER PUMP
B	TAB ON GASKET
37	PUMP ASSEMBLY, WATER
38	GASKET, WATER PUMP
39	COUPLING, WATER PUMP DRIVESHAFT
40	BOLT/SCREW, WATER PUMP
41	BOLT/SCREW, WATER PUMP
63	DRIVESHAFT ASSEMBLY, WATER PUMP
156	SEAL, WATER PUMP DRIVESHAFT

Exploded view of the water pump mounting and related components—F Body 5.7L engines

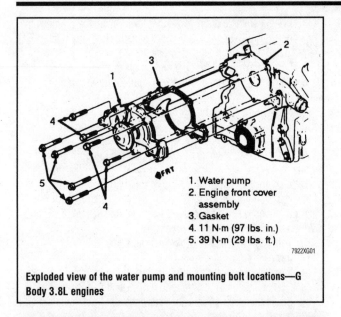

1. Water pump
2. Engine front cover assembly
3. Gasket
4. 11 N·m (97 lbs. in.)
5. 39 N·m (29 lbs. ft.)

7922XG01

Exploded view of the water pump and mounting bolt locations—G Body 3.8L engines

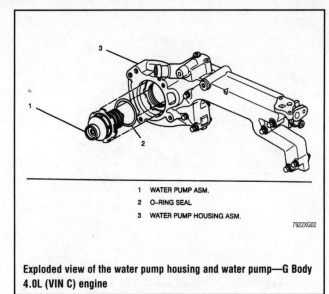

1 WATER PUMP ASM.
2 O-RING SEAL
3 WATER PUMP HOUSING ASM.

7922XG02

Exploded view of the water pump housing and water pump—G Body 4.0L (VIN C) engine

✳✳ CAUTION

Never open, service or drain the radiator or cooling system when hot; serious burns can occur from the steam and hot coolant.

2. Drain the cooling system into a suitable container.
3. Remove the serpentine belt.
4. Disconnect the heater and bypass hoses from the water pump.
5. Remove the water pump pulley bolts. The long bolt can be removed by lining the head of the bolt up with the hole in the frame rail. Remove the pulley.
6. Install an engine support fixture and remove the torque axis mount.
7. Remove the water pump mounting bolts and remove the water pump.

To install:
8. Thoroughly clean all the sealing surfaces.
9. Apply a thin bead of sealer around the outside edge of the water pump and install the gasket on the pump.
10. Install the water pump on the engine and tighten the water pump bolts on the sides of the water inlet and outlet to 29 ft. lbs. (39 Nm) and the remainder of the bolts to 97 inch lbs. (11 Nm).
11. Reinstall the torque axis mount and remove the support fixture.
12. Reinstall the water pump pulley and tighten the bolts finger-tight.
13. Reconnect the hoses to the water pump.
14. Reinstall the serpentine belt.
15. Tighten the water pump pulley bolts to 115 inch lbs. (13 Nm).
16. Refill the cooling system.
17. Reconnect the negative battery cable.
18. Start the vehicle and check for proper operation.

4.0L (VIN C) Engine

1. Disconnect the negative battery cable.

✳✳ CAUTION

Never open, service or drain the radiator or cooling system when hot; serious burns can occur from the steam and hot coolant.

2. Drain the cooling system.
3. Remove the air inlet duct.
4. Remove the water pump drive belt cover.
5. Remove the water pump drive belt and set aside.
6. Disconnect the lower radiator hose and bypass hose.
7. Remove the thermostat housing from the water pump housing.
8. Remove the water pump from the water pump housing by rotating turning the locking ring with J-38816, or an equivalent water pump remover/installer.

To install:
9. Install the water pump and seat the locking ring using J-38816.
10. Reinstall the thermostat housing to the water pump housing.
11. Reconnect the lower radiator hose and coolant bypass hose.
12. Reinstall the drive belt and drive belt cover.
13. Reinstall the air inlet duct.
14. Reconnect the negative battery cable.
15. Refill and bleed the cooling system.

J BODY—CHEVROLET CAVALIER & PONTIAC SUNFIRE

2.2L Engine

✳✳ WARNING

When adding coolant, it is important that you use GM Goodwrench DEX-COOL® coolant meeting GM Specification 6277M.

1. Disconnect the negative battery cable.

✳✳ CAUTION

The EPA warns that prolonged contact with used engine oil may cause a number of skin disorders, including cancer! You should make every effort to minimize your exposure to used engine oil. Protective gloves should be worn when changing the oil. Wash your hands and any other exposed skin areas as soon as possible after exposure to used engine oil. Soap and water, or waterless hand cleaner should be used.

2. Drain the cooling system into a suitable container.
3. Loosen, but do not remove, the water pump pulley bolts.

4. Remove the serpentine belt.

5. Remove the alternator mounting bolts and set the alternator aside.

6. Remove the water pump pulley bolts, then remove the pulley.

7. Remove the 4 water pump mounting bolts, then remove the water pump.

To install:

8. Clean all the gasket surfaces completely.

9. Apply a thin bead of sealer around the outer edge of the water pump gasket seating area and place he gasket on the pump.

10. Install the water pump on the engine and tighten the 4 mounting bolts to 18 ft. lbs. (25 Nm).

11. Install the water pump pulley and tighten the mounting bolts finger-tight.

12. Install the alternator in the mounting bracket.

13. Install the serpentine belt.

14. Tighten the water pump pulley mounting bolts to 22 ft. lbs. (30 Nm).

15. Connect the negative battery cable.

16. Refill and bleed the cooling system.

2.3L and 2.4L Engines

1. Disconnect the negative battery cable

2. Detach the oxygen sensor connector.

❄❄ CAUTION

Never open, service or drain the radiator or cooling system when hot; serious burns can occur from the steam and hot coolant.

3. Properly drain the engine coolant into a suitable container. Remove the heater hose from the thermostat housing for more complete coolant drain.

4. Remove upper exhaust manifold heat shield.

5. Remove the bolt that attaches the exhaust manifold brace to the manifold.

6. Remove the lower exhaust manifold heat shield.

7. Break loose the manifold to exhaust pipe spring loaded bolts using a 13mm box wrench.

8. Raise and safely support the vehicle.

➡**It is necessary to relieve the spring pressure from 1 bolt prior to removing the second bolt. If the spring pressure is not relieved, it will cause the exhaust pipe to twist and bind up the bolt as it is removed.**

9. Unfasten the 2 radiator outlet pipe-to-water pump cover bolts.

10. Remove the manifold to exhaust pipe bolts from the exhaust pipe flange as follows:

   a. Unscrew either bolt clockwise 4 turns.

   b. Remove the other bolt.

   c. Remove the first bolt.

❄❄ WARNING

On the 2.4L engines, DO NOT rotate the flex coupling more than 4 degrees or damage may occur.

11. Pull down and back on the exhaust pipe to disengage it from the exhaust manifold bolts.

12. Remove the radiator outlet pipe from the oil pan and transaxle. If equipped with a manual transaxle, remove the exhaust manifold brace. Leave the lower radiator hose attached and pull down on the outlet pipe to remove it from the water pump. Leave the radiator outlet pipe hang.

13. Carefully lower the vehicle.

14. Unfasten the exhaust manifold-to-cylinder head retaining nuts, then remove the exhaust manifold, seals and gaskets.

15. For the 2.4L engine, remove the front timing chain cover and the chain tensioner.

16. Unfasten the water pump-to-cylinder block bolts. Remove the water pump-to-timing chain housing nuts. Remove the water pump and cover mounting bolts and nuts. Remove the water pump and cover as an assembly, then separate the 2 pieces.

To install:

17. Thoroughly clean and dry all mounting surfaces, bolts and bolt holes. Using a new gasket, install the water pump to the cover and tighten the bolts finger-tight.

18. Lubricate the splines of the water pump with clean grease and install the assembly to the engine using new gaskets. Install the mounting bolts and nuts finger-tight.

19. Lubricate the radiator outlet pipe O-ring with antifreeze and slid the pipe onto the water pump cover. Install the bolts finger-tight.

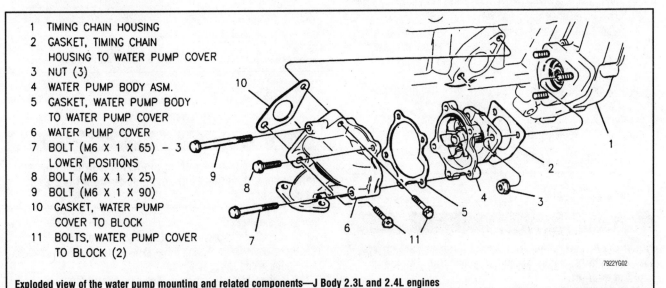

1   TIMING CHAIN HOUSING
2   GASKET, TIMING CHAIN HOUSING TO WATER PUMP COVER
3   NUT (3)
4   WATER PUMP BODY ASM.
5   GASKET, WATER PUMP BODY TO WATER PUMP COVER
6   WATER PUMP COVER
7   BOLT (M6 X 1 X 65) – 3 LOWER POSITIONS
8   BOLT (M6 X 1 X 25)
9   BOLT (M6 X 1 X 90)
10   GASKET, WATER PUMP COVER TO BLOCK
11   BOLTS, WATER PUMP COVER TO BLOCK (2)

7922YG02

Exploded view of the water pump mounting and related components—J Body 2.3L and 2.4L engines

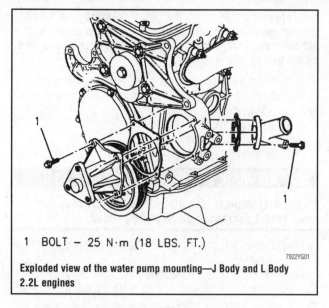

1 BOLT – 25 N·m (18 LBS. FT.)

7922YG01

Exploded view of the water pump mounting—J Body and L Body 2.2L engines

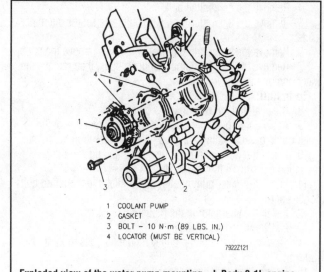

1 COOLANT PUMP
2 GASKET
3 BOLT – 10 N·m (89 LBS. IN.)
4 LOCATOR (MUST BE VERTICAL)

79222121

Exploded view of the water pump mounting—L Body 3.1L engine

20. With all gaps closed, tighten the bolts, in the following sequence, to the proper values:

    a. Pump assembly-to-chain housing nuts—19 ft. lbs. (26 Nm).

    b. Pump cover-to-pump assembly—106 inch lbs. (12 Nm).

    c. Cover-to-block, bottom bolt first—19 ft. lbs. (26 Nm).

    d. Radiator outlet pipe assembly-to-pump cover—125 inch lbs. (14 Nm).

21. Using new gaskets, install the exhaust manifold.

22. Raise and safely support the vehicle.

23. Index the exhaust manifold bolts into the exhaust pipe flange.

24. Connect the exhaust pipe to the manifold. Install the exhaust pipe flange bolts evenly and gradually to avoid binding. Turn the bolts in until fully seated.

25. Connect the radiator outlet pipe to the transaxle and oil pan. Install the exhaust manifold brace, if removed.

26. On the 2.4L engine, install the timing chain tensioner and front cover.

27. Install the lower heat shield.

28. Carefully lower the vehicle.

29. Fasten the bolt that attaches the exhaust manifold brace to the manifold.

30. Tighten the manifold-to-exhaust pipe nuts to 26 ft. lbs. (35 Nm).

31. Install the upper heat shield.

32. Attach the oxygen sensor connector.

33. Fill the radiator with coolant until it comes out the heater hose outlet at the thermostat housing. Then, connect the heater hose. Leave the radiator cap off.

34. Connect the negative battery cable, then start the engine. Run the vehicle until the thermostat opens, fill the radiator and recovery tank to their proper levels, then turn the engine **OFF**.

35. Once the vehicle has cooled, recheck the coolant level.

L BODY—CHEVROLET BERETTA & CORSICA

2.2L and 3.1L Engines

❋❋ CAUTION

Never open, service or drain the radiator or cooling system when hot; serious burns can occur from the steam and hot coolant.

1. Disconnect the negative battery cable. Drain the cooling system.

2. Loosen, but do not remove, the water pump pulley bolts.

3. Remove the serpentine belt.

4. Remove the alternator mounting bolts and set the alternator aside.

5. Remove the water pump pulley bolts and remove the water pump pulley.

6. Remove the 4 water pump mounting bolts and remove the water pump.

To install:

7. Clean all the gasket surfaces completely.

8. Apply a thin bead of sealer around the outer edge of the water pump gasket seating area and place he gasket on the pump.

9. Install the water pump on the engine and tighten the mounting bolts to 18 ft. lbs. (25 Nm) for the 2.2L engine or to 89 inch lbs. (10 Nm) on the 3.1L engine.

10. Install the remaining components in the reverse order of removal. Tighten the water pump pulley mounting bolts to 22 ft. lbs. (30 Nm) for the 2.2L engine or to 18 ft. lbs. (25 Nm) on the 3.1L engine.

11. Connect the negative battery cable. Refill the cooling system with the proper coolant. Bleed the cooling system and check for leaks with the engine running at idle.

L/N BODY—CHEVROLET MALIBU & OLDSMOBILE CUTLASS

2.4L Engine

1. Disconnect the negative battery cable.

❋❋ CAUTION

Never open, service or drain the radiator or cooling system when hot; serious burns can occur from the steam and hot coolant.

2. Drain the cooling system.

3. Detach the oxygen sensor (O_2) electrical connector.

4. Remove the upper and lower exhaust manifold heat shields.

5. Raise and safely support the vehicle.

6. Remove the exhaust manifold brace-to-manifold bolt.

7. Loosen the exhaust pipe-to-manifold spring loaded nuts.

8. Remove the radiator outlet pipe-to-water pump cover bolts.

9. Raise and safely support the vehicle.

10. Disconnect the exhaust pipe from the manifold.

11. Disconnect the radiator outlet pipe from the oil pan and transaxle.

12. If equipped with a manual transaxle, remove the exhaust manifold brace. If equipped with an automatic transaxle, leave the lower radiator hose attached and pull down on the radiator pipe to disconnect it from the water pump.

13. Lower the vehicle.

14. Remove the exhaust manifold.

15. Remove the water pump cover-to-engine bolts.

16. Remove the three water pump-to-timing chain housing nuts.

17. Remove the water pump and cover assembly.

18. Remove the five water pump cover-to-pump bolts and remove the cover.

To install:

19. Install the water pump-to-engine components and tighten the bolts finger-tight.

20. Lubricate the splines of the water pump drive with chassis grease.

21. To complete the installation, reverse the removal procedures.

22. With all the bolts in place tighten in the following sequence:

• Water pump-to-timing chain housing: 19 ft. lbs. (26 Nm)

• Water pump cover-to-water pump: 106–124 inch lbs. (12–14 Nm)

• Water pump cover-to-engine block (bottom first): 19 ft. lbs. (26 Nm)

• Radiator outlet pipe-to-pump cover: 10 ft. lbs. (14 Nm).

23. Refill the cooling system. Connect the negative battery cable. Start the vehicle and verify no leaks.

3.1L Engine

> ❊❊❊ **CAUTION**
>
> **Never open, service or drain the radiator or cooling system when hot; serious burns can occur from the steam and hot coolant.**

1. Disconnect the negative battery cable. Drain the cooling system into a suitable container.

2. Loosen, but do not remove, the water pump pulley bolts. Remove the serpentine belt.

3. Remove the water pump pulley bolts and pulley. Remove the five water pump mounting bolts and pump.

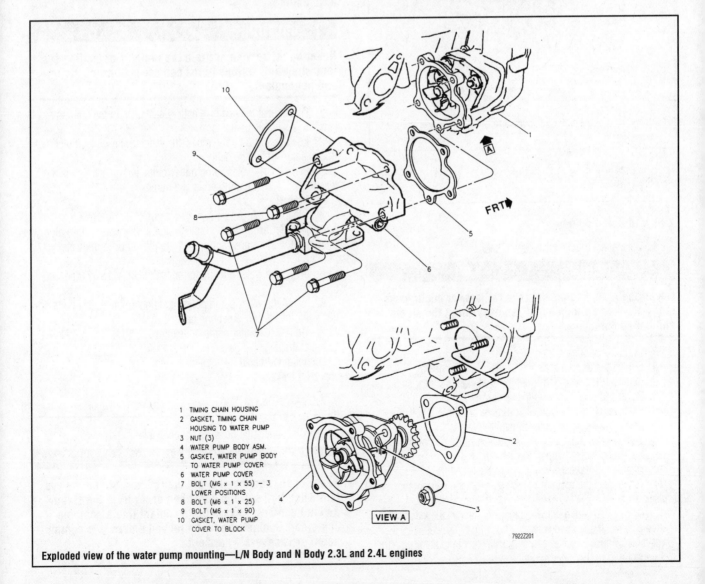

```
1  TIMING CHAIN HOUSING
2  GASKET, TIMING CHAIN
   HOUSING TO WATER PUMP
3  NUT (3)
4  WATER PUMP BODY ASM.
5  GASKET, WATER PUMP BODY
   TO WATER PUMP COVER
6  WATER PUMP COVER
7  BOLT (M6 x 1 x 55) — 3
   LOWER POSITIONS
8  BOLT (M6 x 1 x 25)
9  BOLT (M6 x 1 x 90)
10 GASKET, WATER PUMP
   COVER TO BLOCK
```

VIEW A

7922Z201

Exploded view of the water pump mounting—L/N Body and N Body 2.3L and 2.4L engines

To install:

4. Clean all the gasket surfaces completely. Apply a thin bead of sealer around the outside edge of the water pump along the gasket sealing area and install the gasket onto the water pump.

5. Install the water pump and tighten the bolts to 89 inch lbs. (10 Nm).

6. Install the water pump pulley and tighten the pulley bolts finger-tight. Install the serpentine belt.

7. Tighten the water pump pulley bolts to 18 ft. lbs. (25 Nm). Connect the negative battery cable.

8. Refill and bleed the cooling system. Test run the engine and check for leaks.

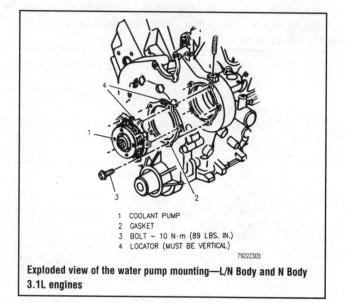

1 COOLANT PUMP
2 GASKET
3 BOLT – 10 N·m (89 LBS. IN.)
4 LOCATOR (MUST BE VERTICAL)

79222303

Exploded view of the water pump mounting—L/N Body and N Body 3.1L engines

N BODY—BUICK SKYLARK, OLDSMOBILE ACHIEVA & PONTIAC GRAND AM

2.3L and 2.4L Engines

1. Disconnect the negative battery cable.

✳✳ CAUTION

Never open, service or drain the radiator or cooling system when hot; serious burns can occur from the steam and hot coolant.

2. Drain the cooling system.
3. Detach the oxygen sensor (O$_2$) electrical connector.
4. Remove the upper and lower exhaust manifold heat shields.
5. Raise and safely support the vehicle.
6. Remove the exhaust manifold brace-to-manifold bolt.
7. Loosen the exhaust pipe-to-manifold spring loaded nuts.
8. Remove the radiator outlet pipe-to-water pump cover bolts.
9. Raise and safely support the vehicle.
10. Disconnect the exhaust pipe from the manifold.
11. Disconnect the radiator outlet pipe from the oil pan and transaxle.
12. If equipped with a manual transaxle, remove the exhaust manifold brace. If equipped with an automatic transaxle, leave the lower radiator hose attached and pull down on the radiator pipe to disconnect it from the water pump.

13. Lower the vehicle.
14. Remove the exhaust manifold.
15. Remove the water pump cover-to-engine bolts.
16. Remove the three water pump-to-timing chain housing nuts.
17. Remove the water pump and cover assembly.
18. Remove the five water pump cover-to-pump bolts and remove the cover.

To install:

19. Install the water pump-to-engine components and tighten the bolts finger-tight.
20. Lubricate the splines of the water pump drive with chassis grease.
21. To complete the installation, reverse the removal procedures.
22. With all the bolts in place tighten in the following sequence:
• Water pump-to-timing chain housing: 19 ft. lbs. (26 Nm)
• Water pump cover-to-water pump: 106–124 inch lbs. (12–14 Nm)
• Water pump cover-to-engine block (bottom first): 19 ft. lbs. (26 Nm)
• Radiator outlet pipe-to-pump cover: 10 ft. lbs. (14 Nm).

23. Refill the cooling system. Connect the negative battery cable. Start the vehicle and verify no leaks.

3.1L Engine

✳✳ CAUTION

Never open, service or drain the radiator or cooling system when hot; serious burns can occur from the steam and hot coolant.

1. Disconnect the negative battery cable. Drain the cooling system into a suitable container.
2. Loosen, but do not remove, the water pump pulley bolts. Remove the serpentine belt.
3. Remove the water pump pulley bolts and pulley. Remove the five water pump mounting bolts and pump.

To install:

4. Clean all the gasket surfaces completely. Apply a thin bead of sealer around the outside edge of the water pump along the gasket sealing area and install the gasket onto the water pump.

5. Install the water pump and tighten the bolts to 89 inch lbs. (10 Nm).

6. Install the water pump pulley and tighten the pulley bolts finger-tight. Install the serpentine belt.

7. Tighten the water pump pulley bolts to 18 ft. lbs. (25 Nm). Connect the negative battery cable.

8. Refill and bleed the cooling system. Test run the engine and check for leaks.

V BODY—CADILLAC CATERA

1. Disconnect the negative battery cable.

✳✳ CAUTION

Never open, service or drain the radiator or cooling system when hot; serious burns can occur from the steam and hot coolant. Always drain coolant into a sealable container. Coolant should be reused unless it is contaminated or is several years old.

2. Drain the coolant into a suitable container.

3. Remove the resonance chamber.

4. Remove the front timing belt cover.

5. Remove the water pump.

To install:

6. Clean the water pump mounting surface on the engine block.

7. Apply silicone grease to the water pump O-ring and install the water pump on the engine block. Tighten the mounting bolts to 18 ft. lbs. (21 Nm).

8. Install the front timing belt cover.

9. Install the resonance chamber.

10. Connect the negative battery cable.

11. Refill the cooling system through the reservoir tank. The cooling system will bleed automatically during warm-up. Use only GM DEX-COOL® or equivalent coolant. When refilling the cooling system, add two crushed engine coolant supplement sealant pellets PN 3634621 or equivalent into the coolant reservoir.

12. Check the coolant level in the reservoir after the engine has cooled and add coolant as needed.

W BODY—BUICK CENTURY, REGAL, CHEVROLET LUMINA, MONTE CARLO, OLDSMOBILE CUTLASS SUPREME, INTRIGUE & PONTIAC GRAND PRIX

3.1L and 3.4L Engines

1. Disconnect the negative battery cable. Drain the cooling system.

2. Remove the coolant reservoir and lay aside.

3. Remove the serpentine belt guard, bolts and nuts.

4. Loosen the water pump pulley bolts. Remove the serpentine belt.

5. Remove the water pulley bolts and pulley.

6. Remove the water pump attaching bolts, then remove the water pump and gasket.

To install:

7. Clean all pump mating surfaces.

8. Inspect the pump. There should be a locator tab to identify the top of the pump. This locator must be in the vertical position when the pump is installed. Install the water pump and gasket. Tighten the attaching bolts to 89 inch lbs. (10 Nm).

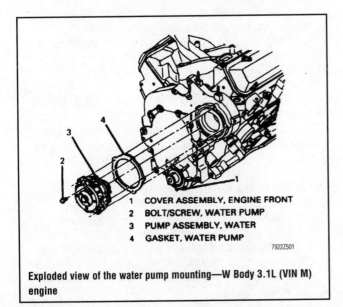

1	COVER ASSEMBLY, ENGINE FRONT
2	BOLT/SCREW, WATER PUMP
3	PUMP ASSEMBLY, WATER
4	GASKET, WATER PUMP

79222501

Exploded view of the water pump mounting—W Body 3.1L (VIN M) engine

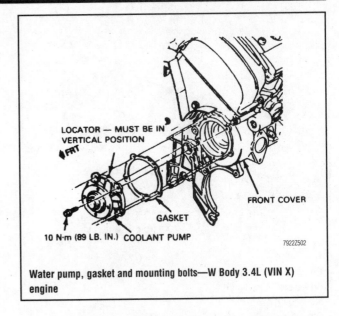

Water pump, gasket and mounting bolts—W Body 3.4L (VIN X) engine

9. Install the remaining components in the reverse order of removal. Tighten the pulley bolts to 18 ft. lbs. (25 Nm).

10. Connect the negative battery cable. Install the air cleaner assembly.

11. Refill the cooling system with the proper coolant. Bleed the cooling system and check for leaks with the engine running at idle.

3.8L Engines

VIN L AND VIN K MODELS

1. Disconnect the negative battery cable. Drain the cooling system.

2. Remove the coolant recovery reservoir.

3. Remove the serpentine belt. If additional access is needed, remove the inner fender electrical cover.

4. For VIN L engines, remove the alternator and position aside, then remove the serpentine belt tensioner pulley.

5. For VIN K engines, remove the power steering pump pulley using pulley remover J 25034-B or the equivalent.

6. Remove the water pump pulley.

7. Remove the water pump attaching bolts. Note that there are different length bolts. Use care to keep them organized for proper assembly. Remove the water pump from the vehicle.

To install:

8. Clean the water pump mating surfaces. Install the water pump using a new gasket.

9. Clean the pump bolt threads well. Install the attaching water pump bolts using care to install the proper bolts in the proper locations. Tighten long bolts to 22 ft. lbs. (30 Nm) and short bolts to 13 ft. lbs. (18 Nm) plus an additional 80 degrees using a torque angle meter.

10. Install the water pump pulley and tighten the bolts to 115 inch lbs. (13 Nm).

11. For VIN L engines, install the serpentine belt tensioner pulley, then install the alternator.

12. For VIN K engines, install the power steering pump pulley using J 25033-B or the equivalent to press the pulley onto the shaft.

13. Install the serpentine belt.

14. Install the coolant recovery reservoir.

15. Refill the cooling system with the correct amount and type of coolant.

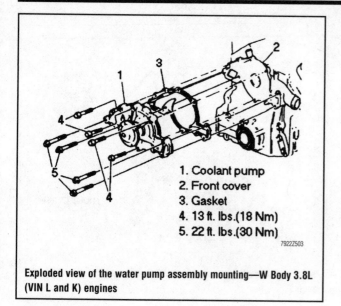

1. Coolant pump
2. Front cover
3. Gasket
4. 13 ft. lbs.(18 Nm)
5. 22 ft. lbs.(30 Nm)

Exploded view of the water pump assembly mounting—W Body 3.8L (VIN L and K) engines

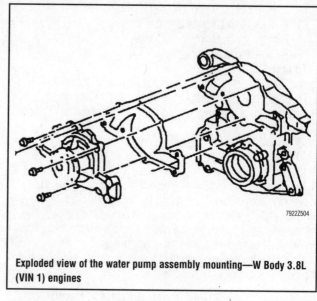

Exploded view of the water pump assembly mounting—W Body 3.8L (VIN 1) engines

16. Connect the negative battery cable. Start the engine and bleed the cooling system using the recommended procedure.

VIN 1 MODELS

1. Disconnect the negative battery cable.
2. Drain the cooling system into an approved container.
3. Remove the accessory drive and supercharger belts.
4. Remove the alternator and brace.
5. Disconnect the hoses and pipes from the water pump.
6. Remove the pulley bracket assembly.
7. Raise and safely support the vehicle.
8. Remove the power steering pump and lines.
9. Lower the vehicle.
10. Support the engine using engine support fixture J 28467-A or equivalent, and remove the front engine mount.
11. Remove the water pump pulley.
12. Remove the water pump mounting bolts and remove the water pump from the vehicle.

To install:
13. Clean all sealing surfaces.
14. Apply a thin bead of sealant to the gasket mating surface of the water pump and install a new gasket to the pump.
15. Install the water pump on the engine. Tighten the pump-to-front cover bolts to 11 ft. lbs. (15 Nm) plus an additional 80° turn with a torque angle meter.
16. Install the water pump pulley and tighten the bolts to 9.5 ft. lbs. (13 Nm).
17. Install the front engine mount.
18. Raise and safely support the vehicle.
19. Install the power steering pump and lines following the proper procedure.
20. Lower the vehicle.
21. Install the pulley bracket assembly.
22. Connect the hoses and pipes from the water pump.
23. Install the alternator and brace.
24. Install the supercharger and accessory drive and belts.
25. Connect the negative battery cable.
26. Fill and bleed the cooling system following the proper procedure.
27. Run the engine and check for leaks.
28. Recheck the coolant level when the engine has cooled.

Y BODY—CHEVROLET CORVETTE

5.7L (VIN P and 5) Engines

1. Disconnect the negative battery cable.
2. Unplug the IAT electrical connection.
3. Remove the air cleaner and air intake duct assembly.
4. Drain the cooling system into a suitable container. Remove the knock sensors from the lower left and right side of the block to assure proper draining.
5. Remove the upper and lower radiator hoses and the heater hose from the water pump. Remove the throttle body hose from the tee fitting.
6. Unplug the coolant sensor electrical connection and remove the sensor wiring harness from the retainer on the front of the coolant pump.
7. Use a box wrench or socket on the tensioner pulley bolt to rotate the tensioner and relieve belt tension, then remove the serpentine drive belt from the alternator pulley. This should create sufficient room to work, if additional room is necessary, the belt can be completely removed.
8. Remove the 6 bolts securing the water pump flanges to the engine block, then remove the water pump from the vehicle.
9. Remove and discard the old gaskets from the mating surfaces.
10. If replacing the pump, remove the coolant sensor from the old pump.

To install:
11. If replacing the pump, install the coolant sensor on the new pump and tighten to 17 ft. lbs. (23 Nm).
12. Thoroughly clean all gasket mating surfaces and apply a light coat of grease to the seals and splines before assembling the coupling to the water pump. The white band on the coupling should be positioned towards the engine.
13. Install the new gaskets with the tabs up, the coolant pump with the drive coupling and the mounting bolts. Tighten the bolts to 30 ft. lbs. (41 Nm).
14. Install the serpentine drive belt.
15. Install the coolant sensor wiring harness to the retainer on the front of the pump, then engage the sensor electrical connection.
16. Connect the heater hose and the upper and lower radiator hoses to the water pump.

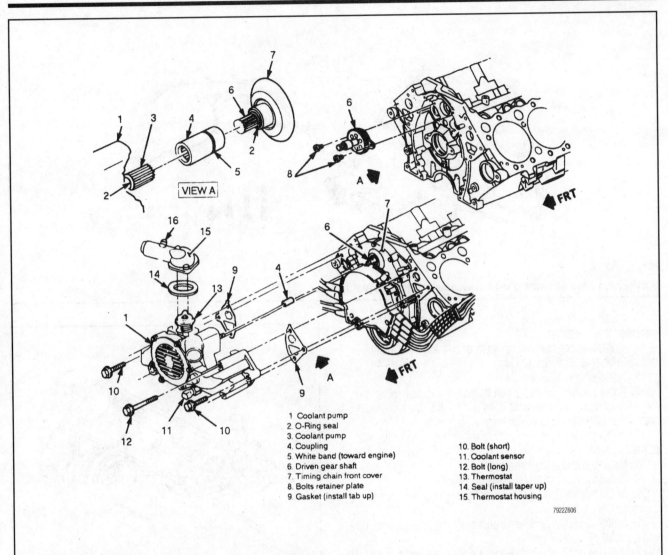

1. Coolant pump
2. O-Ring seal
3. Coolant pump
4. Coupling
5. White band (toward engine)
6. Driven gear shaft
7. Timing chain front cover
8. Bolts retainer plate
9. Gasket (install tab up)
10. Bolt (short)
11. Coolant sensor
12. Bolt (long)
13. Thermostat
14. Seal (install taper up)
15. Thermostat housing

79222606

Exploded view of the water pump mounting—Y Body 5.7L (VIN P and 5) engines

17. Install the throttle body hose at the tee.
18. If removed, install the knock sensors.
19. Open the bleed valves on the thermostat housing and the throttle body. Fill the cooling system through the radiator surge tank until a solid stream of coolant comes out of the bleeds.
20. Close all bleeds and continue to fill the surge tank until the coolant is level at the base of the surge tank neck.
21. Install the radiator pressure cap and check the coolant recovery reservoir for the proper level of coolant, add as necessary.
22. Install the air cleaner and intake duct assembly.
23. Engage the IAT electrical connection and clean any excess coolant from the engine compartment.
24. Connect the negative battery cable, start the engine and check for leaks.
25. If the low coolant indicator lamp is lit, the engine must be cycled from cold to normal operating temperature and back to cold 3 times. If the lamp does not go out after this and coolant is at the proper level, the indicator system must be repaired.

5.7L (VIN J) Engine

1. Disconnect the negative battery cable.
2. Drain engine coolant into a suitable container.

3. Remove the air cleaner and intake duct assembly.
4. Remove the screws attaching the throttle body extension to the throttle body, then remove the throttle body extension and gasket.
5. Remove clamps and hoses from the coolant outlets, radiator inlet and inlet pipe.
6. Remove the inlet pipe assembly and hose from the vehicle.
7. Remove the serpentine drive belt, then remove the tensioner retaining bolt and remove the tensioner from the pump. It is not necessary to remove the water pump pulley.
8. Remove the engine hose clamp, then the hose from the water pump.
9. Remove the alternator lower bracket mounting bolts, then remove the bracket from the vehicle.
10. Remove the water pump attaching bolts (noting the position and size of each bolt) and remove the bolt attaching the air conditioning compressor to the water pump. Remove the water pump from the vehicle.
    To install:
11. Thoroughly clean the pump and front cover sealing surfaces.
12. Install the water pump, new gasket and bolts, finger-tight only.

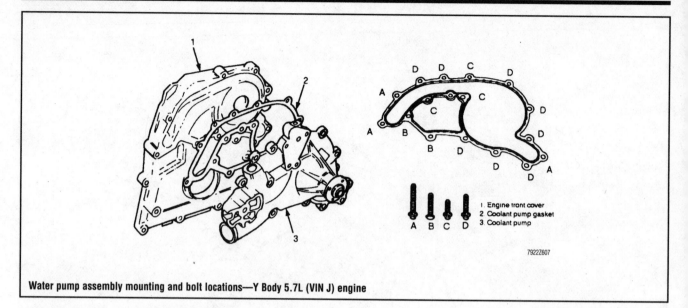

Water pump assembly mounting and bolt locations—Y Body 5.7L (VIN J) engine

13. Install and finger-tighten the bolt attaching air conditioning compressor to the pump.

14. Tighten air conditioning compressor bolt and water pump attaching bolts to 20 ft. lbs. (26 Nm).

15. Install engine hose and clamp.

16. Apply Loctite® 565 to the bolt threads, then install the alternator mounting bolts. Tighten the bolts to 39 ft. lbs. (52 Nm) and the bracket bolts to 20 ft. lbs. (26 Nm).

17. Install the belt tensioner and tighten the retaining bolt to 45 ft. lbs. (60 Nm).

18. Install the serpentine drive belt.

19. Install the hose and inlet pipe assembly.

20. Install throttle body extension and gasket. Tighten bolts to 53 inch lbs. (6 Nm).

21. Install air cleaner and intake duct assembly, then connect the negative battery cable.

22. Refill the cooling system with the proper type and quantity of antifreeze and inspect the system for leaks.

5.7L (VIN G) Engine

1. Disconnect the negative battery cable.

2. Disconnect the wiring harness from the Intake Air Temperature (IAT) and Mass Air Flow (MAF) sensors.

3. Remove the fuel pressure regulator purge line from the intake air duct.

4. Remove the air cleaner assembly.

5. Remove the accessory drive belts.

6. Disconnect the radiator and heater hoses from the water pump.

7. Remove the water pump pulley.

8. Remove the six mounting bolts and the water pump assembly.

To install:

9. Clean the sealing surfaces on the water pump and engine block.

10. Be sure the tabs on the new gaskets are pointing up and install the water pump on the engine. Tighten the mounting bolts to 30 ft. lbs. (41 Nm).

11. Install the water pump pulley. Tighten the bolts first to 89 inch lbs. (10 Nm), then to 18 ft. lbs. (25 Nm).

12. Connect the radiator and heater hoses.

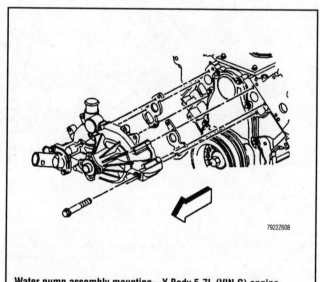

Water pump assembly mounting—Y Body 5.7L (VIN G) engine

13. Install the drive belts.

14. Install the intake air duct and fuel regulator purge line.

15. Attach the IAT and MAF sensor connectors.

16. Refill the cooling system with the proper type of coolant.

17. Connect the negative battery cable.

18. Start the engine and inspect for leakage.

Geo/Chevrolet

METRO

1. Disconnect the negative battery cable.

✽✽✽ CAUTION

Never open, service or drain the radiator or cooling system when hot; serious burns can occur from the steam and hot coolant.

2. Drain the cooling system.

3. Remove the air cleaner.

4. Remove the suction pipe bracket for the A/C compressor, if equipped.

5. Loosen but do not remove the four water pump pulley bolts.

6. Raise and safely support the vehicle.

7. Remove the lower splash shield.

8. Remove the A/C compressor drive belt, if equipped.

9. Remove the alternator drive belt.

10. Remove the crankshaft and water pump pulleys.

11. Remove the timing belt.

12. Remove the oil level dipstick and tube.

13. Remove the alternator adjusting bracket from the water pump.

14. Remove the water pump rubber seals.

15. Remove the water pump mounting bolts and nuts and remove the water pump from the engine.

To install:

16. Clean the gasket mating surfaces thoroughly.

17. Check the water pump by hand for smooth operation. If the pump does not operate smoothly or is noisy, replace it.

18. Install the pump using a new gasket. Tighten the bolts to 115 inch lbs. (13 Nm).

19. Install new rubber seals.

20. Install the upper alternator adjusting bracket and tighten the bolt to 17 ft. lbs. (23 Nm).

21. Install the oil level dipstick and tube.

22. Install the timing belt.

23. Install the water pump and crankshaft pulleys. Leave the water pump pulley bolts hand-tight.

24. Install the alternator drive belt.

25. Install the lower alternator cover plate and tighten the bolts to 89 inch lbs. (10 Nm).

26. Install the A/C compressor drive belt, if equipped.

27. Install the lower splash shield and lower the vehicle.

28. Tighten the water pump pulley mounting bolts to 18 ft. lbs. (24 Nm).

29. Adjust the water pump drive belt tension and tighten the alternator adjustment bolt to 17 ft. lbs. (23 Nm).

30. Install the suction pipe bracket for the A/C compressor, if equipped.

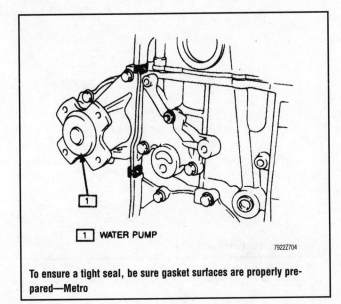

1 WATER PUMP

7922Z704

To ensure a tight seal, be sure gasket surfaces are properly prepared—Metro

31. Install the air cleaner.

32. Refill the cooling system.

33. Connect the negative battery cable.

34. Start the engine and check for leaks.

PRIZM

1. Disconnect the negative battery cable.

�֍֍ CAUTION

Never open, service or drain the radiator or cooling system when hot; serious burns can occur from the steam and hot coolant.

2. Drain the engine coolant into a suitable container.

3. Support the engine using a lifting device.

4. Remove the right engine mount and insulator.

5. Remove the upper and middle timing belt covers.

6. If equipped with power steering, proceed as follows:

  a. Raise and safely support the vehicle.

  b. Remove the plastic cover from the front transaxle mount.

  c. Remove the two mount-to-chassis bolts.

  d. Remove the nut and through-bolt, then remove the front transaxle mount from the vehicle.

  e. Lower the vehicle.

  f. Remove the coolant reservoir.

  g. Disconnect the upper hose from the radiator.

  h. Unclip the wiring connectors from the shroud.

  i. Remove the cooling fan assembly.

7. Remove the two nuts, one bolt and the engine wiring harness retainer.

8. If equipped, unclip the crankshaft position sensor electrical connector from the dipstick tube.

9. Remove the dipstick tube and dipstick. Immediately plug the hole in the block.

10. Disconnect the cooling fan switch electrical connector.

11. Remove the two nuts securing the engine coolant inlet pipe to the cylinder block.

12. Loosen the hose clamps and remove the coolant inlet pipe from the water pump.

13. Loosen the hose clamp and remove the coolant inlet hose from the water pump.

14. Remove the water pump bolts and remove the assembly. Discard the O-ring.

To install:

15. Install a new O-ring. Install the water pump, and tighten the retaining bolts evenly to 10 ft. lbs. (14 Nm).

16. Connect the coolant hose to the water pump and secure the clamp.

17. Install the engine coolant inlet pipe. Secure to the hose with clamps and nuts. Tighten the nuts to 11 ft. lbs. (15 Nm).

18. Connect the cooling fan switch electrical connector.

19. Install the dipstick and tube. Tighten the retaining bolt to 84 inch lbs. If equipped, clip the crankshaft position sensor electrical connector to the dipstick tube.

20. Install the engine wiring harness retainer and tighten the nuts and bolt to 89 inch lbs. (10 Nm).

21. If equipped with power steering, proceed as follows:

  a. Install the cooling fan assembly and tighten the bolts to 52 inch lbs. (6 Nm).

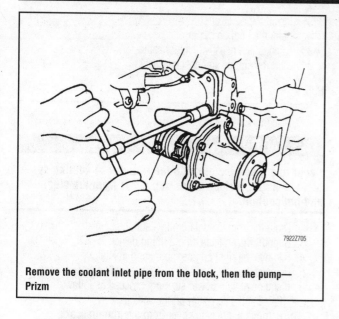

Remove the coolant inlet pipe from the block, then the pump—Prizm

b. Attach the wiring connectors to the shroud.

c. Connect the upper hose to the radiator.

d. Install the coolant reservoir.

e. Raise and safely support the vehicle.

f. Install the front transaxle mount and tighten the through-bolt nut to 64 ft. lbs. (87 Nm).

g. Install the mount-to-chassis bolts and tighten to 47 ft. lbs. (64 Nm).

h. Install the plastic cover to the front transaxle mount.

i. Lower the vehicle.

22. Install the upper and middle timing belt covers.

23. Support the engine using a lifting device. Install the right engine mount and insulator.

24. Remove the engine lifting equipment.

25. Connect the negative battery cable.

26. Fill the cooling system. Start the engine and check for leaks.

Saturn

SC1, SC2, SL, SL1, SL2, SW1 & SW2 MODELS

1. Allow a sufficient amount of time for the engine to cool down.

2. Disconnect the negative battery cable. Drain the engine coolant from the radiator and block drains into a suitable clean container.

✳✳ CAUTION

Never open, service or drain the radiator or cooling system when hot; serious burns can occur from the steam and hot coolant.

3. Remove the serpentine drive belt.

4. Raise and support the vehicle safely. Remove the right front tire and inner wheel well splash shield.

5. If access to the water pump is desired from underhood, remove the air conditioning compressor bolts and position the compressor aside with the refrigerant lines intact.

6. Spray the water pump hub with penetrating oil to loosen any rust or corrosion that might bind the pulley and damage it during removal.

7. Remove the water pump pulley bolts and allow the pulley to hang freely on the hub. A 1 in. (25.4mm) block of wood or a hammer handle may be wedged between the pump and crank-shaft pulleys to hold the assembly while loosening the retaining bolts.

8. Move the pulley outward or remove as necessary for access and remove the 6 water pump flange bolts. Carefully pull the pump and pulley assembly away from the engine and remove the assembly from the vehicle. If necessary, a gasket scraper may be inserted under the flange, but be careful not to damage the machined aluminum block sealing surface.

To install:

9. Thoroughly clean the gasket mating surfaces of all old gasket material. Apply a small amount of gasket sealant at the outer edges of the bolt holes to hold the gasket in place, then install the gasket onto the water pump assembly.

10. Install the pump assembly with the small bump located next to one of the attaching bolts in the 11 o'clock position. Install and tighten the bolts in a crisscross sequence as shown to 22 ft. lbs. (30 Nm).

11. Install or reposition the pump pulley, as applicable and tighten the bolts to 19 ft. lbs. (25 Nm). If the pump hub exposed through the pulley is rusty, clean it with a wire brush and apply a thin coat of primer to prevent the pulley from rusting onto the hub.

12. Install the serpentine drive belt, the right splash shield and right tire assembly.

13. If repositioned, install the air conditioning compressor.

14. Close the radiator drain plug and install the cylinder block drain plug. Tighten the block plug to 26 ft. lbs. (35 Nm).

15. Connect the negative battery cable and properly fill the engine cooling system.

16. Operate the engine and check for coolant leaks.

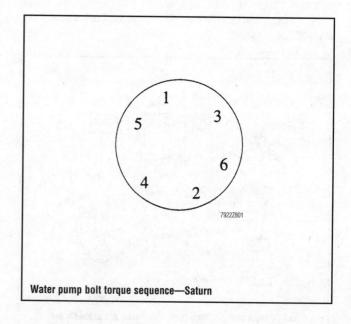

Water pump bolt torque sequence—Saturn

IMPORT CARS

Acura

2.2CL, 2.3CL, 2.5TL, 3.0CL, 3.2TL, 3.5RL, INTEGRA, INTEGRA GSR & LEGEND

➡ **The radio may have a coded theft protection circuit. Obtain the code from the owner before disconnecting the battery, removing the radio fuse, or removing the radio.**

1.8L Engines

1. Disconnect the negative battery cable.
2. If applicable, remove the front under panel.
3. Gradually release the system pressure by slowly and carefully removing the radiator cap. Be sure to protect your hands with gloves or a shop rag.
4. Drain the engine coolant into a sealable container.
5. Remove the timing belt from the engine.
6. Remove the camshaft pulleys and remove the back cover.
7. Remove the five water pump mounting bolts and remove the water pump.
8. Remove and discard the old O-ring.
9. Remove the dowel pins from the oil water pump.
10. Clean the O-ring groove and the water pump mounting surface on the engine.

To install:

11. Install the dowel pins to the new water pump.
12. Position a new O-ring to the new water pump, Apply a small amount of sealant to the O-ring to hold it in position.

13. Place the new water pump on the engine and install the mounting bolts. Tighten the mounting bolts to 8.7 ft. lbs. (12 Nm).
14. Install the back cover and the camshaft pulleys.
15. Install the timing belt.
16. Fill the engine with coolant and bleed the air from the cooling system.
17. Connect the negative battery cable and enter the radio security code.
18. Run the engine and check for cooling system leaks.

2.5L and 3.2L Engines

➡ **Perform this service operation with the engine cold.**

1. Disconnect the negative battery cable.
2. Remove the front splash panel and release the system pressure by slowly removing the radiator cap.
3. Drain the cooling system.
4. Remove the timing belt. Inspect the timing belt for any signs of damage or oil and coolant contamination. Replace the timing belt if there is any doubt about its condition.
5. If extra clearance is required, remove the camshaft pulleys and the timing belt rear cover.
6. Remove the two mounting bolts from the thermostat housing.
7. Remove the water pump bolts. Then, remove the water pump and sprocket assembly from the engine block. Remove the O-rings from the water passage.

To install:

8. Before installation, be sure all gasket and O-ring groove surfaces are clean.

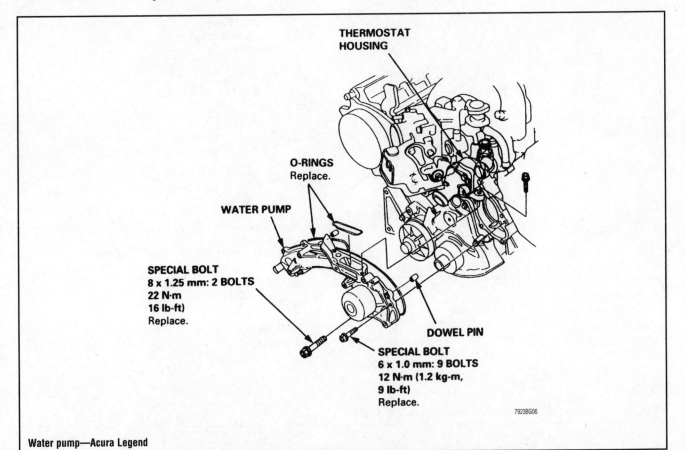

Water pump—Acura Legend

7923BG06

THERMOSTAT HOUSING

O-RINGS
Replace.

WATER PUMP

SPECIAL BOLT
8 x 1.25 mm: 2 BOLTS
22 N·m
16 lb-ft)
Replace.

DOWEL PIN

SPECIAL BOLT
6 x 1.0 mm: 9 BOLTS
12 N·m (1.2 kg-m,
9 lb-ft)
Replace.

9. Install the water pump with a new O-ring. Use liquid gasket if it was present on the sealing surfaces of the water pump that was removed. Use new 6mm mounting bolts and evenly tighten them to 9 ft. lbs. (12 Nm). Use new 8mm bolts and tighten them to 16 ft. lbs. (22 Nm).

10. Install the timing belt rear cover and camshaft pulleys. Tighten the pulley bolts for 2.5L (G25A1, G25A4) engines to 33 ft. lbs. (44 Nm). For 3.2L (C32A1) engines, tighten the pulley bolts to 23 ft. lbs. (32 Nm).

11. Install the thermostat housing and the two mounting bolts. Use a new O-ring.

12. Install the timing belt and timing belt covers.

13. Install and adjust the tension of the accessory drive-belts.

14. Close the cooling system drain plug. Refill and bleed the cooling system.

15. Connect the negative battery cable.

16. Start the engine, allow it to reach normal operating temperature, and check for leaks.

3.0L Engine

1. Remove the timing belt.
2. Remove the timing belt tensioner.
3. Remove the five water pump mounting bolts, then remove the pump and seal.

To install:

4. Clean the seal groove and mating surfaces.
5. Using a new seal, install the water pump. Tighten the bolts to 8.7 ft. lbs. (12 Nm).

6. Install the timing belt tensioner.
7. Install the timing belt.
8. Refill the cooling system.
9. Start the engine and check for leaks.
10. Top off the cooling system if necessary after the engine has cooled.

3.5L Engine

1. Disconnect the negative battery cable.

✳✳ CAUTION

Never open, service or drain the radiator or cooling system when hot; serious burns can occur from the steam and hot coolant.

2. Drain and recycle the engine coolant.
3. Remove the timing belt.
4. Remove the left camshaft pulley and rear cover.
5. Remove the water pump.

To install:

6. Clean the mounting surface and the O-ring grooves.
7. Install the water pump using new O-rings. Tighten the bolts to the specification in the diagram.
8. Install the rear cover and camshaft pulley.
9. Install the timing belt and remaining components.
10. Refill the cooling system.
11. Connect the negative battery cable.
12. Start the engine and check for leaks.

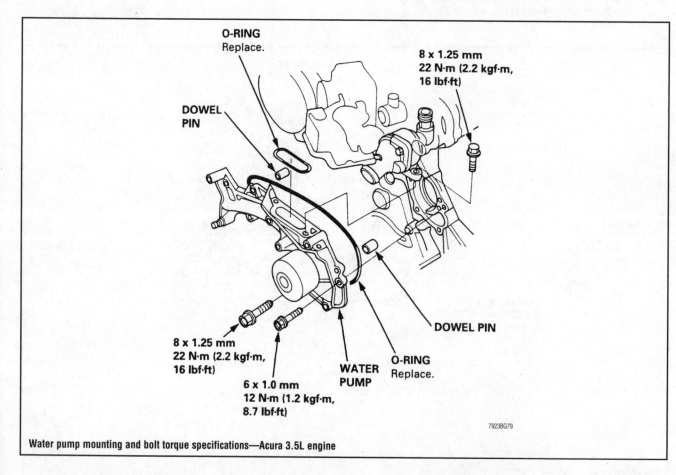

O-RING
Replace.

8 x 1.25 mm
22 N·m (2.2 kgf·m,
16 lbf·ft)

DOWEL
PIN

DOWEL PIN

8 x 1.25 mm
22 N·m (2.2 kgf·m,
16 lbf·ft)

6 x 1.0 mm
12 N·m (1.2 kgf·m,
8.7 lbf·ft)

WATER
PUMP

O-RING
Replace.

7923BG79

Water pump mounting and bolt torque specifications—Acura 3.5L engine

Audi

90, A4, A6, S6 & CABRIOLET

4-Cylinder Engine

➡**The coolant pump is bolted to the brackets for the generator, power steering pump, and cooling fan.**

1. Lock the carrier into service position as follows:
   a. Remove the front bumper.
   b. Tag and remove any wiring or connector that would inhibit locking the carrier.
   c. Remove the three quick-release screws on the front noise insulation panel.
   d. Unbolt the air guide between the lock carrier and the air filter.
   e. If installed, remove the retaining clamps for the wiring harness at the left side of the radiator frame.
   f. Remove the No. 2 bolts and install Support tool 3369 or equivalent.
   g. Remove the remaining bolts and pull the lock carrier out to the stop.
   h. To secure the lock carrier, install the appropriate M6 bolts into the rear of the lock carrier and fender.
2. Turn the ignition switch to the **OFF** position, then disconnect the negative battery cable.
3. Remove the accessory drive belt, then the engine driven cooling fan.
4. Drain and recycle the engine coolant.
5. Loosen the clamps for the coolant hoses at the water pump.
6. Remove the intake air duct between the intake manifold and the charge air cooler.
7. Remove the generator mounting bolts and slide it forward.
8. Disconnect the wiring from the generator once it is removed.
9. Unbolt the following supports and brackets for the generator, power steering pump, and engine cooling fan:
   • Intake manifold support
   • Support for the engine torque bracket
   • Brace to the cylinder block (remove completely)
10. Remove the brackets for the generator, power steering pump, and engine cooling fan.
11. Position the brackets for the generator, power steering pump, and engine cooling fan to the left side using a piece of wire.
12. Pull off the coolant hoses from the pump and thermostat housing.
13. Unbolt the coolant pump housing from the timing belt cover.
14. Remove the coolant pump mounting bolts, then the pump.
15. Unbolt the impeller housing from the pump housing.
16. Clean all gasket and O-ring sealing surfaces.

To install:

17. Using a new gasket, mount the new coolant pump to the pump housing and tighten the mounting bolts to 7 ft. lbs. (10 Nm).
18. Using a new gasket and O-ring, install the coolant pump and tighten the mounting bolts in numerical sequence to 18 ft. lbs. (25 Nm).
19. Tighten the coolant pump housing to the timing belt cover to 7 ft. lbs. (10 Nm).
20. Install the coolant hoses to the pump and thermostat housing.
21. Install the brackets that were removed and tighten the mounting bolts to 18 ft. lbs. (25 Nm).

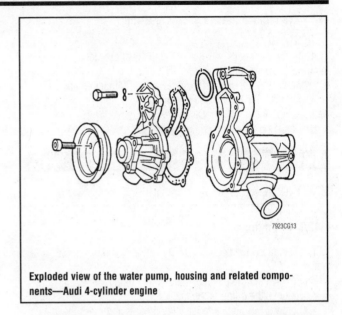

Exploded view of the water pump, housing and related components—Audi 4-cylinder engine

22. Connect the wires to the generator, then install the generator.
23. Install the air intake duct between the intake manifold and the charge air cooler.
24. The remaining steps are the reverse of the removal procedure noting the following items:
   • Fill the engine with coolant
   • Verify that the key is in the **OFF** position before connecting the battery
   • Fully close all power windows to stop, operate all window switches for at least one second in the close direction to activate the one-touch opening/closing function
   • Set the clock to the correct time
   • After installing the lock carrier, check the wiring for proper routing near the cooling fan

Except 4-Cylinder Engine

1. Drain the cooling system.
2. Remove the V-belts and the timing belt covers. Remove the timing belt from the water pump.
3. Remove the water pump mounting bolts, then the pump.

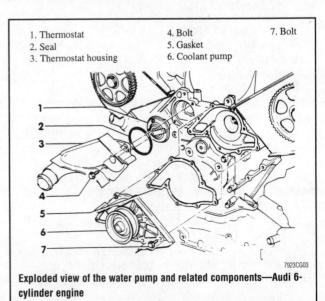

1. Thermostat	4. Bolt	7. Bolt
2. Seal	5. Gasket	
3. Thermostat housing	6. Coolant pump	

Exploded view of the water pump and related components—Audi 6-cylinder engine

➡️**Always replace the old gasket or O-ring.**

To install:

4. Installation is the reverse of the removal procedure. Tighten the water pump retaining bolts to 15 ft. lbs. (20 Nm) on the 5-Cylinder engines and 7 ft. lbs. (10 Nm) for the 6-Cylinder engines.

5. Reinstall the timing belt and properly tension the belt with the water pump. Refer to the necessary service procedures.

6. Refill and bleed the cooling system.

BMW

M3, Z3, 3, 5, 7 & 8 SERIES

M42/M44 Engines

1. Disconnect the negative battery cable. Drain the cooling system.

2. Remove the drive belt and the water pump pulley.

3. Remove the pump mounting bolts.

4. Screw 2 M6 bolts into the tapped bores and press the water pump out of the cover uniformly.

To install:

5. Lubricate and install a new O-ring.

6. Install the water pump and tighten the bolts to 6 ft. lbs. (9 Nm).

7. The remaining components are installed in the reverse order from which they were removed.

8. Start the engine and check for coolant leaks.

M50/M52/S50 and M70/M73/S70 Engines

1. Disconnect the negative battery cable. Drain the cooling system.

2. Remove the fan cowl and fan, if necessary.

3. Remove the drive belt and the pulley. Disconnect the bracket, if necessary.

4. Remove the air cleaner with the air flow sensor, if needed.

5. Disconnect the cooling hoses and remove the water pump.

6. The installation is the reverse of the removal procedure. Tighten the M8 bolts to 16 ft. lbs. (22 Nm) and the M6 bolts to 6.5 ft. lbs. (9 Nm).

M60/M62 Engines

1. Disconnect the negative battery cable.

2. Drain the cooling system.

3. Remove the heat shields at the left and right-hand sides of the front axle carrier.

4. Remove the front and rear engine splash guards.

5. Remove the fan. The fan must be held stationary with tool 11–5–030 or some sort of flat blade cut to fit over the hub and drilled to fit over 2 of the studs on the front of the pulley. Remove the fan coupling nut; left-hand thread-turn clockwise to remove.

6. Remove the drive belt tensioner, and the serpentine drive belt.

7. Remove the vibration damper and hub. There are eight mounting bolts, and a central bolt. Use a suitable holding tool 11–2–230 or equivalent for the central bolt.

8. Disconnect the coolant hose at the cover of the thermostat housing. Remove the thermostat.

9. Remove the water pump pulley by counterholding the pulley with the drive belt and removing the four pulley mounting bolts.

10. Disconnect the hoses from the water pump. Remove the 6 mounting bolts and remove the water pump.

To install:

11. Clean the gasket surfaces and use a new gasket.

12. Check for the correct seating of the dowel sleeves. Install the water pump in position. Tighten the M8 mounting bolts to 16 ft. lbs. (22 Nm), and the M6 mounting bolts to 7 ft. lbs. (10 Nm). Connect the hoses.

13. Install the vibration damper and hub. Tighten the eight mounting bolts to 17 ft. lbs. (24 Nm). Tighten the central mounting bolt in four steps as follows:

- An initial torque of 74–81 ft. lbs. (100–110 Nm)
- Add an additional 60 degrees torque
- Add an additional 60 degrees torque
- Then, add an additional 30 degrees torque.

14. Install the thermostat. Connect the coolant hose.

15. Install the drive belt tensioner and the drive belt.

16. Install the pulley and tighten the bolts to 6–7 ft. lbs. (8–10 Nm). Install the belt and tighten. Install the fan.

17. Install the heat shields and splash guards.

18. Connect the negative battery terminal.

19. Refill and bleed the cooling system.

Chrysler Imports

DODGE/PLYMOUTH COLT, EAGLE SUMMIT, SUMMIT WAGON & MITSUBISHI EXPO

1.5L and 1.8L Engines

1. Disconnect the negative battery cable.

2. Rotate the engine and position the No. 1 piston to TDC of its compression stroke.

3. Drain the cooling system.

4. Remove the engine undercover.

5. Disconnect the clamp bolt from the power steering hose.

6. Support the engine with the appropriate equipment and remove the engine mount bracket.

7. Remove the timing belt from the front of the engine.

8. Remove the timing belt rear cover.

9. Remove the power steering pump bracket.

10. Remove the alternator brace if necessary.

➡️**The water pump mounting bolts are different in length, note their positioning for reassembly.**

11. Remove the water pump mounting bolts and remove the pump.

To install:

12. Thoroughly clean and dry both mating surfaces of the water pump and block.

13. Apply a 0.09–0.12 in. (2.5–3.0mm) continuous bead of sealant to water pump and install the pump assembly.

➡️**Install the water pump within 15 minutes of the application of the sealant. Wait 1 hour after installation of the water pump to refill the cooling system or starting the engine.**

14. Properly position the bolts and tighten the bolts to 18 ft. lbs. (24 Nm).

15. The remaining components are installed in the reverse order from which they were removed.

16. Fill the system with coolant.

17. Connect the negative battery cable, run the vehicle until the thermostat opens and fill the radiator completely.

18. Once the vehicle has cooled, recheck the coolant level.

2.4L Engine

1. Disconnect the negative battery cable.
2. Drain the cooling system.
3. Remove the engine undercover.
4. Disconnect the clamp bolt from the power steering hose.
5. Support the engine with the appropriate equipment and remove the engine mount bracket.
6. Remove the timing belt from the front of the engine.
7. Disconnect the coolant hoses from the pump, if equipped.
8. Remove the alternator brace.
9. Remove the water pump, gasket and O-ring where the water inlet pipe(s) joins the pump.

To install:

10. Thoroughly clean and dry both gasket surfaces of the water pump and block.
11. Install a new O-ring into the groove on the front end of the water inlet pipe. Do not apply oils or grease to the O-ring. Wet with clean antifreeze only.
12. Install the gasket and pump assembly and tighten the bolts.
13. Connect the hoses to the pump.
14. Reinstall the timing belt and related parts.
15. Install the engine drive belts and adjust.
16. Fill the system with coolant.
17. Connect the negative battery cable, run the vehicle until the thermostat opens and fill the radiator completely.
18. Once the vehicle has cooled, recheck the coolant level.

Honda

ACCORD, CIVIC, DEL SOL & PRELUDE

1.5L, 1.6L, 2.2L and 2.3L Engines

➡**The original radio contains a coded anti-theft circuit. Obtain the security code number before disconnecting the battery cables.**

1. Disconnect the negative battery cable.
2. Drain the cooling system.
3. Remove the accessory drive belts, the valve cover, and the upper timing belt cover.
4. Set the timing at TDC/compression for No. 1 piston.
5. Remove the crankshaft pulley and lower timing belt cover.
6. Remove the timing belt. Replace the timing belt if it is contaminated with oil or coolant or shows any signs of wear and damage.
7. If equipped with a Crankshaft Speed Fluctuation (CKF) sensor at the crankshaft sprocket, unbolt the sensor bracket and move the sensor out of the way. Cover the sensor with a shop towel to keep coolant off of it.
8. Unbolt the water pump and remove it from the engine block. On 1.5L and 1.6L engines, the top right water pump mounting bolt also secures the alternator adjusting bracket. Leave the bracket attached to the alternator.

To install:

9. Clean the water pump and O-ring mating surfaces before installation.
10. Install the water pump with a new O-ring. Coat only the bolt threads with liquid gasket and tighten them to 9 ft. lbs. (12 Nm). On 1.5L and 1.6L engines, tighten the bracket bolt to 33 ft. lbs. (44 Nm).
11. Install the timing belt. Be sure it is fitted and adjusted properly.

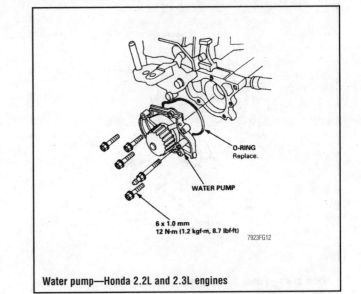

Water pump—Honda 2.2L and 2.3L engines

12. If equipped, install the CKF sensor and tighten the bracket bolts to 9 ft. lbs. (12 Nm).
13. Install the lower belt cover and crankshaft pulley.
14. Install the upper timing belt cover, the valve cover, and the accessory drive belts.
15. Be sure the cooling system drain plug is closed. Refill and bleed the cooling system.
16. Connect the negative battery cable and enter the radio security code.
17. Start the engine, allow it to reach normal operating temperature, and check for coolant leaks Check the tensions of the accessory belts.
18. If equipped with 4WS, turn the steering wheel lock-to-lock to reset the 4WS control unit.

2.7L Engine

1. Disconnect the negative battery cable.
2. Drain the coolant into a sealable container.
3. Remove the timing belt covers and the timing belt.
4. Remove the timing belt tensioner.
5. Remove the nine water pump bolts, take note of their locations for reinstallation.
6. Remove the water pump from the engine and discard the O-ring. Remove the dowel pins.

To install:

7. Clean the water pump mounting surface and O-ring groove, then install the dowel pins to the engine.
8. Install a new O-ring to the engine, then install the water pump, be careful not to pinch the O-ring. Install the mounting bolts to their original locations and when tightening the bolts, be sure that the O-ring does not bulge out of the groove. Tighten the six 1.0mm bolts to 9 ft. lbs. (12 Nm), and tighten the eight 1.25mm bolts to 16 ft. lbs. (22 Nm).
9. Inspect the water pump, making sure that the pump turns freely.
10. Install the timing belt tensioner.
11. Install the timing belt and timing belt covers.
12. Refill and bleed the air from the cooling system.
13. Connect the negative battery cable and enter the radio security code.

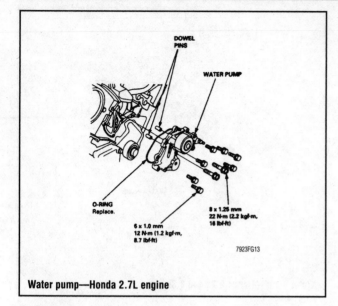

Water pump—Honda 2.7L engine

3.0L Engine

1. Remove the timing belt.
2. Remove the timing belt tensioner.
3. Remove the five water pump mounting bolts, then remove the pump and seal.
   To install:
4. Clean the seal groove and mating surfaces.
5. Using a new seal, install the water pump. Tighten the bolts to 8.7 ft. lbs. (12 Nm).
6. Install the timing belt tensioner.
7. Install the timing belt.
8. Refill the cooling system.
9. Start the engine and check for leaks.
10. Top off the cooling system if necessary after the engine has cooled.

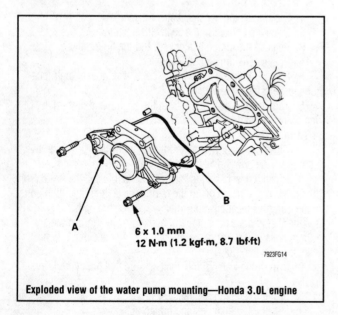

Exploded view of the water pump mounting—Honda 3.0L engine

ACCENT, ELANTRA, SCOUPE, SONATA & TIBURON

1. Disconnect the negative battery cable
2. Remove the water pump pulley bolts.
3. Remove the drive belt.
4. Drain the engine coolant.
5. Remove the timing belt covers.
6. Rotate the crankshaft clockwise and align the timing marks so the No. 1 piston will be at TDC of the compression stroke.
7. Remove the timing belt and tensioner.
8. Remove the water pump mounting bolts.
9. As required, remove the alternator brace.

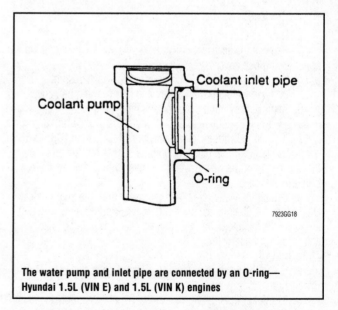

Exploded view of the water pump assembly—Hyundai 1.5L (VIN E) and 1.5L (VIN K) engines

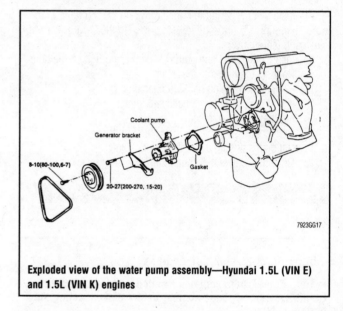

The water pump and inlet pipe are connected by an O-ring—Hyundai 1.5L (VIN E) and 1.5L (VIN K) engines

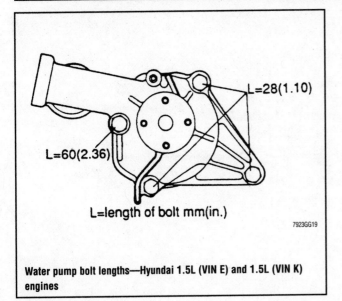

L=28(1.10)

L=60(2.36)

L=length of bolt mm(in.)

7923GG19

Water pump bolt lengths—Hyundai 1.5L (VIN E) and 1.5L (VIN K) engines

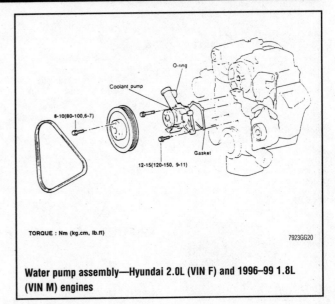

O-ring

Coolant pump

8-10(80-100,6-7)

12-15(120-150, 9-11)

Gasket

TORQUE : Nm (kg.cm, lb.ft)

7923GG20

Water pump assembly—Hyundai 2.0L (VIN F) and 1996–99 1.8L (VIN M) engines

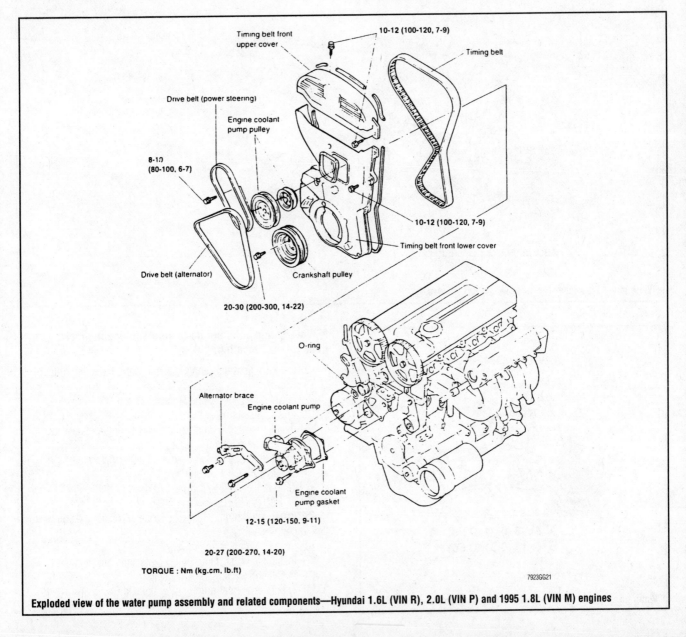

Timing belt front upper cover

10-12 (100-120, 7-9)

Timing belt

Drive belt (power steering)

Engine coolant pump pulley

8-10 (80-100, 6-7)

10-12 (100-120, 7-9)

Timing belt front lower cover

Drive belt (alternator)

Crankshaft pulley

20-30 (200-300, 14-22)

O-ring

Alternator brace

Engine coolant pump

Engine coolant pump gasket

12-15 (120-150, 9-11)

20-27 (200-270, 14-20)

TORQUE : Nm (kg.cm, lb.ft)

7923GG21

Exploded view of the water pump assembly and related components—Hyundai 1.6L (VIN R), 2.0L (VIN P) and 1995 1.8L (VIN M) engines

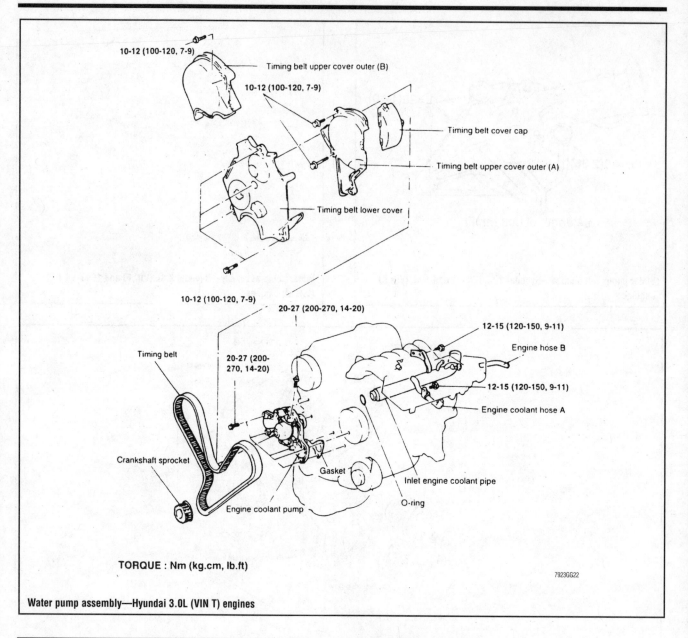

10-12 (100-120, 7-9) — Timing belt upper cover outer (B)

10-12 (100-120, 7-9)

Timing belt cover cap

Timing belt upper cover outer (A)

Timing belt lower cover

10-12 (100-120, 7-9)

20-27 (200-270, 14-20)

12-15 (120-150, 9-11)

Engine hose B

12-15 (120-150, 9-11)

Engine coolant hose A

Timing belt

20-27 (200-270, 14-20)

Crankshaft sprocket

Gasket

Inlet engine coolant pipe

O-ring

Engine coolant pump

TORQUE : Nm (kg.cm, lb.ft)

7923GG22

Water pump assembly—Hyundai 3.0L (VIN T) engines

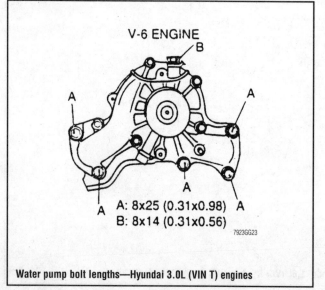

V-6 ENGINE

B

A   A

A   A

A   A

A: 8x25 (0.31x0.98)
B: 8x14 (0.31x0.56)

7923GG23

Water pump bolt lengths—Hyundai 3.0L (VIN T) engines

➡**Water pump bolts are three different lengths. Make a note of length and location.**

10. Remove the water pump, disconnecting the water outlet pipe.
To install:
11. Clean all gasket mating surfaces thoroughly.
12. Install the alternator brace.
13. Install the water pump using a new O-ring and gaskets.
14. Tighten water pump bolts as follows:
• 9–11 ft. lbs. (12–15 Nm)—except 2.0L (VIN F), 3.0L (VIN T) and 1996–98 1.8L (VIN M) engines
• 14–20 ft. lbs. (20–27 Nm)—2.0L (VIN F), 3.0L (VIN T) and 1996–98 1.8L (VIN M) engines
15. Install the timing belt and tensioner. Properly tension the timing belt.
16. Install the timing belt covers.
17. Install the water pump pulley bolts and hand-tighten.
18. Install and tension the drive belts.
19. Tighten the water pump pulley bolts to 6–7 ft. lbs. (8–10 Nm).

20. Fill the cooling system.
21. Start the engine and allow it to reach operating temperature. Check for leaks.
22. Once the vehicle has cooled, recheck the coolant level.

Infiniti

G20, I30, J30 & Q45

2.0L Engine

1. Disconnect the negative battery cable.
2. Drain the coolant from the radiator and engine block. The drain plug in the engine block is located at the left front of the cylinder block.
3. Remove the right wheel and the engine side cover.
4. Remove the drive belts.
5. Remove the front engine mount.
6. Loosen the water pump attaching bolts and remove the water pump. Take care not to drip coolant on the drive belts.

To install:
7. Clean all mating surfaces and place a 2–3mm bead of liquid gasket on the water pump mating surface.
8. Install the water pump and tighten the bolts to 12–15 ft. lbs. (16–21 Nm).
9. Install and tighten drive belts.
10. Install the front engine mount.
11. Install the engine side cover and the right wheel.
12. Using a radiator tester or equivalent, check the system for leaks.
13. Connect the negative battery cable.
14. Refill with coolant and bleed the system of air.

3.0L (VG30DE) Engine

1. Disconnect the negative battery cable.
2. Drain the coolant from the radiator and from the drain plugs on both sides of the cylinder block.
3. Remove the cooling fan assembly. Remove the timing belt covers.

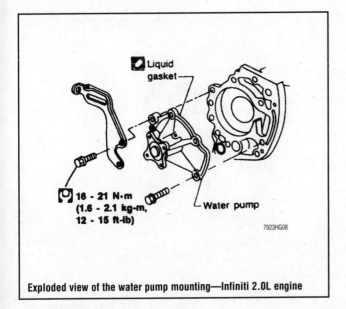

Exploded view of the water pump mounting—Infiniti 2.0L engine

➡️**Use the proper precautions to avoid getting coolant on the timing belt.**

4. Remove the water pump mounting bolts and remove the pump from the engine.

To install:
5. Thoroughly clean and dry the mating surfaces, bolts and bolt holes.
6. Apply liquid gasket to the water pump and install to the engine. Torque the bolts to 12–15 ft. lbs. (16–21 Nm).
7. Open the air release plug, as required. Fill the cooling system and check for leaks using a pressure tester before continuing.
8. Install the timing belt covers and all related parts.
9. Connect the negative battery cable, run the vehicle until the thermostat opens and fill the radiator completely. Recheck for coolant leaks.
10. Once the vehicle has cooled, recheck the coolant level.

3.0L (VQ30DE) Engine

1. Disconnect the negative battery cable.
2. Drain the coolant from the plugs on the radiator and both sides of the engine block.
3. Position a jack under the oil pan for support. Be sure to place a block of wood on the jack for protection to the engine parts.
4. Remove the right side engine mount and engine mounting bracket.
5. Remove the drive belts and the idler pulley bracket.
6. Remove the chain tensioner cover and the water pump cover.
7. Push the timing chain tensioner sleeve and apply a stopper pin so it does not return.
8. Remove the timing chain tensioner assembly.
9. Remove the three bolts that secure the water pump.
10. Rotate the crankshaft 20 degrees counterclockwise to provide timing chain slack.
11. Put the two grade M8 bolts in the two M8 threaded holes of the water pump.
12. Tighten each bolt by turning alternately ½ turn until they reach the timing chain rear case. Be sure to turn each bolt ½ turn at a time to prevent damage.
13. Lift up the water pump and remove it.
14. When removing the water pump, do not allow the water pump gear to hit the timing chain.
15. Remove and discard the O-rings from the water pump.
16. Clean all traces of liquid gasket from the water pump and covers.

To install:
17. Using new O-rings, install the water pump to the engine block.
18. Tighten the three water pump mounting bolts evenly to 62–86 inch lbs. (7–10 Nm).
19. Rotate the crankshaft pulley to its original position by turning it 20 degrees clockwise.
20. Install the timing chain tensioner and tighten the mounting bolts to 75–96 inch lbs. (9–10 Nm).
21. Remove the stopper pin from the timing chain tensioner.
22. Apply a continuous 0.091–0.130 in. (2.3–3.3mm) bead of liquid sealant to the mating surfaces of the timing chain tensioner and water pump covers.
23. Install the timing chain tensioner and water pump covers to the engine block. Tighten the cover mounting bolts to 84–108 inch lbs. (10–13 Nm).

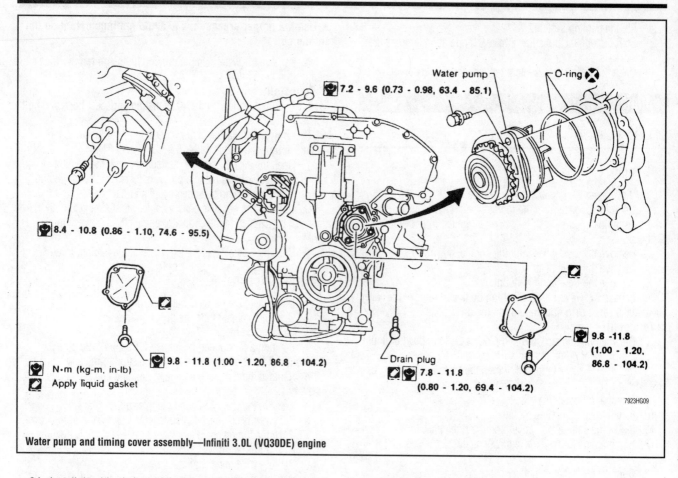

🔧 7.2 - 9.6 (0.73 - 0.98, 63.4 - 85.1)

Water pump

O-ring ⊗

🔧 8.4 - 10.8 (0.86 - 1.10, 74.6 - 95.5)

🔧 9.8 - 11.8 (1.00 - 1.20, 86.8 - 104.2)

Drain plug

📄🔧 7.8 - 11.8
(0.80 - 1.20, 69.4 - 104.2)

🔧 9.8 -11.8
(1.00 - 1.20,
86.8 - 104.2)

🔧 N·m (kg-m, in-lb)

📄 Apply liquid gasket

7923HG09

Water pump and timing cover assembly—Infiniti 3.0L (VQ30DE) engine

24. Install the drive belts and the idler pulley bracket.
25. Install the right side engine mounting bracket and the engine mount.
26. Remove the jack from under the engine and install the drain plugs to the cylinder block.
27. Connect the negative battery cable and refill the cooling system.
28. Start the engine, bleed the cooling system, and check for leaks.

4.1L and 4.5L Engines

1. Disconnect the negative battery cable.
2. Drain the coolant from the radiator and from the drain cocks on both sides of the cylinder block.
3. Remove the drive belts.
4. Unbolt the fan shroud and move it backward in order to remove the fan and coupling. Remove the fan to water pump bolts and remove the fan, coupling, water pump pulley and shroud.

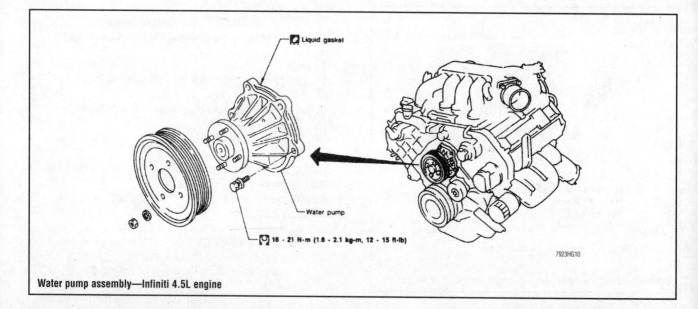

📄 Liquid gasket

Water pump

🔧 16 - 21 N·m (1.6 - 2.1 kg-m, 12 - 15 ft-lb)

7923HG10

Water pump assembly—Infiniti 4.5L engine

5. Remove all necessary accessories to gain access to the water pump.

6. Note the positioning of the clamp and disconnect the hose from the water pump.

7. Remove the water pump mounting bolts and remove the pump from the engine.

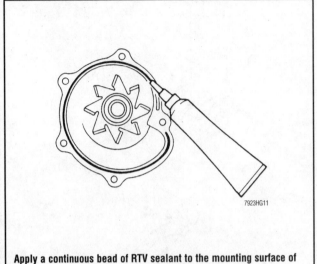

Apply a continuous bead of RTV sealant to the mounting surface of the water pump assembly—Infiniti 4.1L engine

To install:

8. Thoroughly clean and dry the mating surfaces, bolts and bolt holes.

9. Apply liquid gasket to the water pump and install it to the engine. Torque the bolts to 10–13 (14–18 Nm) on the 4.1L or 12–15 ft. lbs. (16–21 Nm) on the 4.5L engine.

10. Connect the hose and install the clamp in the same position as when it was removed. Fill the cooling system and check for leaks using a pressure tester before continuing.

11. Install all removed accessories.

12. Install the shroud, pulley, coupling and fan. Torque the water pump pulley nuts to 7 ft. lbs. (10 Nm). Install and adjust all the belts.

13. Connect the negative battery cable, run the vehicle until the thermostat opens and fill the radiator completely. Recheck for coolant leaks.

14. Once the vehicle has cooled, recheck the coolant level.

Kia

SEPHIA

1. Disconnect the negative battery cable. Drain the cooling system.

2. Remove the timing belt covers, and remove the timing belt.

3. Disconnect the coolant inlet pipe and gasket.

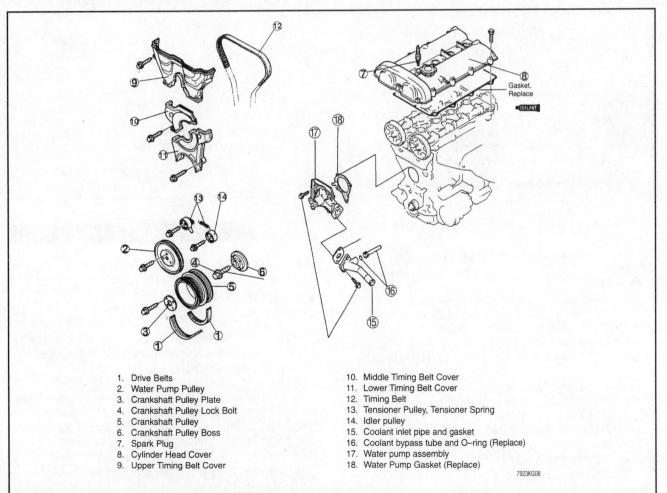

1. Drive Belts	10. Middle Timing Belt Cover
2. Water Pump Pulley	11. Lower Timing Belt Cover
3. Crankshaft Pulley Plate	12. Timing Belt
4. Crankshaft Pulley Lock Bolt	13. Tensioner Pulley, Tensioner Spring
5. Crankshaft Pulley	14. Idler pulley
6. Crankshaft Pulley Boss	15. Coolant inlet pipe and gasket
7. Spark Plug	16. Coolant bypass tube and O–ring (Replace)
8. Cylinder Head Cover	17. Water pump assembly
9. Upper Timing Belt Cover	18. Water Pump Gasket (Replace)

Exploded view of the water pump assembly—Kia DOHC engine

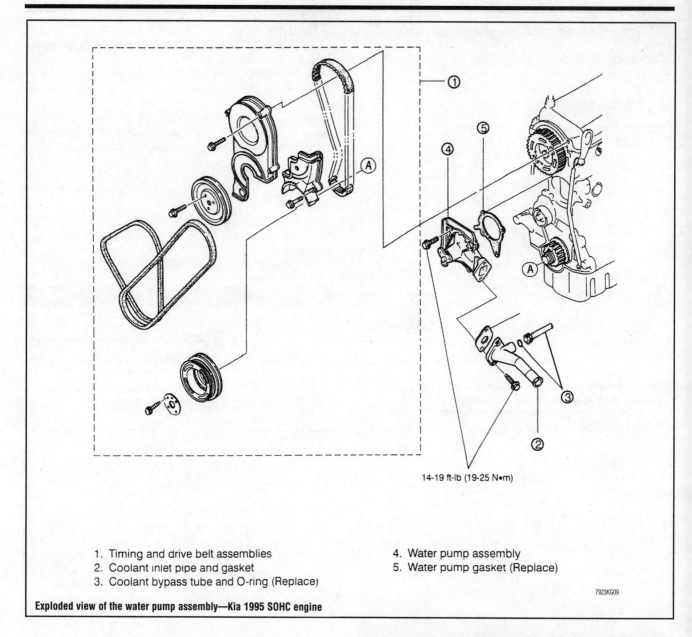

14-19 ft-lb (19-25 N•m)

1. Timing and drive belt assemblies
2. Coolant inlet pipe and gasket
3. Coolant bypass tube and O-ring (Replace)
4. Water pump assembly
5. Water pump gasket (Replace)

7923KG09

Exploded view of the water pump assembly—Kia 1995 SOHC engine

4. Remove the timing belt idler pulleys still attached to the water pump.

5. Remove the water pump mounting bolts, and remove the water pump.

To install:

6. Clean all gasket mating surfaces.

7. Install a new rubber seal on the water pump.

8. Using a new gasket, install the water pump on the engine. Tighten the mounting bolts to 14–18 ft. lbs. (19–25 Nm). Tighten the bolt from the water pump to the alternator bracket to 28–38 ft. lbs. (38–51 Nm).

9. Install the timing belt idler pulleys that were removed.

10. Install the coolant inlet pipe, using a new gasket. Tighten the bolts to 14–18 ft. lbs. (19–25 Nm).

11. Install the timing belt and the timing belt covers.

12. Fill and bleed the cooling system. Connect the negative battery cable, start the engine and bring to normal operating temperature. Check for leaks.

Lexus

ES300, GS300, GS400, LS400, SC300 & SC400

3.0L (1MZ-FE) Engine

1. With the ignition switch in the **LOCK** position, disconnect the negative battery terminal. If equipped with an air bag system, wait at least 90 seconds or longer before performing any other work.

2. Drain the engine coolant.

3. Remove the timing belt.

4. Mark the left and right camshaft pulleys with a touch of paint. Using SST tools 09249–63010 and 09960–1000 or equivalents, remove the bolts to the right and left camshaft pulleys. Remove the pulleys from the engine. Be sure not to mix up the pulleys.

5. Remove the No. 2 idler pulley by removing the bolt.

6. Disconnect the three clamps and engine wire from the rear timing belt cover.

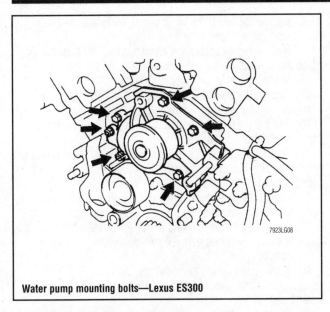

Water pump mounting bolts—Lexus ES300

7. Remove the six bolts holding the rear timing belt cover to the engine block.

8. Remove the four bolts and two nuts to the water pump.

9. Remove the water pump.

10. Remove all the old packing (sealant) and gasket material from the water pump and clean the mounting surfaces.

11. Scrape and clean all gasket material from the upper inner timing belt cover.

To install:

12. Check that the water pump turns smoothly. Also check the air hole for coolant leakage.

13. Using a new gasket, apply liquid sealer to the gasket, water pump and engine block.

14. Install the gasket and pump to the engine and install the four bolts and two nuts. Tighten the nuts and bolts to 53 inch lbs. (6 Nm).

15. Install the rear timing belt cover and tighten the six bolts to 74 inch lbs. (9 Nm).

16. Connect the engine wire with the three clamps to the rear timing belt cover.

17. Install the No. 2 idler pulley with the bolt. Tighten the bolt to 32 ft. lbs. (43 Nm). After tightening the bolt, be sure the idler pulley moves smoothly.

18. With the flange side **outward**, install the right-hand camshaft pulley to the engine. Be sure to align the knock pin hole on the camshaft pulley with the knock pin on the camshaft. Using the same tools as removal, tighten the camshaft bolt to 65 ft. lbs. (88 Nm).

19. With the flange side **inward**, install the left-hand camshaft pulley to the engine. Be sure to align the knock pin hole on the camshaft pulley with the knock pin on the camshaft. Using the same tools as removal, tighten the camshaft bolt to 94 ft. lbs. (125 Nm).

20. Install the timing belt to the engine.

21. Fill the engine coolant.

22. Connect the negative battery cable to the battery and start the engine.

23. Top off the engine coolant and check for leaks.

3.0L (2JZ-GE) Engine

GS300

1. Disconnect the negative battery cable. Wait at least 90 seconds before performing any other work.

2. Drain the cooling system.

3. Remove the lower engine cover.

4. Remove the nut, then remove the air cleaner duct.

5. Disconnect the high tension lead from the ignition coil.

6. Disconnect the high tension lead from the clamp on the VAF meter.

7. Disconnect the VAF meter electrical connector and disconnect the harness from the meter.

8. Disconnect the power steering air hose from the timing belt cover.

9. Disconnect the PCV hose from the cylinder head cover.

10. Loosen the hose clamp bolt securing the intake air connector pipe to the throttle body.

11. Remove the three bolts, air cleaner, VAF meter and intake air connector pipe assembly.

12. Remove the drive belt.

13. Place matchmarks on the cooling fan clutch and the cooling fan pulley. Remove the cooling fan by removing the four nuts.

14. Remove the water pump pulley.

15. Remove the radiator.

16. Remove the two nuts and disconnect the water inlet from the water pump.

17. Remove the thermostat.

18. Remove the distributor.

19. On all except California vehicles, remove the four nuts and exhaust manifold heat insulator.

20. Remove the two bolts, the No. 1 water bypass outlet and pipe. Discard the O-rings.

21. Remove the timing belt.

22. Remove the idler pulley.

23. Remove the mounting bolt and disconnect the engine wire bracket.

24. Remove the alternator mounting bolt and disconnect the alternator from the water pump.

25. Remove the two nuts and disconnect the No. 2 water bypass pipe from the water pump.

26. Remove the water pump mounting bolts. Note the position and type of each bolt for correct installation.

27. Lift out the water pump. If prying is necessary, use a protected blade and take great care not to damage the mating surfaces.

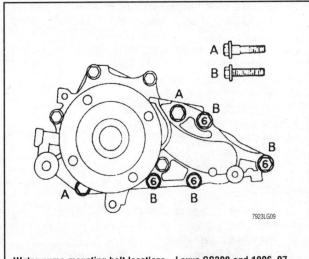

Water pump mounting bolt locations—Lexus GS300 and 1996-97 SC300

28. Remove all the old sealant and gasket material; clean the mounting surfaces.

To install:

29. Install a new O-ring to the cylinder block and a new gasket to the water pump.

30. Connect the water bypass pipe to the water pump. Do not install the nuts yet.

31. Install the water pump. Fit the bolts and hand-tighten them, then tighten the mounting bolts to 15 ft. lbs. (21 Nm).

➠**Hand-tighten the A bolts prior to hand-tightening the B bolts.**

32. Install the nuts to the No. 2 water bypass pipe and tighten them to 15 ft. lbs. (21 Nm).

33. Install the alternator mounting bolt and nut and tighten them to 27 ft. lbs. (37 Nm).

34. Connect the engine wiring harness. Install and tighten the bolts.

35. Install the idler pulley.

36. Install the timing belt.

37. Install new O-rings to the No. 1 water bypass pipe and outlet. Install the bolts and tighten them to 78 inch lbs. (9 Nm).

38. Install the exhaust manifold heat insulator if it was removed and tighten the nuts to 13 ft. lbs. (18 Nm).

39. Install the distributor.

40. Install the thermostat and align the jiggle valve with the protrusion on the water inlet housing.

41. Install the water inlet housing; tighten the bolts to 78 inch lbs. (9 Nm).

42. Install the radiator.

43. Align the matchmarks and install the water pump pulley and fan assembly. Do not tighten the nuts at this time.

44. Install the drive belt and tighten the fan assembly nuts to 12 ft. lbs. (16 Nm).

45. Connect the intake air connector pipe to the throttle body.

46. Install the air cleaner, VAF meter and intake air connector pipe assembly with the three (3) bolts.

47. Install the hose clamp.

48. Connect the power steering air hose to the timing belt cover.

49. Connect the PCV hose to the cylinder head cover.

50. Connect the VAF meter harness and electrical connector.

51. Connect the high tension lead to the VAF meter and ignition coil.

52. Connect the air cleaner duct to the air cleaner.

53. Install the air cleaner duct.

54. Refill the cooling system.

55. Connect the negative battery cable.

56. Start the engine and check for leaks.

57. Check the fluid level on vehicles with A/T.

58. Inspect the engine ignition timing.

59. Install the lower engine cover.

60. Road test the vehicle.

1995 SC300

1. Disconnect the negative battery cable. Wait at least 90 seconds before performing any other work.

2. Drain the cooling system.

3. Remove the timing belt.

4. Remove the idler pulley.

5. Remove the thermostat.

6. Remove the two bolts, water bypass outlet and the No. 1 water bypass pipe. Discard the three O-rings.

7. Remove the mounting bolt and disconnect the engine wiring harness bracket above the alternator.

8. Remove the alternator mounting nut.

9. Remove the alternator mounting bolt and disconnect the alternator from the water pump.

10. Remove the two nuts and disconnect the No. 2 water bypass pipe from the water pump.

11. Remove the six water pump mounting bolts. The bolts are of different lengths and styles; note the correct position of each bolt during removal.

12. Lift out the water pump by carefully prying between the pump and the cylinder head.

13. Remove all the old packing and clean the mounting surfaces.

14. Remove the O-ring from the cylinder block.

To install:

15. Install a new O-ring to the cylinder block.

16. Apply a thin layer of liquid sealant to the water pump and install a new gasket.

17. Connect the No. 2 water bypass pipe to the water pump. Do not install the nuts yet.

18. Install the water pump. Install the bolts in the correct positions and tighten them finger-tight. Tighten the mounting bolts to 15 ft. lbs. (21 Nm).

19. Install the nuts to the No. 2 water bypass pipe and tighten them to 15 ft. lbs. (21 Nm).

20. Install the alternator mounting bolt and nut and tighten them to 27 ft. lbs. (37 Nm).

21. Connect the engine wiring harness. Install and tighten the bolts.

22. Install new O-rings to the No. 1 water bypass pipe and outlet. Install the bolts and tighten them to 78 inch lbs. (9 Nm).

23. Install the thermostat.

24. Install the idler pulley.

25. Install the timing belt.

26. Refill the cooling system.

27. Connect the negative battery cable.

28. Start the engine, bleed the cooling system and check for leaks.

29. Recheck the fluid levels and the ignition timing.

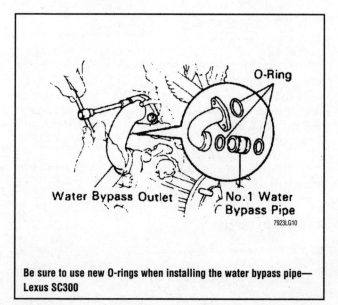

Be sure to use new O-rings when installing the water bypass pipe— Lexus SC300

1996–99 SC300

1. Disconnect the negative battery cable. Wait at least 90 seconds before performing any work.
2. Drain the engine coolant and remove the radiator assembly.
3. Remove the air cleaner, MAF meter and the intake air connector pipe assembly.
4. Remove the timing belt.
5. Remove the idler pulley.
6. Remove the water inlet and the thermostat.
7. Remove the 2 bolts, the water bypass outlet and the No. 1 water bypass pipe. Remove the three O-rings from the water bypass outlet and the No. 1 water bypass pipe.
8. Remove the generator.
9. Remove the bolt and disconnect the engine wire bracket. Remove the bolt and disconnect the clamp bracket for the crankshaft position sensor connector. Remove the nuts and disconnect the No. 2 water bypass pipe from the water pump. Remove the six bolts and the water pump and gasket.
10. Remove the drain hose and the O-ring from the cylinder block.

To install:

11. Install a new O-ring to the cylinder block.
12. Install the drain hose.
13. Install a new gasket to the water pump. Connect the water pump to the water bypass pipe. Do not install the nut yet. Install the water pump with the two bolts (A) and the four bolts (B).

➡**Hand-tighten the (A) bolts first. Tighten all six bolts to 15 ft. lbs. (21 Nm).**

14. Install the two nuts holding the No. 2 water bypass pipe to the water pump. Tighten the nuts to 15 ft. lbs. (21 Nm). Install the clamp bracket for the crankshaft position sensor connector.
15. Install the engine wire bracket.
16. Install the generator.
17. Install new O-rings to the No. 1 water bypass pipe. Install a new O-ring and the water bypass outlet with the two bolts and tighten them to 78 inch lbs. (9 Nm).
18. Install the thermostat and the water inlet.
19. Install the idler pulley.
20. Install the timing belt.
21. Install the air cleaner, the MAF meter and the intake air connector pipe assembly.
22. Install the radiator assembly.
23. Reconnect the negative battery cable. Refill the cooling system. Start the engine, check for leaks and bleed the cooling system.
24. Road test for proper operation.

4.0L Engine

LS400

1. Disconnect the negative battery cable.
2. Drain the cooling system.
3. Remove the timing belt and the No. 2 idler pulley.
4. Remove the right side ignition coil.
5. Remove the two water inlet housing to water pump bolts.
6. Disconnect the IAC valve bypass hose from the water inlet housing.
7. Remove the water inlet housing and discard the O-ring.
8. Remove the mounting bolts, studs and nut. Lift out the water pump by carefully prying between the pump and the cylinder head.
9. Remove all the old packing and clean all mounting surfaces. Remove the O-ring from the water bypass pipe.

To install:

10. Install new seal packing to the water pump groove and a new O-ring to the water bypass pipe.
11. Connect the water pump to the water bypass pipe end.
12. Install the water pump and tighten the mounting bolts to 13 ft. lbs. (18 Nm).
13. Apply sealant to the groove of the water inlet housing.
14. Install a new O-ring to the water inlet housing.
15. Push the water inlet housing end into the water pump hole.
16. Connect the IAC valve bypass hose to the water inlet housing.
17. Install the water inlet and housing assembly with the two bolts. Alternately tighten the bolts to 13 ft. lbs. (18 Nm).
18. Install the ignition coil.
19. Install the No. 2 idler pulley for the timing belt.
20. Install the timing belt.
21. Refill the cooling system.
22. Connect the negative battery cable.
23. Start the engine and check for leaks.

SC400

1. Disconnect the negative battery cable.
2. Drain the cooling system.
3. Remove the timing belt.
4. Remove the No. 2 idler pulley.
5. Remove the right side ignition coil by removing the ignition coil connector and two bolts.
6. Disconnect the bypass hose(s) from the water inlet housing.
7. Remove the two bolts holding the water inlet housing to water pump.
8. Remove the water inlet housing and discard the gasket.
9. Remove the mounting bolts, studs and the nut to the water pump. Remove the water pump by carefully prying between the pump and the cylinder head.
10. Remove all the old gasket and clean all mounting surfaces.

To install:

11. Install new seal packing to the water pump groove and a new O-ring to the water bypass pipe end.
12. Connect the water pump to the water bypass pipe end.
13. Install the water pump and tighten the mounting bolts and nut to 13 ft. lbs. (18 Nm).

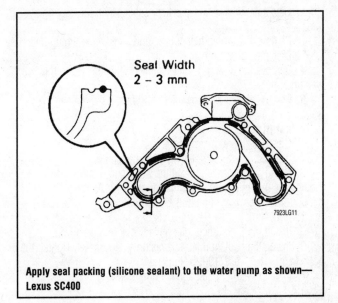

Apply seal packing (silicone sealant) to the water pump as shown—Lexus SC400

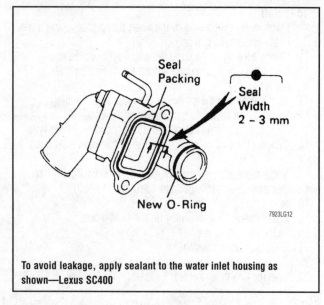

To avoid leakage, apply sealant to the water inlet housing as shown—Lexus SC400

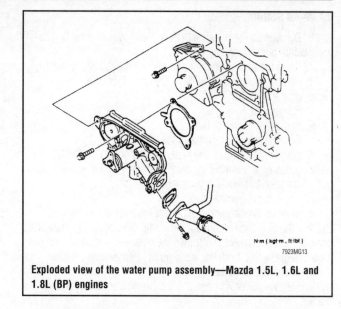

Exploded view of the water pump assembly—Mazda 1.5L, 1.6L and 1.8L (BP) engines

14. Apply sealant to the groove of the water inlet housing.

15. Install a new O-ring to the water inlet housing.

16. Push the water inlet housing end into the water pump hole.

17. Install the water inlet and housing assembly with the two bolts. Alternately tighten the bolts to 13 ft. lbs. (18 Nm).

18. Connect the bypass hose(s) to the water inlet housing.

19. Replace the ignition coil by installing the two bolts and coil connector.

20. Install the No. 2 idler pulley.

21. Install the timing belt.

22. Refill the cooling system.

23. Connect the negative battery cable.

24. Start the engine, bleed the cooling system and check for leaks.

Mazda

323, 626, MX3, MX6, MIATA, MILLENIA & PROTEGÉ

1.5L, 1.6L and 1.8L (BP) Engines

1. Disconnect the negative battery cable. Drain the cooling system.

2. Remove the timing belt covers, and remove the timing belt.

3. Disconnect the coolant inlet pipe and gasket.

4. Remove the timing belt idler pulleys still attached to the water pump.

5. Remove the water pump mounting bolts, and remove the water pump.

To install:

6. Clean all gasket mating surfaces.

7. Install a new rubber seal on the water pump.

8. Using a new gasket, install the water pump on the engine. Tighten the mounting bolts to 14–18 ft. lbs. (19–25 Nm). Tighten the bolt from the water pump to the alternator bracket to 28–38 ft. lbs. (38–51 Nm).

9. Install the timing belt idler pulleys that were removed.

10. Install the coolant inlet pipe, using a new gasket. Tighten the bolts to 14–18 ft. lbs. (19–25 Nm).

11. Install the timing belt and the timing belt covers.

12. Fill and bleed the cooling system. Connect the negative battery cable, start the engine and bring to normal operating temperature. Check for leaks.

2.0L Engines

1. Disconnect the negative battery cable. Drain the cooling system.

2. Remove the timing belt.

3. Remove the power steering oil pump adjuster.

4. Remove the five water pump mounting bolts and remove the water pump.

To install:

5. Clean all gasket mating surfaces.

6. Install a NEW gasket on the water pump and install the water pump on the engine. Install the mounting bolts and tighten to 14–18 ft. lbs. (19–25 Nm).

7. Install the power steering oil pump adjuster, torque the mounting bolts to 12–16 ft. lbs. (16–22 Nm).

8. Install the timing belt.

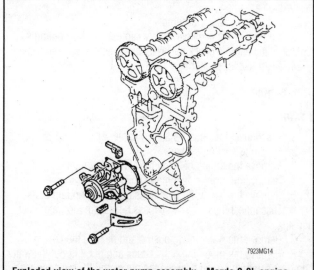

Exploded view of the water pump assembly—Mazda 2.0L engine

9. Connect the negative battery cable. Fill and bleed the cooling system.

10. Start the engine and bring to normal operating temperature. Check for leaks.

1.8L (K8) and 2.5L Engines

1. Disconnect the negative battery cable. Drain the cooling system.

2. Remove the timing belt.

3. Remove the No.3 engine mount bracket.

4. Position a drain pan under the water pump.

5. Remove the five water pump mounting bolts, and remove the water pump.

To install:

6. Clean the mating surfaces of the water pump and the engine block.

7. Install a NEW rubber seal onto the water pump.

8. Install the water pump and torque the bolts 14–18 ft. lbs. (19–25 Nm).

9. Install the engine mount bracket, and tighten the mounting bolt to 32–44 ft. lbs. (44–60 Nm).

10. Install the timing belt.

11. Connect the negative battery cable. Fill and bleed the cooling system.

12. Start the engine and bring to normal operating temperature. Check for leaks.

2.3L Engine

1. Disconnect the negative battery cable. Drain the cooling system.

2. Remove the timing belt covers and the timing belt.

3. Use a pulley removal tool to hold the water pump pulley and remove the bolts. Remove the water pump pulley.

4. Position a drain pan under the water pump.

5. Remove the water pump mounting bolts, and remove the water pump.

To install:

6. Clean the mating surfaces of the water pump and the engine block.

7. Install a new O-ring onto the water pump.

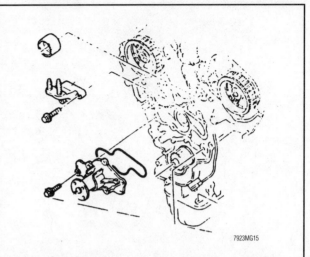

Exploded view of the water pump assembly—Mazda 1.8L (K8) and 2.5L engines, 2.3L engine is similar

7923MG15

8. Install the water pump and torque the bolts 18 ft. lbs. (25 Nm).

9. Install the water pump pulley with the bolts. Hold the pulley with the tool and tighten the bolts to 88 inch lbs. (10 Nm).

10. Install the timing belt and timing covers.

11. Connect the negative battery cable. Fill and bleed the cooling system.

12. Start the engine and bring to normal operating temperature. Check for leaks.

Mercedes-Benz

C, CLK, E, S, SL & SLK CLASSES

1. Disconnect the negative battery cable.

❊❊ CAUTION

Never open, service or drain the radiator or cooling system when hot; serious burns can occur from the steam and hot coolant.

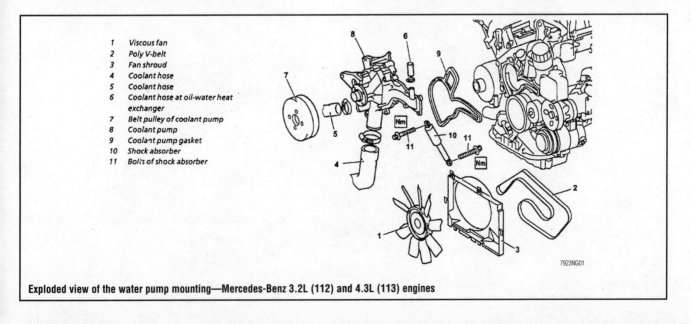

1	Viscous fan
2	Poly V-belt
3	Fan shroud
4	Coolant hose
5	Coolant hose
6	Coolant hose at oil-water heat exchanger
7	Belt pulley of coolant pump
8	Coolant pump
9	Coolant pump gasket
10	Shock absorber
11	Bolts of shock absorber

7923NG01

Exploded view of the water pump mounting—Mercedes-Benz 3.2L (112) and 4.3L (113) engines

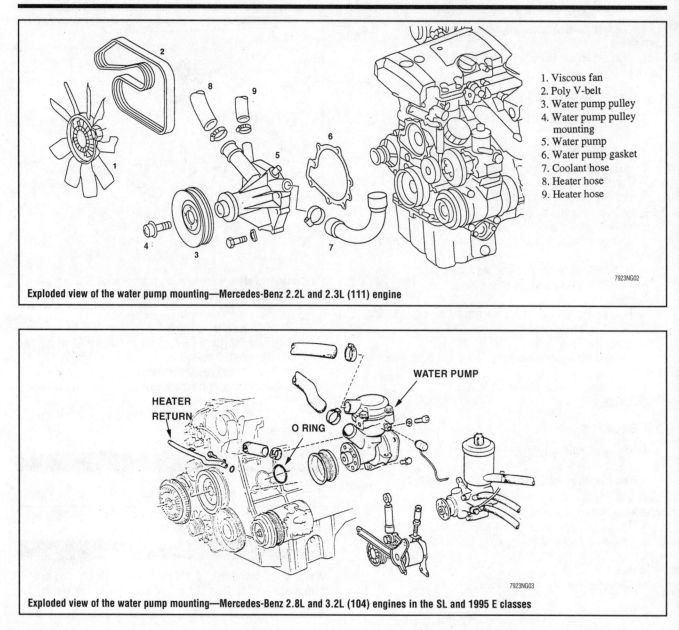

1. Viscous fan
2. Poly V-belt
3. Water pump pulley
4. Water pump pulley mounting
5. Water pump
6. Water pump gasket
7. Coolant hose
8. Heater hose
9. Heater hose

7923NG02

Exploded view of the water pump mounting—Mercedes-Benz 2.2L and 2.3L (111) engine

WATER PUMP

HEATER RETURN

O RING

7923NG03

Exploded view of the water pump mounting—Mercedes-Benz 2.8L and 3.2L (104) engines in the SL and 1995 E classes

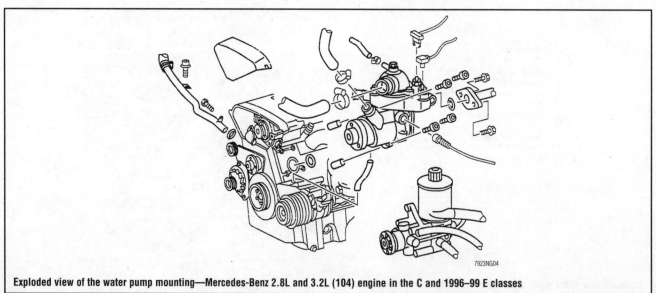

7923NG04

Exploded view of the water pump mounting—Mercedes-Benz 2.8L and 3.2L (104) engine in the C and 1996–99 E classes

2. Drain and recycle the engine coolant.

3. Remove the engine cooling fan and clutch, then the fan shroud.

4. Drain and recycle the engine coolant.

5. If equipped, remove the engine cover.

6. Remove the accessory drive belt.

7. If necessary, unbolt the power steering pump and position it aside leaving the hoses attached.

8. Disconnect the coolant hoses from the water pump.

9. If equipped, disconnect the coolant hoses from the oil-to-water heat exchanger.

10. Remove the belt pulley.

11. Remove the water pump mounting bolts, then the water pump.

12. Clean and dry the gasket mating surface for the water pump.

To install:

13. Install the water pump and gasket, and tighten the M6 bolts to 88 inch lbs. (10 Nm) and the M8 bolts to 177 inch lbs. (20 Nm).

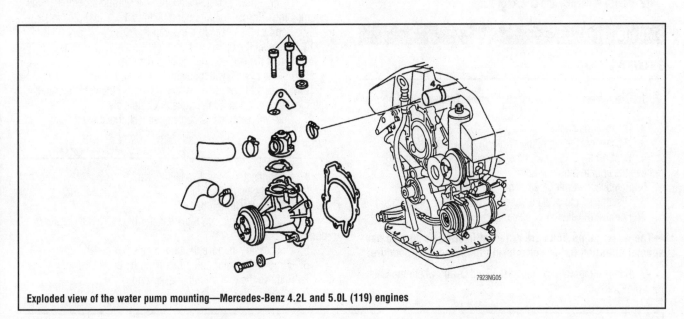

Exploded view of the water pump mounting—Mercedes-Benz 4.2L and 5.0L (119) engines

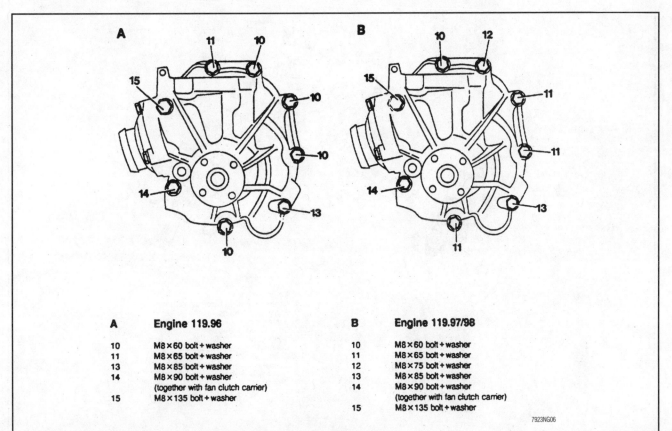

A	Engine 119.96	B	Engine 119.97/98
10	M8 × 60 bolt + washer	10	M8 × 60 bolt + washer
11	M8 × 65 bolt + washer	11	M8 × 65 bolt + washer
13	M8 × 85 bolt + washer	12	M8 × 75 bolt + washer
14	M8 × 90 bolt + washer	13	M8 × 85 bolt + washer
	(together with fan clutch carrier)	14	M8 × 90 bolt + washer
15	M8 × 135 bolt + washer		(together with fan clutch carrier)
		15	M8 × 135 bolt + washer

Water pump mounting bolt identification—Mercedes-Benz 4.2L and 5.0L (119) engines

14. Install the water pump belt pulley and tighten the mounting bolts to 88 inch lbs. (10 Nm).
15. Connect the coolant hoses to the water pump.
16. Install the power steering pump.
17. Install the accessory drive belt.
18. Install the engine cover.
19. Install the fan shroud and fan.
20. Fill the engine with coolant.
21. Connect the negative battery cable.
22. Start the vehicle and check for leaks.

Mitsubishi

3000GT & DIAMANTE

1. Disconnect the negative battery cable. Drain the cooling system.
2. Remove the engine undercover.
3. Disconnect the clamp bolt from the power steering hose.
4. Support the engine with the appropriate equipment and remove the engine mount bracket.
5. Remove the timing belt from the front of the engine.
6. Disconnect the coolant hoses from the pump, if equipped.
7. Remove the alternator brace.

➡**The water pump bolts are different in size. Be sure to pay special attention to the bolts during the removal procedure.**

8. Remove the water pump, gasket and O-ring where the water inlet pipe joins the pump.

To install:

9. Thoroughly clean and dry both gasket surfaces of the water pump and block.
10. Install a new O-ring into the groove on the front end of the water inlet pipe. Do not apply oils or grease to the O-ring. Wet with water only.

➡**Use care when aligning the water pump with the inlet water pipe.**

11. Using a new gasket, install the water pump assembly to the engine block. Torque the mounting bolts to 17 ft. lbs. (24 Nm).
12. Connect the hoses to the pump.
13. Reinstall the timing belt and related parts.
14. Install the engine drive belts and adjust.
15. Fill the system with coolant.
16. Connect the negative battery cable, run the vehicle until the thermostat opens and fill the radiator completely.
17. Once the vehicle has cooled, recheck the coolant level.

ECLIPSE

1. Disconnect the negative battery cable.
2. Drain the engine coolant.
3. Remove the timing belt.
4. If necessary, remove the alternator brace from the water pump.
5. If necessary, remove the timing belt rear cover.
6. Remove the water pump mounting bolts.
7. Remove the water pump, gasket and O-ring.

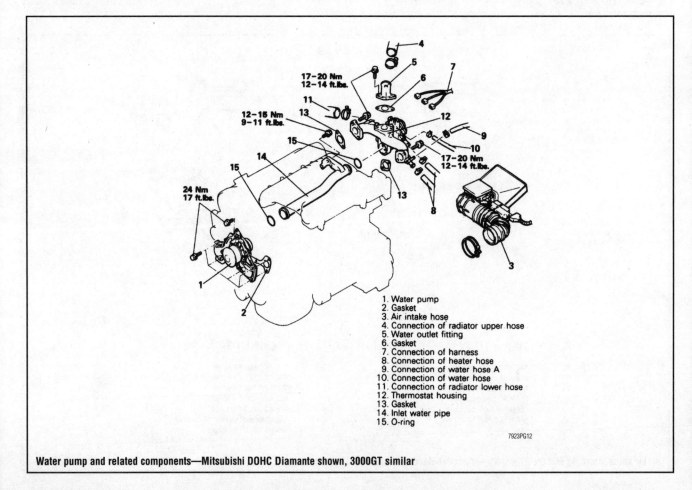

1. Water pump
2. Gasket
3. Air intake hose
4. Connection of radiator upper hose
5. Water outlet fitting
6. Gasket
7. Connection of harness
8. Connection of heater hose
9. Connection of water hose A
10. Connection of water hose
11. Connection of radiator lower hose
12. Thermostat housing
13. Gasket
14. Inlet water pipe
15. O-ring

Water pump and related components—Mitsubishi DOHC Diamante shown, 3000GT similar

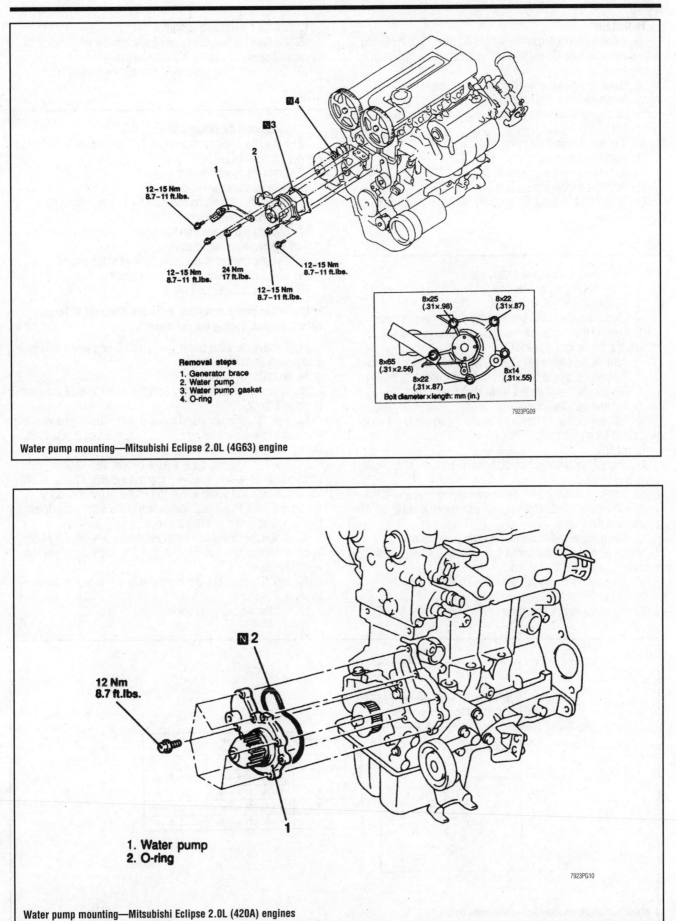

Removal steps
1. Generator brace
2. Water pump
3. Water pump gasket
4. O-ring

12–15 Nm
8.7–11 ft.lbs.

12–15 Nm
8.7–11 ft.lbs.

24 Nm
17 ft.lbs.

12–15 Nm
8.7–11 ft.lbs.

12–15 Nm
8.7–11 ft.lbs.

8x25
(.31x.98)

8x22
(.31x.87)

8x65
(.31x2.56)

8x22
(.31x.87)

8x14
(.31x.55)

Bolt diameter x length: mm (in.)

7923PG09

Water pump mounting—Mitsubishi Eclipse 2.0L (4G63) engine

12 Nm
8.7 ft.lbs.

1. Water pump
2. O-ring

7923PG10

Water pump mounting—Mitsubishi Eclipse 2.0L (420A) engines

To install:

8. Install a new O-ring on the water inlet pipe. Coat the O-ring with water or coolant. Do not allow oil or other grease to contact the O-ring.

9. Use a new gasket and install the water pump to the engine block. Torque the mounting bolts to 8.7–11 ft. lbs. (12–15 Nm). Install the alternator brace on the water pump. Torque the brace pivot bolt to 17 ft. lbs. (24 Nm).

10. If removed, install the timing belt rear cover.

11. Install the timing belt.

12. Install the remaining components.

13. Refill the engine with coolant.

14. Connect the negative battery cable, start the engine and check for leaks.

GALANT

1. Disconnect the negative battery cable.

2. Drain the cooling system.

3. Remove the engine undercover.

4. Disconnect the clamp bolt from the power steering hose.

5. Support the engine with the appropriate equipment and remove the engine mount bracket.

6. Remove the engine drive belts and the A/C tensioner bracket.

7. Remove the timing belt covers from the front of the engine.

8. Remove the camshaft and silent shaft timing belts.

9. Remove the alternator brace.

10. Remove the water pump, gasket and O-ring where the water inlet pipe(s) joins the pump.

To install:

11. Thoroughly clean and dry both gasket surfaces of the water pump and block.

12. Install a new O-ring into the groove on the front end of the water inlet pipe and wet with clean antifreeze only. Do not apply oils or grease to the O-ring.

13. Using a new gasket, install the water pump assembly. Tighten bolts with the head mark **4** to 10 ft. lbs. (14 Nm) and bolts with the head mark **7** to 18 ft. lbs. (24 Nm).

14. Reinstall the timing belt and related parts.

15. Install the engine drive belts and adjust.

16. Install the engine undercover.

17. Fill the system with coolant.

18. Connect the negative battery cable, run the vehicle until the thermostat opens and fill the radiator completely.

19. Once the vehicle has cooled, recheck the coolant level.

MIRAGE

1. Disconnect the negative battery cable.

2. Rotate the engine and position the No. 1 piston to TDC of its compression stroke.

3. Drain the cooling system.

4. Remove the engine undercover.

5. Disconnect the clamp bolt from the power steering hose.

6. Remove the engine drive belts.

7. Support the engine with the appropriate equipment and remove the engine mount bracket.

8. Remove the timing belt from the front of the engine.

9. Remove the power steering pump bracket.

10. Remove the alternator brace.

➡️**The water pump mounting bolts are different in length, note their positioning for reassembly.**

11. Remove the water pump, gasket and O-ring where the water inlet pipe(s) joins the pump.

To install:

12. Thoroughly clean and dry both gasket surfaces of the water pump and block.

13. For 1.5L engines, install a new O-ring into the groove on the front end of the water inlet pipe. Do not apply oils or grease to the O-ring. Wet the O-ring with water only.

14. For 1.8L engines, apply a 0.09–0.12 in. (2.5–3.0mm) continuous bead of sealant to water pump and install the pump assembly. Install the water pump within 15 minutes of the application of the sealant. Wait 1 hour after installation of the water pump to refill the cooling system or starting the engine.

15. Install the gasket and pump assembly and tighten the bolts to specifications. Use care when aligning the water pump with the water inlet pipe.

16. Install the remaining components in the reverse order of removal.

17. Fill the system with coolant.

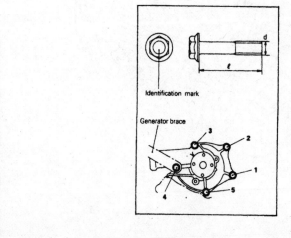

No.	Identification mark	Bolt diameter (d) x length (ℓ) mm (in.)	Torque Nm (ft.lbs.)
1	4	8 x 14 (.31 x .55)	
2	4	8 x 22 (.31 x .87)	12–15 (9–10)
3	4	8 x 30 (.31 x 1.18)	
4	7	8 x 65 (.31 x 2.56)	20–27 (15–19)
5	4	8 x 28 (.31 x 1.10)	12–15 (9–10)

7923PG11

Water pump bolt identification—Mitsubishi Galant

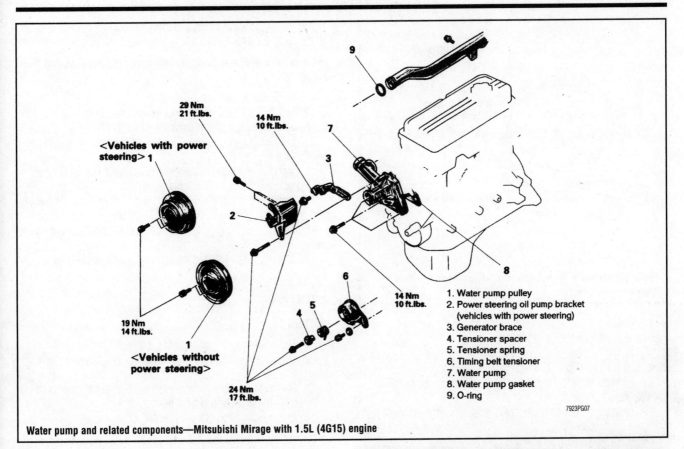

1. Water pump pulley
2. Power steering oil pump bracket (vehicles with power steering)
3. Generator brace
4. Tensioner spacer
5. Tensioner spring
6. Timing belt tensioner
7. Water pump
8. Water pump gasket
9. O-ring

29 Nm 21 ft.lbs.
14 Nm 10 ft.lbs.
<Vehicles with power steering> 1
19 Nm 14 ft.lbs.
<Vehicles without power steering>
24 Nm 17 ft.lbs.
14 Nm 10 ft.lbs.

7923PG07

Water pump and related components—Mitsubishi Mirage with 1.5L (4G15) engine

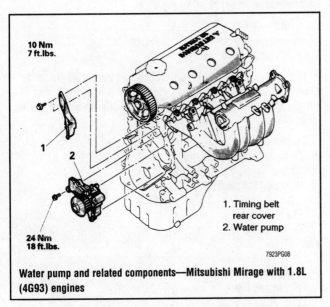

10 Nm 7 ft.lbs.
24 Nm 18 ft.lbs.

1. Timing belt rear cover
2. Water pump

7923PG08

Water pump and related components—Mitsubishi Mirage with 1.8L (4G93) engines

18. Connect the negative battery cable, run the vehicle until the thermostat opens and fill the radiator completely.

19. Once the vehicle has cooled, recheck the coolant level.

Nissan

200SX, 240SX, 300ZX, ALTIMA, MAXIMA & SENTRA

1.6L Engine

1. Disconnect the negative battery cable.

❊❊ CAUTION

Never open, service or drain the radiator or cooling system when hot; serious burns can occur from the steam and hot coolant. Always drain coolant into a sealable container. Coolant should be reused unless it is contaminated or is several years old.

2. Drain the cooling system.

3. Remove the cylinder head front mounting bracket.

4. Loosen the water pump pulley bolts.

5. Remove the engine drive belts from the A/C compressor, power steering pump and alternator.

6. Remove the belt pulley from the water pump.

7. Detach electrical connectors and coolant hoses from the thermostat housing.

8. Unbolt and remove the water pump and thermostat housing from the engine.

➡**Remove the thermostat housing with water pump assembly.**

9. Remove the bolts that secure the thermostat housing to the water pump.

10. Remove all traces of gasket material from sealing surfaces.

To install:

11. Apply a continuous bead of liquid sealer to the sealing surface of the thermostat housing. The sealant should be 0.079–0.118 in. (2–3mm) diameter.

12. Install the thermostat housing to the water pump and tighten mounting bolts to 56–73 inch lbs. (7–8 Nm).

13. Apply a continuous bead of liquid sealer to the sealing surface of the water pump. The sealant should be 0.079–0.118 in. (2–3mm) diameter.

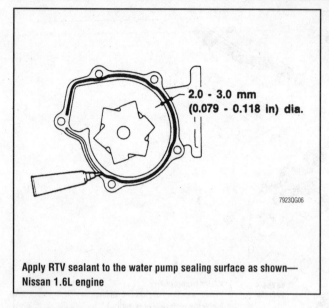

2.0 - 3.0 mm
(0.079 - 0.118 in) dia.

79230G06

Apply RTV sealant to the water pump sealing surface as shown—Nissan 1.6L engine

14. Install the water pump on the engine and tighten mounting bolts to 56–73 inch lbs. (7–8 Nm).

15. Install the pulley to the water pump and tighten the mounting bolts to 56–73 inch lbs. (7–8 Nm).

16. Connect electrical connectors and coolant hoses to the thermostat housing.

17. Install and adjust the alternator, power steering and A/C compressor drive belts.

18. Refill the cooling system and connect the negative battery cable.

19. Start the engine, bleed the cooling system, warm the engine to full operating temperature, and check for leaks.

20. If necessary, refill the cooling system when the engine has cooled.

2.0L Engine

1. Disconnect the negative battery cable.
2. Drain the radiator coolant.
3. Remove the cylinder block drain plug located at the left front of the engine and drain coolant.
4. Loosen the water pump pulley bolts.
5. Remove the power steering pump, alternator and A/C compressor drive belts (if equipped).
6. Remove the water pump pulley.
7. Note positioning of power steering pump adjusting bracket and remove the power steering pump adjusting bracket from the water pump. If necessary, remove the power steering pump for access to bracket.

➡ **When removing the power steering pump, it is not necessary to disconnect the pressure hoses or drain the system. Position or tie the pump aside.**

8. Support the engine and remove the front engine mount.
9. Remove the mounting bolts from the water pump and remove the water pump.
10. Remove all traces of liquid gasket material from sealing surfaces.

To install:

11. Apply a continuous bead of liquid sealer to the mating surface of the water pump. Sealer should be 0.079–0.118 in. (2–3mm) wide.

12. Install the water pump assembly and tighten mounting bolts to 12–15 ft. lbs. (16–21 Nm).

➡ **Be sure to properly position the adjusting bracket that was noted during removal.**

13. Install the front engine mount.

14. Install the power steering pump adjusting bracket and install the power steering pump if removed.

15. Install the water pump pulley and tighten mounting bolts to 55–73 inch lbs. (6–8 Nm).

16. Install and adjust the power steering pump, alternator and A/C compressor drive belts (if equipped).

17. Install the cylinder block drain plug located at the left front of the engine and tighten drain plug to 70–104 inch lbs. (8–12 Nm).

18. Refill the cooling system and connect the negative battery cable.

19. Start the engine, bleed the cooling system, and check for leaks.

2.4L Engine

ALTIMA

1. Disconnect the negative battery cable.
2. Drain the cooling system and water pipe, using the drain plugs.
3. Remove the upper radiator hose to provide working room and remove the drive belt(s) from the pulleys.
4. Remove the alternator and the A/C compressor.

➡ **Do not disconnect the A/C compressor lines. Unbolt the compressor and lay it off to the side.**

5. Remove the water pump pulley.
6. Remove the mounting bolts and remove the water pump from the engine.

➡ **The mounting bolts are different sizes and must be reinstalled in the correct location, therefore it is a good idea to arrange the bolts so that they can be easily identified during installation.**

To install:

7. Be sure all gasket surfaces are clean and properly apply a continuous bead of silicone sealer to the pump.

8. Install the pump to the engine and tighten the 6mm bolts to 57–66 inch lbs. (6–8 Nm) and the 8mm bolts to 12–14 ft. lbs. (16–19 Nm).

9. Install the water pump pulley and tighten the bolts to 57–66 inch lbs. (6–8 Nm).

10. Install the remaining components in the opposite order from which they were removed.

11. Fill and bleed the cooling system.

12. Start the engine and check for leaks.

240SX

1. Disconnect the negative battery cable.
2. Drain the cooling system and the engine block, using the radiator petcock and cylinder block drain plug.
3. Remove the upper radiator hose to provide working room and remove the drive belt(s) from the pulleys.
4. Remove the retaining screws and lift the fan shroud from the engine.
5. While holding the pulley, remove the nuts retaining the cooling fan and pulley to the water pump.

6. Remove the mounting bolts and remove the water pump from the engine.

To install:

7. Be sure all gasket surfaces are clean and properly apply liquid sealer to the pump.

8. Install the pump to the engine and tighten the bolts to 12–14 ft. lbs. (16–19 Nm).

9. Install the remaining components in the reverse order of removal.

10. Tighten the fan clutch, fan, and pulley mounting nuts to 66 inch lbs. (8 Nm).

11. Start the engine and check for leaks.

3.0L (VG30DE and VG30DETT) Engines

1. Disconnect negative battery cable.
2. Drain coolant from radiator and engine block.
3. Remove the undercover and the radiator.
4. Remove cooling fan assembly, water inlet, outlet and drive belts.
5. Remove crankshaft pulley and timing belt cover.
6. Remove water pump.

To install:

7. Remove all traces of gasket material.

8. Apply a continuous bead of liquid gasket to water pump mating surface.

9. Install water pump to engine block.

10. Tighten water pump bolts to 12–15 ft. lbs. (16–21 Nm).

11. Reinstall timing belt cover and tighten the cover bolts to 26–43 inch lbs. (3–5 Nm).

12. Replace crankshaft pulley and tighten the mounting bolt to 159–174 ft. lbs. (216–235 Nm).

13. Reinstall the drive belts, water inlet, water outlet, and cooling fan assembly.

14. Tighten cooling fan nuts to 51–86 inch lbs. (6–10 Nm).

15. Reinstall the radiator and undercover.

16. Refill and bleed the cooling system.

17. Reconnect the negative battery cable

18. Start engine and check for proper operation.

3.0L (VQ30DE) Engine

1. Disconnect the negative battery cable.

2. Drain the coolant from the plugs on the radiator and both sides of the engine block.

3. Position a jack under the oil pan for support. Be sure to place a block of wood on the jack for protection to the engine parts.

4. Remove the right side engine mount and engine mounting bracket.

5. Remove the drive belts and the idler pulley bracket.

6. Remove the chain tensioner cover and the water pump cover.

7. Push the timing chain tensioner sleeve and apply a stopper pin so it does not return.

8. Remove the timing chain tensioner assembly.

9. Remove the three bolts that secure the water pump.

10. Rotate the crankshaft 20 degrees counterclockwise to provide timing chain slack.

11. Put M8 bolts to two M8 threaded holes of the water pump.

12. Tighten each bolt by turning alternately ½ turn until they reach the timing chain rear case. Be sure to turn each bolt ½ turn at a time to prevent damage.

13. Lift up the water pump and remove it.

14. When removing the water pump, do not allow the water pump gear to hit the timing chain.

15. Remove and discard the O-rings from the water pump.

16. Clean all traces of liquid gasket from the water pump and covers.

To install:

17. Using new O-rings, install the water pump to the engine block.

18. Tighten the three water pump mounting bolts evenly to 62–86 inch lbs. (7–10 Nm).

19. Rotate the crankshaft pulley to its original position by turning it 20 degrees clockwise.

20. Install the timing chain tensioner and tighten the mounting bolts to 75–96 inch lbs. (9–10 Nm).

21. Remove the stopper pin from the timing chain tensioner.

22. Apply a continuous 0.091–0.130 in. (2.3–3.3mm) bead of liquid sealant to the mating surfaces of the timing chain tensioner and water pump covers.

23. Install the timing chain tensioner and water pump covers to the engine block. Tighten the cover mounting bolts to 84–108 inch lbs. (10–13 Nm).

24. Install the drive belts and the idler pulley bracket.

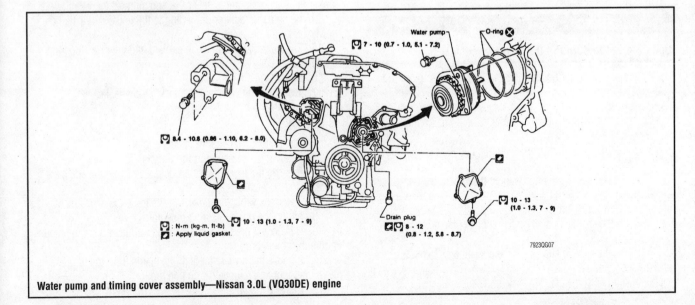

Water pump and timing cover assembly—Nissan 3.0L (VQ30DE) engine

25. Install the right side engine mounting bracket and the engine mount.

26. Remove the jack from under the engine and install the drain plugs to the cylinder block.

27. Connect the negative battery cable and refill the cooling system.

28. Start the engine, bleed the cooling system, and check for leaks.

Saab

900, 9000 & 9-3

2.0L and 2.3L Engines

1. Disconnect the negative battery cable.
2. Loosen the expansion tank pressure cap.
3. Raise and safely support the vehicle.
4. Remove the center air deflector and properly drain the coolant.
5. Remove the right front wheel assembly and remove the front section of the inner fender panel.
6. Lower the vehicle.
7. Remove the expansion tank, then remove the coolant hoses and electrical connector. Pull hard on the drive belt and lock the tensioner using tool 83–94–488 or equivalent. Remove the belt from the water pump and the air conditioning compressor.
8. Protect the engine oil cooler and place a protective cover over the upper radiator crossmember.
9. Unplug the electrical connector to the air conditioning compressor and detach the compressor with lines attached and position it to the side. Remove the air conditioning compressor bracket.
10. Disconnect the coolant hoses from the water pump.
11. Remove the oxygen sensor wire from the clips.
12. Remove the coolant pipe from the turbocharger.
13. Raise and safely support the vehicle.
14. Remove the coolant pipe from the water pump and lower the vehicle.
15. Remove the 3 retaining screws securing the water pump to the timing cover.
16. Carefully pry the water pump loose. Start at the sleeve in the cylinder block and remove the water pump.

✳✳ WARNING

Use care not to damage the oxygen sensor.

17. Remove the water pump pulley and take the pump out of its housing.
   To install:
18. Clean the sealing surfaces and install the water pump in the pump housing with a new gasket.
19. Install the water pump pulley and tighten the retaining bolts to 72 inch lbs. (8 Nm).
20. Lubricate the new O-rings with acid free petroleum jelly and fit the sleeve together with the water pump. Tighten the water pump bolts to 15 ft. lbs. (20 Nm).
21. Raise the vehicle and attach the coolant pipe to the water pump.
22. Lower the vehicle and attach the coolant pipe to the turbocharger.
23. Install the oxygen sensor wire in the clips.

24. Connect the hoses to the water pump.
25. Install the air conditioning compressor bracket. Install the air conditioning compressor and attach the electrical connector.
26. Remove the protective covers from the engine oil cooler and the upper radiator crossmember.
27. Install the belt, pull on it firmly and remove the locking pin from the tensioner. Check for proper belt alignment and tension.
28. Install the coolant hoses and electrical connector to the expansion tank, then install the expansion tank.
29. Check that the radiator drain plug is tightened and install the center air deflector.
30. Install the front section of the inner fender panel and install the wheel.
31. Properly fill the cooling system.
32. Connect the negative battery cable. Start the engine and check for proper cooling system operation.

2.5L Engine

1. Disconnect the negative battery cable.
2. Safely raise and support the vehicle.
3. Remove the lower center air deflector. Drain the engine coolant into a catch pan.
4. Lower the vehicle and remove the top engine covers, air filter assembly and mass air flow sensor.
5. Loosen the power steering pump and water pump pulley bolts. Remove the drive belt and tensioner. Remove the power steering pump, water pump pulley and timing cover.
6. Check the timing belt for damage and replace if necessary.
7. Remove the water pump.
   To install:
8. Clean water pump mounting surface. Coat the water pump O-ring and sealing surface with acid-free petroleum jelly.
9. Install water pump with new O-ring and tighten the water pump bolts to 18 ft. lbs. (25 Nm)
10. Install the timing cover.
11. Install the water pump pulley. Tighten the water pump pulley bolts to 6 ft. lbs. (8 Nm).
12. Install power steering pump, drive belt tensioner and drive belt.
13. Install air filter assembly, mass air flow sensor. Install the upper engine covers and safely raise and support the vehicle.
14. Be sure the radiator drain plug is tightened and install the lower center air deflector.
15. Fill the radiator with clean coolant and check the cooling system for leaks
16. Bleed the cooling system.
17. Test drive the vehicle and check for any leaks.

3.0L Engine

1. Disconnect the negative battery cable.
2. Safely raise and support the vehicle.
3. Remove the lower center air deflector. Drain the engine coolant into a catch pan.
4. Lower the vehicle and remove the top engine covers.
5. Lift up the power steering reservoir.
6. Disconnect the connection on the torque arm engine mount and remove the torque arm.
7. Remove the power steering line clamp from the torque arm engine mount and remove the engine mount.
8. Remove the upper coolant hose.

9. Disconnect the hose from the coolant expansion tank and remove the upper alternator air intake.

10. Loosen the power steering pump and water pump pulley bolts. Remove the drive belt and tensioner. Remove the power steering pump, water pump pulley and timing cover.

11. Check the timing belt for damage and replace if necessary.

12. Remove the water pump.

To install:

13. Clean water pump mounting surface. Coat the water pump O-ring and sealing surface with acid-free petroleum jelly.

14. Install water pump with new O-ring and tighten the water pump bolts to 18 ft. lbs. (25 Nm)

15. Install the timing cover.

16. Install the water pump pulley. Tighten the water pump pulley bolts to 6 ft. lbs. (8 Nm).

17. Install power steering pump, drive belt tensioner and drive belt.

18. Install the upper alternator air intake.

19. Install the torque arm engine mount and attach the power steering line clamp.

20. Install the torque arm and bolt the connection to the torque arm engine mount.

21. Install the upper coolant hose.

22. Connect the upper hose to the coolant expansion tank and set the power steering reservoir back into position.

23. Install the upper engine covers and safely raise and support the vehicle.

24. Be sure the radiator drain plug is tightened and install the lower center air deflector.

25. Fill the radiator with clean coolant and check the cooling system for leaks

26. Bleed the cooling system.

27. Test drive the vehicle and check for any leaks.

Subaru

LEGACY, LEGACY OUTBACK, IMPREZA, IMPREZA OUTBACK, IMPREZA OUTBACK SPORT & SVX

All Engines

✷✷ CAUTION

Never open, service or drain the radiator or cooling system when hot; serious burns can occur from the steam and hot coolant.

1. Disconnect the negative battery cable.

2. If equipped, remove the engine undercover.

3. Drain the coolant into a suitable container.

4. Disconnect the radiator outlet hose.

5. Remove the radiator fan motor assembly.

6. Remove the accessory drive belts.

7. Remove the timing belt, tensioner and camshaft angle sensor.

8. Remove the left side camshaft pulley(s) and left side rear timing belt cover. Remove the tensioner bracket.

9. Disconnect the radiator hose and heater hose from the water pump.

10. Remove the water pump retainer bolts.

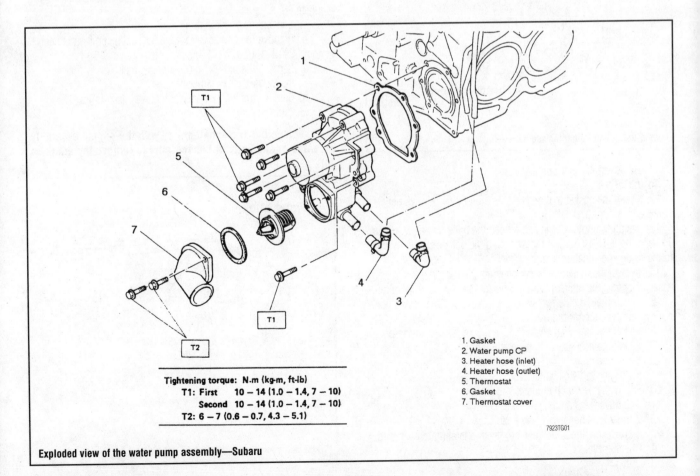

Tightening torque: N.m (kg-m, ft-lb)
T1: First   10 – 14 (1.0 – 1.4, 7 – 10)
    Second 10 – 14 (1.0 – 1.4, 7 – 10)
T2: 6 – 7 (0.6 – 0.7, 4.3 – 5.1)

1. Gasket
2. Water pump CP
3. Heater hose (inlet)
4. Heater hose (outlet)
5. Thermostat
6. Gasket
7. Thermostat cover

7923TG01

Exploded view of the water pump assembly—Subaru

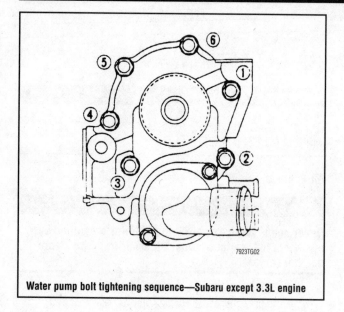

Water pump bolt tightening sequence—Subaru except 3.3L engine

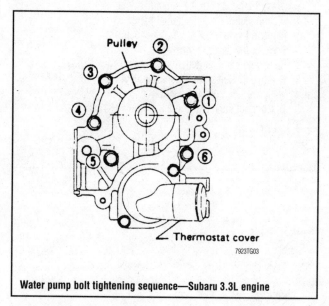

Water pump bolt tightening sequence—Subaru 3.3L engine

11. Remove the water pump.

To install:

12. Clean the gasket mating surfaces thoroughly. Always use new gaskets during installation.

13. Install the water pump and tighten the bolts, in sequence, to 7–10 ft. lbs. (10–14 Nm). After tightening the bolts once, retighten to the same specification again.

14. Inspect the radiator hoses for deterioration and replace as necessary. Connect the radiator hose and heater hose to the water pump.

15. Install the left side rear timing belt cover, left side camshaft pulley(s) and tensioner bracket.

16. Install the camshaft angle sensor, tensioner and timing belt.

17. Install the accessory drive belts.

18. Install the radiator ran motor assembly.

19. Install the radiator outlet hose.

20. Fill the system with coolant.

21. If removed, install the engine undercover.

22. Connect the negative battery cable.

23. Start the engine and allow it to reach operating temperature.

24. Check for leaks.

Suzuki

ESTEEM & SWIFT

1.0L and 1.3L Engines

1. Disconnect the negative battery cable.

2. Drain the cooling system into a suitable container and tighten the drain plug.

3. Remove the air cleaner assembly and the MAF sensor and outlet hose.

4. Remove the air cleaner bracket.

5. Raise and safely support the vehicle.

6. Remove the right side fender apron clips by pushing the center pin.

➡**Do not push the center pin too far in, or it will fall off into the fender.**

7. If equipped, remove the power steering and air conditioning belt.

8. Loosen the water pump pulley bolts.

9. Remove the alternator drive belt.

10. Remove the water pump pulley.

11. To remove the crankshaft pulley perform the following:

a. If equipped with a manual transaxle, insert a suitable flat bladed tool into the hole in the bell housing next to the exhaust pipe. This will lock the crankshaft in place.

b. If equipped with a automatic transaxle, hold a suitable flat bladed tool in line with the oil pan and insert into the teeth of the drive plate. This will lock the crankshaft in place.

c. Loosen the crankshaft pulley bolts.

d. Remove the crankshaft timing belt pulley bolt with special tool 09919–16020 or a 17mm socket.

e. Remove the pulley from the crankshaft.

f. Install the crankshaft bolt.

g. Remove the flat bladed tool that was used to lock the crankshaft in place.

➡**To remove the crankshaft pulley with the engine assembly mounted on the body, it is necessary to remove the crank-**

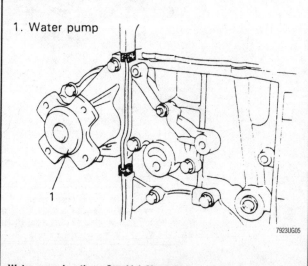

1. Water pump

Water pump location—Suzuki 1.3L engine

shaft timing belt pulley bolt. If the engine assembly is dis-mounted, the bolt does not need to be removed.

12. Remove the resonator and the timing belt outside cover.
13. Loosen the right engine mounting bolt.
14. Remove the timing belt.

✲✲ CAUTION

After the timing belt is removed never turn the camshafts or the crankshaft. Interference may occur between the pistons and the valves causing component damage.

15. Remove the timing belt inside cover.
16. Remove the water pump belt adjusting arm.
17. Carefully remove the rubber seal between the water and oil pumps, and remove the seal between the water pump and the cylinder head.
18. Remove the water pump bolts and remove the water pump.

To install:

19. Clean the water pump mounting surface of old gasket material.
20. Install a new water pump gasket to the cylinder block.
21. Install the water pump to the cylinder block and tighten the bolts to 7–9 ft. lbs. (10–13 Nm).
22. Install the rubber seal between the water pump and the oil pump. Install the seal between the water pump and the cylinder head.
23. Install the water pump belt adjusting arm.
24. Install the timing belt inside cover.
25. With the crankshaft locked in position, remove the crankshaft bolt and install the crankshaft pulley. Tighten the crankshaft pulley bolts to 10–13 ft. lbs. (14–18 Nm). Using special tool 09919–16020 or a 17mm socket, tighten the crankshaft timing belt pulley bolt to 76–83 ft. lbs. (105–115 Nm).
26. Install the timing belt.
27. Install the water pump pulley and drive belt. Tighten the water pump pulley bolts to 7–8 ft. lbs. (9–12 Nm).
28. Install the remaining components.
29. Fill the cooling system.
30. Connect the negative battery cable.
31. Start the engine and top off the coolant as necessary.
32. Check the cooling system for leaks.
33. Check the ignition timing.

1.6L Engine

1. Disconnect the negative battery cable.
2. Drain the cooling system into a re-sealable container and tighten the drain plug.
3. Remove the timing belt.
4. Remove the alternator adjusting shim.
5. Remove the oil dipstick guide and dipstick.
6. Remove the water pump bolts, gasket, the water pump and rubber seal.

To install:

7. Clean the water pump mounting surface of old gasket material.
8. Install a new water pump gasket to the cylinder block.
9. Install the water pump to the cylinder block and tighten the bolts to 7–9 ft. lbs. (10–13 Nm).
10. Install the rubber seal between the water pump and the oil pump. Install the seal between the water pump and the cylinder head.

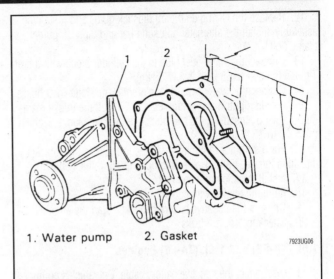

1. Water pump   2. Gasket   7923UG06

Exploded view of the water pump mounting—Suzuki 1.6L engine

11. Install the timing belt.
12. Install the alternator adjusting arm.
13. Using a new O-ring, install the oil dipstick guide and dipstick.
14. Lower the vehicle.
15. Fill the cooling system with engine coolant.
16. Connect the negative battery cable.
17. Start the engine and top off the coolant as necessary.
18. Check the ignition timing.

Toyota

AVALON, CAMRY, CELICA, COROLLA, PASEO, SUPRA & TERCEL

1.5L (5E-FE) Engine

1. Disconnect the negative battery cable. On vehicles equipped with an air bag, wait at least 90 seconds before proceeding.
2. Drain the engine coolant.
3. Remove the alternator.
4. For engines with distributorless ignition, remove the intake manifold stay bracket by disconnecting the wire clamps and removing the two nuts.
5. For engine with distributor ignition, remove the intake manifold stay bracket by removing the two nuts and two bolts.
6. Remove the water inlet pipe as follows:
    a. Disconnect the water inlet hose.
    b. Disconnect the heater hose.
    c. Disconnect the bypass hose.
    d. Remove the bolt, water inlet pipe, and O-ring.
7. Remove the oil dipstick guide.
8. Remove the alternator adjusting bar.
9. Remove the water pump attaching bolt and nuts. Remove the water pump assembly.

To install:

10. Scrape any remaining gasket material off the pump mating surface. Apply a 2–3mm (0.08–0.12 in.) bead of sealant to the groove in the pump.
11. Replace the O-ring on the water inlet pipe and lubricate the O-ring with a little soap and water. Install the pump assembly. Tighten the bolts to 13 ft. lbs. (17 Nm).

12. Replace the O-ring on the oil dipstick guide and install the assembly. Install the alternator adjusting bar and dipstick guide clamp bolt.

13. Connect the water inlet pipe to the cylinder block with a bolt. Tighten the bolt to 65 inch lbs. (7.5 Nm).

14. Connect the water bypass, heater inlet and water inlet hoses.

15. For distributorless ignition engines, install the intake manifold bracket by installing the two bolts and the wire clamp. Tighten the bolts to 15 ft. lbs. (20 Nm).

16. For distributor ignition, install the intake manifold bracket by installing the two bolts. Tighten the bolts to 15 ft. lbs. (20 Nm).

17. Install the alternator and belt.

18. Refill the engine with coolant.

19. Connect the negative battery cable and start the engine.

20. Check for coolant leaks.

1.6L (4A-FE) and 1.8L (7A-FE) Engines

1. Disconnect the negative battery cable. On vehicles equipped with an air bag, wait at least 90 seconds before proceeding.

2. Drain the engine coolant into a suitable container.

3. Remove the RH engine mounting insulator.

4. Remove No. 2 and No. 3 timing belt covers.

5. If equipped with power steering, safely raise and support the engine. Remove the hole cover and remove the two mounting bolts from the front engine mount insulator. Remove the nut and the through-bolt and remove the insulator.

6. If equipped with power steering, remove the electric cooling fan.

7. Remove the bolt and two nuts and remove the engine wire.

8. On the 7A-FE engine, disconnect crankshaft position sensor connector from the dipstick guide.

9. Remove the mounting bolt and pull out the dipstick guide and the dipstick.

10. Disconnect the water temperature sender gauge connector.

11. Remove the two nuts and the No. 2 water inlet from the water inlet hose.

12. Remove the three water pump bolts, the water pump, and the O-ring from the block.

To install:

13. Install a new O-ring on the block and install the water pump with the three bolts. Tighten the bolts to 10 ft. lbs. (14 Nm).

14. Connect the inlet hose to the water pump and install the water inlet No. 2 to the cylinder head with the two nuts. Tighten the nuts to 11 ft. lbs. (15 Nm).

15. Connect the water temperature sender gauge connector.

16. After applying a small amount of oil to the O-ring, install a new O-ring on the oil dipstick guide. Install the guide mounting bolt and tighten it to 82 inch lbs. (9 Nm).

17. On the 7A-FE engine, connect the crankshaft position sensor connector.

18. Connect the engine wire with the two nuts and the bolt.

19. If equipped with power steering, install the electric cooling fan.

20. If equipped with power steering, install the front mounting insulator through-bolt and nut. Tighten the nut to 64 ft. lbs. (87 Nm).

21. Install the engine insulator mounting bolts and tighten the two bolts to 47 ft. lbs. (64 Nm). Install the hole cover and safely lower the engine.

22. Install the No. 2 and No. 3 timing belt covers.

23. Install the RH engine mounting insulator.

24. Refill the cooling system with coolant and connect the negative battery cable. Start the engine and bleed the cooling system. Check for cooling system leaks and proper system operation.

1.8L (1ZZ-FE) Engine

1. Remove the right-hand engine under cover.

2. Drain the engine coolant.

3. Turn the tensioner bolt clockwise to loosen the belt tension, then remove the belt. Slowly release the tensioner.

4. Remove the water pump mounting bolts, then remove the pump.

To install:

5. Place a new o-ring on the timing belt cover and install the water pump. Tighten the bolts marked A (short) to 80 inch lbs. (9 Nm) and the bolts marked B (long) to 8 ft. lbs. (11 Nm).

6. Install the drive belt.

7. Install the right engine under cover.

8. Refill the engine with coolant.

9. Start the engine and check for leaks.

10. Allow the engine to cool and recheck the coolant level.

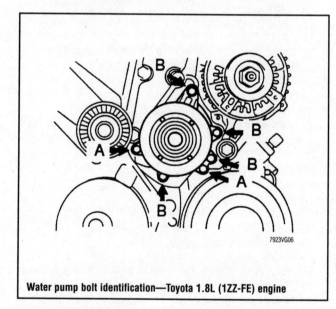

Water pump bolt identification—Toyota 1.8L (1ZZ-FE) engine

2.2L (5S-FE) Engine

1. Disconnect the negative battery cable. On vehicles equipped with an air bag, wait at least 90 seconds before proceeding.

2. Raise and safely support the vehicle.

3. Remove the right engine undercover.

4. Drain the engine coolant into a suitable container. Disconnect the lower radiator hose from the water outlet.

5. Remove the timing belt, timing belt tension spring, and the No. 2 idler pulley.

6. Remove the alternator, drive belt and the adjusting bar if necessary.

7. Remove the two nuts holding the water pump to the water bypass pipe and remove the three bolts in sequence.

8. Disconnect the water pump cover from the water bypass pipe and remove the water pump cover assembly.

9. Remove the gasket and two O-rings from the water pump and the bypass pipe.

10. Remove the water pump from the water pump cover by removing the three bolts in sequence.

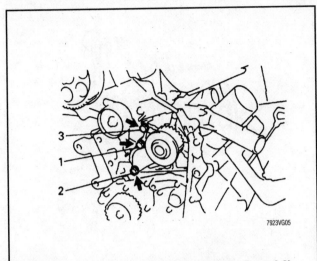

Install the three water pump bolts in this sequence—Toyota 2.2L (5S-FE) engine

To install:

11. Cleaned the gasket mating surfaces.

12. Install a new gasket and assemble the water pump to the water pump cover. Tighten the bolts to 78 inch lbs. (9 Nm) in proper sequence.

13. Install a new O-ring and gasket to the water pump cover and install a new O-ring on the water bypass pipe. Connect the water pump cover to the water bypass pipe, but do not install the nuts yet.

14. Install the water pump and tighten the three bolts in sequence. Tighten the bolts to 78 inch lbs. (9 Nm). Install the two nuts holding the water pump cover to the water bypass pipe and tighten them to 82 inch lbs. (9 Nm).

15. Install the alternator drive belt adjusting bar with the bolt and tighten the bolt to 13 ft. lbs. (18 Nm).

16. Install the No. 2 idler pulley and the timing belt tension spring.

17. Connect the lower radiator hose.

18. Install the timing belt.

19. Install the right undercover and safely lower the vehicle.

20. Fill the cooling system with coolant and connect the negative battery cable. Start the engine and bleed the cooling system. Check the cooling system for leaks and proper operation.

3.0L (1MZ-FE) Engine

1. Disconnect the negative battery cable from the battery. On vehicles equipped with an air bag, wait at least 90 seconds before proceeding.

2. Drain the engine coolant.

3. Remove the timing belt.

4. Mark the left and right camshaft pulleys with a touch of paint. Using SST tools 09249–63010 and 09960–10000 or equivalents, remove the bolts to the right and left camshaft pulleys. Remove the pulleys from the engine. Be sure not to mix up the pulleys.

5. Remove the No. 2 idler pulley by removing the bolt.

6. Disconnect the three clamps and engine wire from the rear timing belt cover.

7. Remove the six bolts holding the rear timing belt cover to the engine block.

8. Remove the four bolts and two nuts to the water pump.

9. Remove the water pump and the gasket from the engine.

To install:

10. Check that the water pump turns smoothly. Also check the air hole for coolant leakage.

11. Using a new gasket, apply liquid sealer to the gasket, water pump and engine block.

12. Install the gasket and pump to the engine and install the four bolts and two nuts. Tighten the nuts and bolts to 53 inch lbs. (6 Nm).

13. Install the rear timing belt cover and tighten the six bolts to 74 inch lbs. (9 Nm).

14. Connect the engine wire with the three clamps to the rear timing belt cover.

15. Install the No. 2 idler pulley with the bolt. Tighten the bolt to 32 ft. lbs. (43 Nm). After tightening the bolt, be sure the idler pulley moves smoothly.

16. With the flange side **outward**, install the right-hand camshaft pulley to the engine. Be sure to align the knock pin hole on the camshaft pulley with the knock pin on the camshaft. Using the same tools as removal, tighten the camshaft bolt to 65 ft. lbs. (88 Nm).

17. With the flange side **inward**, install the left-hand camshaft pulley to the engine. Be sure to align the knock pin hole on the camshaft pulley with the knock pin on the camshaft. Using the same tools as removal, tighten the camshaft bolt to 94 ft. lbs. (125 Nm).

18. Install the timing belt to the engine.

19. Fill the engine coolant.

20. Connect the negative battery cable to the battery and start the engine.

21. Top off the engine coolant and check for leaks.

3.0L (2JZ-GE and 2JZ-GTE) Engines

1. Disconnect the negative battery cable from the battery. On vehicles equipped with an air bag, wait at least 90 seconds before proceeding.

2. Remove the air cleaner and MAF meter assembly.

3. Remove the radiator assembly from the vehicle.

4. If equipped with manual transmission, remove the drive belt tensioner damper by removing the two nuts.

5. Loosen the four nuts holding the fan clutch to the water pump.

6. Loosen the drive belt tension by turning the drive belt tensioner clockwise. Remove the drive belt from the engine.

7. Remove the four nuts, the fan, fan clutch, and the water pump pulley.

8. Remove the water inlet, lower radiator hose assembly, and the thermostat.

9. Remove the timing belt.

10. Remove the alternator from the engine.

11. On turbo models disconnect the turbo water hoses from the water outlet.

12. Except for the California vehicles, remove the exhaust manifold heat insulator.

13. Remove the water outlet and No. 1 water bypass pipe.

14. Disconnect the No. 2 water bypass from the water pump by disconnecting the two nuts.

15. Disconnect the No. 3 turbo water hose from the water pump.

16. Remove the six bolts securing the water pump and remove the water pump from the engine. Be sure to replace the each bolt to its original position.

17. Clean the surface of the engine and remove the O-ring from the cylinder block.

To install:

18. Install the O-ring to the cylinder block.

19. Apply a thin layer of liquid sealant to the engine and water pump. Install a new gasket to the water pump.

20. Connect the water pump to the water bypass pipe. Do not install the nut at this time.

21. Install the water pump with the six bolts. Be sure to replace the bolts to their original positions. Tighten the bolts to 15 ft. lbs. (21 Nm).

22. Install the two nuts holding the No. 2 water bypass pipe to the water pump. Tighten the nuts to 15 ft. lbs. (21 Nm).

23. Connect the No. 3 turbo water hose to the water pump.

24. Install the water bypass outlet and No. 1 water bypass pipe.

25. Connect the turbo water hoses to the water outlet.

26. Install the alternator to the engine.

27. Except for California vehicle, install the exhaust manifold heat insulator.

28. Install the engine wire bracket with the bolt.

29. Install the timing belt.

30. Install the thermostat, water inlet, and the lower radiator hose assembly.

31. Install the water pump pulley, fan, fluid clutch assembly, and the drive belt. Tighten the fan nuts to 12 ft. lbs. (16 Nm).

32. If equipped with manual transmission, install the drive belt tensioner damper.

33. Install the radiator assembly to the vehicle.

34. Install the air cleaner and MAF meter assembly.

35. Install the No. 1 air hose.

36. Connect the negative battery cable to the battery.

37. Fill and bleed the cooling system.

38. Start the engine and check for leaks.

Volkswagen

CABRIO, GOLF, GTI, JETTA & PASSAT

Except 2.8L Engine

1. To drain the cooling system, remove the thermostat housing from under the water pump housing.

2. Raise and safely support the vehicle. Loosen but don't remove the bolts holding the pulley to the water pump.

3. Remove the timing belt cover.

4. Loosen the alternator and/or steering pump as required to remove the water pump drive belt.

5. Remove the water pump pulley. On some vehicles, the crankshaft pulley must also be removed by removing the bolts holding the pulley to the timing belt sprocket.

6. All the bolts are now accessible and the water pump can be removed from its housing.

To install:

7. Be sure to clean the housing before installing the new gasket. Install the pump into the housing and torque the pump-to-housing bolts to 7 ft. lbs. (10 Nm).

8. Install the water pump drive pulley and torque the bolts to 15 ft. lbs. (20 Nm). If the crankshaft drive pulley was removed, install it and torque the bolts to 15 ft. lbs. (20 Nm).

9. Adjust drive belt tension and install the thermostat and housing. Torque the bolts to 7 ft. lbs. (10 Nm).

2.8L Engine

1. Obtain the security code for the radio.

2. Disconnect the negative battery cable.

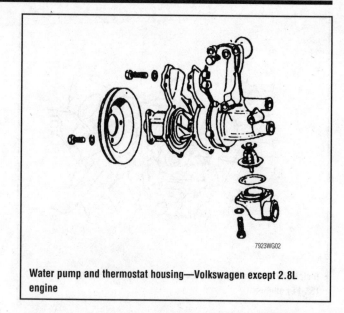

Water pump and thermostat housing—Volkswagen except 2.8L engine

3. Disconnect the front exhaust pipe from the catalytic converter.

> ❉❉ CAUTION

Never open, service or drain the radiator or cooling system when hot; serious burns can occur from the steam and hot coolant.

4. Drain the engine coolant.

5. Remove the accessory drive belt.

6. Remove the air intake duct.

7. Disconnect the ignition wires from the coils and unclip them from the retainers.

8. Remove the ignition wire guide above coil assembly.

9. Disconnect the vacuum hose from the fuel pressure regulator.

10. Remove the Intake Air Temperature (IAT) sensor from the upper intake manifold.

11. Without disconnecting the hoses, remove and place the coolant expansion tank to the side.

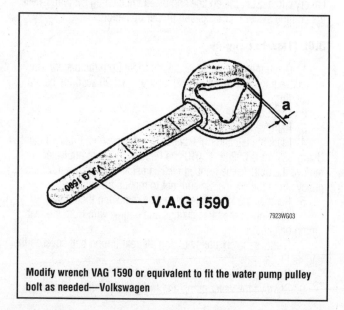

Modify wrench VAG 1590 or equivalent to fit the water pump pulley bolt as needed—Volkswagen

12. Install an engine support fixture to the lifting eyes on the left and right sides of the cylinder head. Lift the engine slightly to remove the weight from the mounts.

13. Remove the right and left rear engine/transaxle mount center bolts.

14. Remove the front engine mounting center bolts.

15. Carefully raise the engine to gain access to the water pump pulley mounting bolts.

16. Remove the water pump pulley using wrench VAG 1590 or equivalent. Modify the wrench as shown in necessary to fit the bolt.

17. Remove the mounting bolts and the water pump.

To install:

18. Using a new O-ring, install the water pump. Tighten the mounting bolts to 15 ft. lbs. (20 Nm).

19. Install the water pump pulley. Tighten the bolt to 18 ft. lbs. (25 Nm).

20. Lower the engine and install the engine/transaxle mount bolts. Tighten the mounting bolts to 44 ft. lbs. (60 Nm). Tighten the front and right rear mounts first, then the left rear mount.

21. Install the expansion tank. Tighten the bolts to 7 ft. lbs. (10 Nm).

22. Install the IAT sensor in the upper intake manifold.

23. Connect the vacuum hose to the fuel pressure regulator.

24. Install the ignition wires and the wire guide.

25. Install the air duct and accessory drive belt.

26. Refill the engine with coolant.

27. Start the engine and check for leaks.

28. Recheck the coolant level after the engine has cooled and add if necessary.

Volvo

850, 940, 960, C70, S70, V70, S90 & V90

4-Cylinder Engines

1. Disconnect the negative battery cable.
2. Set the heater control to MAX heat.

✳✳ CAUTION

Never open, service or drain the radiator or cooling system when hot; serious burns can occur from the steam and hot coolant.

3. Remove the expansion tank cap. Open the draincock on the right-hand side of the engine block and on the radiator and drain the coolant into a suitable container.

4. Close the draincocks when the coolant is completely drained.

5. Remove the radiator shroud and fan

6. Remove the lower radiator hose at the water pump. If required, remove the retaining bolt for the coolant pipe beneath the exhaust manifold and pull the pipe rearward.

7. Remove the drive belts and water pump pulleys.

8. Remove the water pump bolts, washers and nuts. Remove the water pump assembly.

To install:

9. Clean the gasket contact surfaces thoroughly and use a new gasket and O-rings.

10. Install the water pump and tighten the bolts to 11–15 ft. lbs. (14–19 Nm). Install the coolant pipe and lower radiator hose. Install the accessory drive belts and water pump pulley.

11. Install the fan and shroud. Connect the negative battery cable.

12. Fill the coolant system with coolant. Start the engine and allow it to reach normal operating temperature. Check for leaks. Add coolant as necessary.

5-Cylinder Engines

1. Disconnect the negative battery cable.

2. Raise and safely support vehicle. Remove the splash guard from below the engine.

✳✳ CAUTION

Never open, service or drain the radiator or cooling system when hot; serious burns can occur from the steam and hot coolant.

3. Drain the cooling system.

4. Remove the following:

- Spark plug cover
- Fuel line clips
- Expansion tank
- Front timing cover
- Accessory belts

5. Remove the timing belt.

6. Remove the water pump and clean the cylinder block where the two mate.

To install:

7. Install the new water pump and gasket, and tighten the bolts to 15 ft. lbs. (20 Nm).

8. Install the timing belt.

9. Install the following:

- The two fuel line clips
- Front timing cover
- Accessory belts
- Spark plug cover
- Vibration damper guard
- Wheel well panel
- Wheel

10. Connect the negative battery cable.

11. Fill the cooling system. Run the engine to normal operating temperature. Top off as necessary and check for leaks.

6-Cylinder Engines

1. Disconnect the negative battery cable.

✳✳ CAUTION

Never open, service or drain the radiator or cooling system when hot; serious burns can occur from the steam and hot coolant.

2. Drain the cooling system by opening the draincock on the right side of the cylinder block.

3. Remove the timing belt.

4. Remove the water pump retaining bolts (7) and remove the water pump.

To install:

5. Before installing the water pump, clean the mating surfaces.

6. Install the water pump, using a new gasket. Tighten the mounting bolts to 15 ft. lbs. (20 Nm).

7. Install the timing belt.

8. Fill the cooling system. Connect the negative battery cable.

9. Start the engine and check for leaks.

TRUCKS, VANS AND SPORT UTILITY VEHICLES

Acura

SLX

1. Disconnect the negative battery cable.
2. Drain the engine coolant into a clean container.
3. Remove the upper radiator hose.
4. Remove the timing belt and the idler pulley.

➡The timing belt must be replaced if it has been contaminated by oil or coolant.

5. Unbolt and remove the water pump.
6. Remove the water pump gasket. Clean any gasket material or sealant residue from the water pump mating sealing surfaces.

To install:

7. Install the water pump using a new gasket. Tighten the mounting bolts to 13 ft. lbs. (18 Nm) in a two-step crisscross sequence.

8. Install the idler pulley. Tighten the mounting bolt to 31 ft. lbs. (42 Nm).
9. Install and tension the timing belt.
10. Install the upper radiator hose.
11. Refill and bleed the cooling system.
12. Connect the negative battery cable. Start the engine and check for coolant leaks.

Chrysler Corporation

CHRYSLER TOWN & COUNTRY, DODGE CARAVAN & PLYMOUTH VOYAGER

2.4L Engine

1. Disconnect the negative battery cable.

➡This procedure requires removing the engine timing belt and the auto-tensioner. The factory specifies that the timing

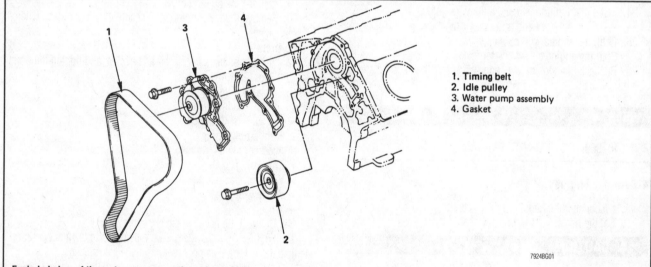

1. Timing belt
2. Idle pulley
3. Water pump assembly
4. Gasket

Exploded view of the water pump mounting—Acura SLX

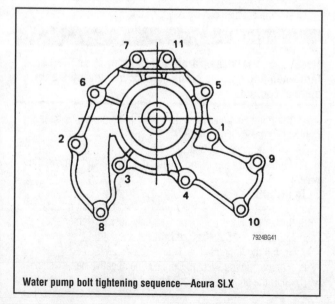

Water pump bolt tightening sequence—Acura SLX

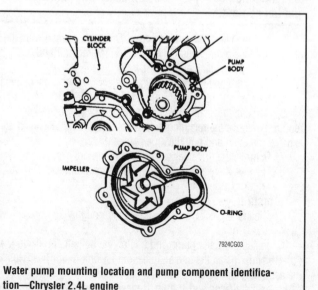

Water pump mounting location and pump component identification—Chrysler 2.4L engine

marks should always be aligned before removing the timing belt. Set the piston in the No. 1 cylinder to TDC on the compression stroke. This should align all timing marks on the crankshaft sprocket and both camshaft sprockets.

2. Raise and safely support the vehicle.
3. Remove the right inner splash shield.
4. Remove the accessory drive belts.
5. Place a drain pan under the radiator drain plug. Drain the cooling system.
6. Support the engine and remove the right motor mount.
7. Remove the power steering pump mounting bracket bolts and place the pump/bracket assembly off to one side. Do not disconnect the power steering fluid lines.
8. Remove the right engine mount bracket.
9. Remove the front timing belt upper and lower covers.
10. Loosen the timing belt tensioner bolts and remove the belt tensioner and timing belt.

❄❄ WARNING

With the timing belt removed, DO NOT rotate the camshaft or crankshaft or damage to the engine could occur.

11. Remove the camshaft sprockets. With the timing belt removed, remove both camshaft sprocket bolts. Do not allow the camshafts to turn when the camshaft sprockets are being removed.
12. Remove the rear timing belt cover to access the water pump.
13. Remove the water pump attaching bolts.
14. Remove the water pump.

To install:

15. Thoroughly clean all parts. Replace the water pump if there are any cracks, signs of coolant leakage from the shaft seal, loose or rough turning bearing, damaged impeller or sprocket, or the sprocket flange is loose or damaged.
16. Clean all sealing surfaces. Install a new rubber O-ring into the water pump O-ring groove.

➡Be sure the O-ring is properly seated in the water pump groove before tightening the bolts. An improperly located O-ring may cause damage to the O-ring and cause a coolant leak.

17. Install the water pump to the engine and tighten the bolts to 105 inch lbs. (12 Nm).
18. Pressurize the cooling system to 15 psi (103 kPa) and check for leaks. If okay, release the pressure and continue the assembly process.
19. Install the rear timing belt cover.
20. Install the camshaft sprockets and tighten the attaching bolts to 75 ft. lbs. (101 Nm). To maintain timing mark alignment, DO NOT allow the camshafts to turn while the sprocket bolts are being tightened.

❄❄ WARNING

Do not attempt to compress the tensioner plunger with the tensioner assembly installed in the engine. This will cause damage to the tensioner and other related components. The tensioner MUST be compressed in a vise.

21. Install the timing belt tensioner and timing belt. Be sure to properly tension the timing belt.
22. Install the front upper and lower timing belt covers.

23. Install the right engine mount bracket and engine mount.
24. Install the crankshaft damper and tighten the center bolt to 105 ft. lbs. (142 Nm).
25. Install the right inner splash shield.
26. Lower the vehicle.
27. Install the power steering pump bracket and power steering pump. Tighten the bracket mounting bolts to 40 ft. lbs. (54 Nm).
28. Install and adjust the drive belts.
29. Refill the cooling system using a ⁵⁰⁄₅₀ mixture of water and ethylene glycol antifreeze. Bleed the cooling system.
30. Start the engine and check for proper operation.
31. Check and top off the cooling system, if necessary.

2.5L Engine

1. Disconnect the negative battery cable.
2. Drain the cooling system.
3. Remove the drive belts.
4. If equipped with air conditioning, remove the compressor from the bracket and position it aside. Do not disconnect the refrigerant lines.
5. Remove the alternator and mounting bracket and set it aside.
6. Raise and safely support the vehicle, if necessary. Remove the pulley from the water pump.
7. Disconnect the lower radiator hose and heater hose from the water pump.
8. Remove the water pump housing attaching bolts and remove the assembly from the vehicle. Discard the O-ring.
9. Remove the water pump from the housing.

To install:

10. Using a new gasket or silicone sealer, install the water pump on the housing.
11. Install a new O-ring to the housing and install on the engine. Tighten the top three bolts to 21 ft. lbs. (30 Nm). Tighten the lower bolt to 50 ft. lbs. (68 Nm).
12. Install the water pump pulley. Tighten the water pump pulley bolts to 20 ft. lbs. (28 Nm). Connect the radiator hose and heater hose to the water pump.
13. Install the mounting bracket, alternator and A/C compressor, if removed. Install the drive belts and adjust the belts to the proper tension, if necessary.

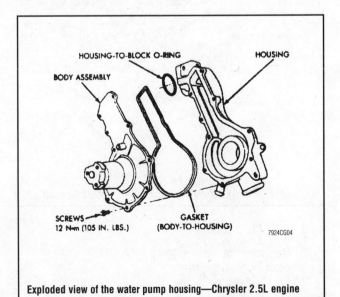

Exploded view of the water pump housing—Chrysler 2.5L engine

14. Remove the hex head plug or vacuum switching valve on the top of the thermostat housing. Fill the radiator with coolant until the coolant comes out the plug hole. Install the plug or valve and continue to fill the radiator. Tighten the vent plug to 15 ft. lbs. (20 Nm).

15. Reconnect the negative battery cable, run the vehicle until the thermostat opens, and fill the overflow tank. Check for leaks.

16. Once the vehicle has cooled, recheck the coolant level.

3.0L Engine

1. Disconnect the negative battery cable.
2. Drain the cooling system.
3. Remove the drive belts. Remove the timing belt covers and the timing belt.
4. Remove the water pump mounting bolts.
5. Separate the water pump from the water inlet pipe and remove the water pump.
6. Inspect the water pump and replace as necessary.

To install:

7. Clean all gasket and O-ring surfaces on the water pump and water pipe inlet tube.
8. Wet a new O-ring with water and install it on the water inlet pipe.
9. Install a new gasket on the water pump.
10. Install the pump inlet opening over the water pipe and press until the pipe is completely inserted into the pump housing.
11. Install the water pump-to-block mounting bolts and tighten to 20 ft. lbs. (27 Nm).
12. Install the timing belt and timing belt covers. Install and adjust the drive belts.
13. Reconnect the negative battery cable. Fill the cooling system to the proper level with a 50/50 mixture of clean, ethylene glycol antifreeze and water.
14. Run the engine and check for leaks. Top off the coolant level, if necessary.

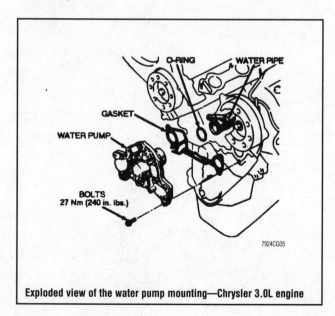

Exploded view of the water pump mounting—Chrysler 3.0L engine

3.3L and 3.8L Engines

1. Disconnect the negative battery cable.
2. Drain the cooling system.
3. Remove the serpentine belt.

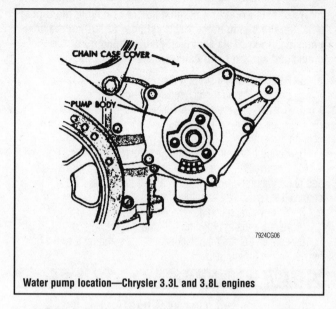

Water pump location—Chrysler 3.3L and 3.8L engines

4. Raise and safely support the vehicle. Remove the right front wheel and lower fender shield.
5. Remove the water pump pulley.
6. Remove the five mounting bolts and remove the pump from the engine.
7. Discard the O-ring. Clean the O-ring sealing surface and inspect the water pump for damage, cracks, seal leaks, and loose or rough turning bearings.

To install:

8. Install a new O-ring into the water pump groove. Install the pump onto the engine. Tighten the mounting bolts to 108 inch lbs. (12 Nm).
9. Install the water pump pulley. Tighten the water pump pulley bolts to 20 ft. lbs. (28 Nm).
10. Install the fender shield and wheel. Tighten the wheel lug nuts, in sequence, to 95 ft. lbs. (129 Nm). Lower the vehicle.
11. Install the serpentine belt.
12. Refill the cooling system to the correct level with a 50/50 mixture of clean, ethylene glycol antifreeze and water. Bleed the cooling system.
13. Reconnect the negative battery cable, run the vehicle until the thermostat opens, fill the overflow tank and check for leaks.
14. Once the vehicle has cooled, recheck the coolant level.

DODGE DAKOTA, DURANGO, RAM TRUCKS & RAM VANS

2.5L (SOHC) Engine

1. Disconnect the negative battery cable.

➡**When removing or installing the constant tension hose clamps from vehicles so equipped, use only the correct clamp tool, such as the Snap-On No. HPC-20, or equivalent.**

2. If the vehicle is equipped with air conditioning, remove the compressor from the bracket and position it to the side.
3. Raise the vehicle and support safely, if necessary and remove the alternator and bracket. Remove the pulley from the water pump.
4. Disconnect the lower radiator hose and heater hose from the water pump.

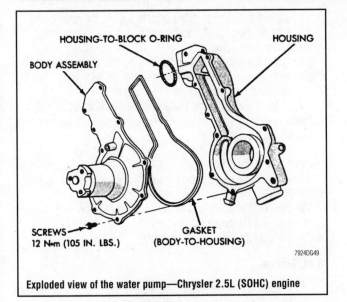

Exploded view of the water pump—Chrysler 2.5L (SOHC) engine

5. Remove the water pump housing attaching screws and remove the assembly from the vehicle. Discard the O-ring.

6. Remove the water pump from the housing.

To install:

7. Clean the mating surfaces prior to sealing the water pump.

➡ **This component is subjected to constant high pressure from hot fluid and must be sealed correctly or it will leak.**

8. Using a new gasket or silicone sealer, install the water pump to the housing.

9. Install a new O-ring to the housing and install to the engine. Tighten the bolts to 21 ft. lbs. (30 Nm).

10. Install the water pump pulley. Connect the radiator hose and heater hose to the water pump.

11. Install all items removed to gain access to the water pump and adjust the belt(s).

12. Remove the hex-head plug on the top of the thermostat housing. Fill the radiator with coolant until the coolant comes out the plug hole. Install the plug or valve and continue to fill the radiator.

13. Connect the negative battery cable, run the vehicle until the thermostat opens, fill the radiator completely and check for leaks.

14. Once the vehicle has cooled, recheck the coolant level.

2.5L (OHV) Engine

➡ **Be aware that on the 2.5L (OHV) engine, the impeller rotates in a counterclockwise direction. Check on the impeller for the letter R stamped on the blade. The use of a water pump from previous year engines will cause overheating.**

1. Disconnect the negative battery cable.
2. Drain the coolant into a suitable container.
3. Remove the drive belt.
4. Remove the power steering pump.
5. Remove the lower radiator hose and heater hose from the water pump connections.

➡ **When removing or installing the constant tension hose clamps from vehicles so equipped, use only the correct clamp tool, such as the Snap-On No. HPC-20, or equivalent.**

6. Remove the four mounting bolts and remove the pump from the engine. Note that one of the mounting bolts is longer than the others.

To install:

7. Clean the mating surfaces of all dirt and gasket material.

8. If the pump is being replaced, remove the heater hose tube from the old pump, wrap the threads with Teflon® tape and install the pipe in the new pump.

9. Install the water pump and new gasket. Tighten the bolts to 22 ft. lbs. (30 Nm).

10. Install the water pump pulley.

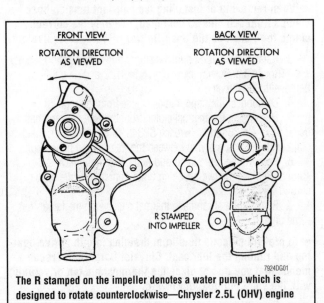

The R stamped on the impeller denotes a water pump which is designed to rotate counterclockwise—Chrysler 2.5L (OHV) engine

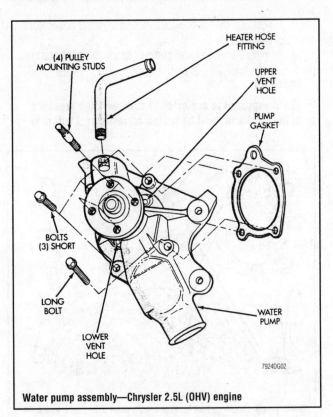

Water pump assembly—Chrysler 2.5L (OHV) engine

11. Connect the lower radiator hose and heater hose to the water pump.

12. Install the power steering pump.

13. Install the drive belt.

14. Properly fill the cooling system.

15. Connect the negative battery cable.

3.9L, 5.2L and 5.9L Engines

1. Disconnect the negative battery cable.

2. Drain the coolant into a suitable container.

➡**When removing or installing the constant tension hose clamps from vehicles so equipped, use only the correct clamp tool, such as the Snap-On No. HPC-20, or equivalent.**

3. Relax the tension on the tensioner pulley by rotating it clockwise. (Rotate it counterclockwise on 5.9L HDC engine only). Remove the drive belt.

4. Disconnect the upper radiator hose from the radiator.

5. Disconnect the thermal clutch from the water pump shaft using Snap-On 36mm fan wrench SP346, or equivalent on the thermal clutch nut and a prybar between the water pump pulley bolts.

6. Remove the four fan shroud bolts. Ram vans have a 2-piece shroud. Remove the two attaching bolts from the middle of the shroud.

7. Remove the shroud and thermal clutch with the fan at the same time.

➡**To prevent silicone fluid from draining into the drive bearing and ruining the lubricant, Chrysler Corporation recommends that you do not place the thermostatic fan drive unit with the shaft pointing downward.**

8. Remove the four bolts attaching the water pump pulley to the pump.

9. Disconnect the lower radiator hose and heater hoses from the water pump.

10. Remove the seven water pump bolts attaching the pump to the engine. Remove the pump.

To install:

➡**This component is subjected to constant high pressure from hot fluid and must be sealed correctly or it will leak.**

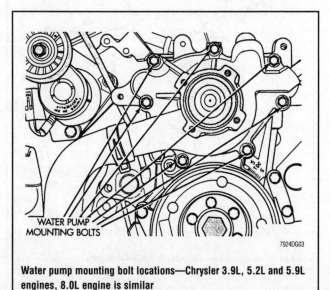

Water pump mounting bolt locations—Chrysler 3.9L, 5.2L and 5.9L engines, 8.0L engine is similar

WATER PUMP MOUNTING BOLTS

7924DG03

11. Clean the mating surfaces prior to sealing the water pump.

12. Transfer the coolant return tube, with a new O-ring installed, to the replacement pump.

13. Install the water pump on the engine. Use a new gasket coated with sealer. Tighten the bolts to 30 ft. lbs. (41 Nm).

14. Reconnect all hoses to the water pump.

15. Install the water pump pulley. tighten the bolts to 20 ft. lbs. (27 Nm).

16. Install the drive belt.

17. Install the shroud and fan assembly. Tighten the thermal clutch nut 42 ft. lbs. (57 Nm).

18. Position the shroud and tighten the mounting bolts 50 inch lbs. (6 Nm).

19. Connect the upper radiator hose to the radiator.

20. Properly fill the cooling system.

21. Connect the negative battery cable.

8.0L Engine

1. Disconnect the negative battery cable.

2. Drain the coolant into a suitable container.

3. Remove the washer bottle from the fan shroud and disconnect the fan shroud but do not remove it from the radiator.

4. Remove the upper radiator hose from the radiator.

➡**When removing or installing the constant tension hose clamps from vehicles so equipped, use only the correct clamp tool, such as the Snap-On No. HPC-20, or equivalent.**

5. Disconnect the thermal clutch from the water pump shaft using Snap-On 36mm fan wrench SP346, or equivalent on the thermal clutch nut and a prybar between the water pump pulley bolts.

➡**To prevent silicone fluid from draining into the drive bearing and ruining the lubricant, Chrysler Corporation recommends that you do not place the thermostatic fan drive unit with the shaft pointing downward.**

6. Relax the tension on the tensioner pulley by rotating it counterclockwise. Remove the drive belt.

7. Remove the four water pump pulley-to-water pump hub bolts and remove the pulley from the vehicle.

8. Remove the lower radiator hose from the water pump.

9. Remove the heater hose at the water pump fitting.

10. Remove the seven water pump mounting bolts.

11. Loosen the clamp at the water pump end of the bypass hose. Slip the hose from the water pump while removing the pump from the vehicle. Do not remove the clamp from the bypass hose.

12. Discard the water pump-to-timing chain/case/cover O-ring seal.

13. Remove the heater hose fitting from the water pump if the pump replacement is necessary. Note the position (direction) of the fitting before removal. The fitting must be installed in the same position.

➡**Do not disconnect any refrigerant lines from the compressor.**

To install:

❊❊❊ WARNING

This component is subjected to constant high pressure from hot fluid and must be sealed correctly or it will leak.

14. Clean the mating surfaces prior to sealing the water pump.
15. Install the water pump on the engine. Use a new gasket coated with sealer.
16. Install the water-pump-to-compressor front bracket mounting bolts. Tighten the bolts to 30 ft. lbs. (41 Nm).
17. If a new pump has been installed, then install the heater hose fitting to the pump. Tighten the fitting to 144 inch lbs. (16 Nm). After the fitting is tightened, position it as shown in the drawing. When positioning the fitting, do not back it off (rotate counterclockwise). Use a suitable Teflon® containing thread sealant. Refer to the directions on the package.
18. Clean the O-ring groove and install a new O-ring.
19. Apply a small amount of petroleum jelly to the O-ring to help it stay in place on the water pump.
20. Install the water pump to the engine as follows:
    a. Guide the pump fitting bypass hose as the hose is being installed.
    b. Install the water pump and tighten them to 30 ft. lbs. (40 Nm).
21. Position the bypass clamp on the hose.
22. Spin the water pump to be sure that the pump impeller does not rub against the timing chain case/cover.
23. Connect the radiator lower hose to the to the water pump.
24. Connect the heater hose and hose clamp to the heater hose fitting.
25. Install the water pump pulley. Tighten the bolts to 16 ft. lbs. (22 Nm). Place a prybar between the water pump pulley bolts to prevent the pulley from relaxing.
26. Install the serpentine belt.
27. Position the fan shroud assembly and fan blade/viscous fan drive assembly as a complete unit.
28. Install the fan shroud to the radiator. Tighten the bolts to 50 inch lbs. (6 Nm).
29. Install the fan blade and viscous fan drive to the water pump shaft.
30. Fill the cooling system, connect the negative battery cable and check for leaks.

JEEP CHEROKEE, GRAND CHEROKEE & WRANGLER

2.5L and 4.0L Engines

➡**Some vehicles use a serpentine drive belt and have a reverse rotating water pump coupled with a viscous fan drive assembly. The components are identified by the words REVERSE stamped on the cover of the viscous drive and on the inner side of the fan. The word REV is also cast into the body of the water pump.**

1. Disconnect the negative battery cable.
2. Drain the cooling system.
3. Disconnect the hoses at the pump.
4. Remove the drive belts.
5. Remove the power steering pump bracket.
6. Remove the fan and shroud.
7. If equipped, remove the idler pulley to gain clearance for pump removal.
8. Unbolt and remove the pump.
To install:
9. Clean the mating surfaces thoroughly.
10. Using a new gasket, install the pump and tighten the bolts to 22 ft. lbs. (30 Nm).

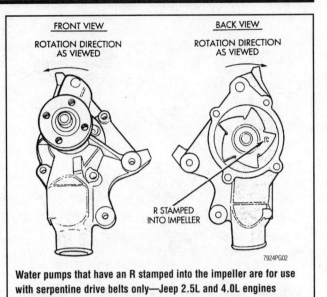

FRONT VIEW
ROTATION DIRECTION
AS VIEWED

BACK VIEW
ROTATION DIRECTION
AS VIEWED

R STAMPED
INTO IMPELLER

7924PG02

Water pumps that have an R stamped into the impeller are for use with serpentine drive belts only—Jeep 2.5L and 4.0L engines

11. If removed, install the idler pulley.
12. Reconnect the hoses at the pump and install accessory drive belt.
13. Install the power steering pump bracket. Install the fan and shroud.
14. Adjust the belt tension and fill the cooling system to the correct level.
15. Operate the engine with the heater control valve in the **HEAT** position until the thermostat opens to purge air from the system. Check coolant level and fill as required.

5.2L and 5.9L Engines

1. Disconnect the negative battery cable.

✳✳ **CAUTION**

Never open, service or drain the radiator or cooling system when hot; serious burns can occur from the steam and hot coolant.

2. Open the radiator valve and drain the cooling system.
3. Remove the cooling fan and shroud as an assembly.
4. Remove the accessory drive belt.
5. Remove the water pump pulley from the hub.
6. Disconnect the hoses from the water pump.
7. Loosen the heater hose coolant return tube mounting bolt and nut and remove the tube. Discard the O-ring.
8. Remove the water pump mounting bolts.
9. Loosen the clamp at the water pump end of the bypass hose. Slip the bypass hose from the water pump while removing the pump from the engine. Discard the gasket.
To install:
10. Clean all gasket mating surfaces.
11. Guide the water pump and new gasket into position while connecting the bypass hose to the pump. Tighten the water pump bolts to 30 ft. lbs. (40 Nm).
12. Install the bypass hose clamp.
13. Spin the water pump to ensure the pump impeller does not rub against the timing chain cover.
14. Coat a new O-ring with coolant and install it to the heater hose coolant return tube.

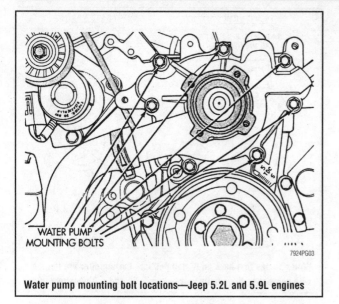

Water pump mounting bolt locations—Jeep 5.2L and 5.9L engines

15. Install the coolant return tube to the engine. Ensure the slot in the tube bracket is bottomed to the mounting bolt. This will properly position the return tube.

16. Connect the radiator hose to the water pump.

17. Connect the heater hose and clamp to the return tube.

18. Install the water pump pulley and tighten the bolts to 20 ft. lbs. (27 Nm).

19. Install the accessory drive belt.

20. Install the cooling fan and shroud.

21. Fill the cooling system.

22. Connect the negative battery cable.

23. Start the engine and check for leaks.

Ford Motor Company

FORD AEROSTAR, EXPLORER, MOUNTAINEER, RANGER & MAZDA B SERIES PICK-UPS

✳ CAUTION

Never open, service or drain the radiator or cooling system when hot; serious burns can occur from the steam and hot coolant.

2.3L Engine

1. Disconnect the negative battery cable.

2. Drain the cooling system.

3. Remove the two bolts that retain the fan shroud and position the shroud back over the fan.

4. Remove the four bolts that retain the cooling fan. Remove the fan and shroud.

5. Loosen and remove the accessory drive belt. Earlier models may have two drive belts, remove them both.

6. Remove the water pump pulley and, if necessary, the vent hose to the emissions canister.

7. Remove the heater hose at the water pump.

8. Remove the timing belt cover. Remove the lower radiator hose from the water pump.

9. Remove the water pump mounting bolts and the water pump. Clean all gasket mounting surfaces.

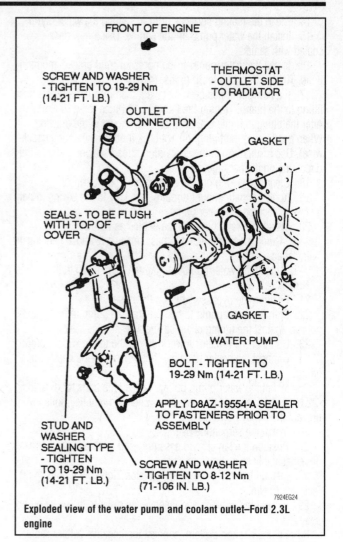

Exploded view of the water pump and coolant outlet–Ford 2.3L engine

10. Install the water pump in the reverse order of removal. Coat the threads of the mounting bolts with sealer before installation.

3.0L Engine

1. Disconnect the negative battery cable.

2. Drain the cooling system.

3. Remove the engine air cleaner outlet tube.

✳ WARNING

The following procedures for removing the fan clutch gives the factory recommended loosening and tightening directions for the fan hub nut. However, it has been our experience that certain aftermarket parts manufacturers have changed this to enable use of universal fit parts. We recommend trying the factory direction first, then, if the nut doesn't seem to be moving, reverse the direction. Placing too much load on the water pump snout will break it.

4. Remove the engine fan and radiator shield.

5. Loosen, but do not remove at this time, the four water pump pulley bolts.

6. Remove the accessory drive belts.

7. Remove the four water pump pulley retaining screws, then remove the pulley itself.

8. Disconnect the engine wiring harness from the alternator.

9. Remove the oil fill tube retaining nut at the alternator stud, then lift the tube from the stud.

10. Remove the alternator adjusting arm and throttle body brace.

11. Using a Torx® 50 driver, remove the engine drive belt tensioner assembly.

12. If equipped with an auxiliary heater, remove the screw retaining the auxiliary heater tube bracket at the power steering pump support bracket.

13. Remove the lower radiator hose.

14. Disconnect the heater hose at the water pump.

15. For vehicles equipped with power steering, remove the 5 screws retaining the power steering pump support bracket to the engine, then secure the power steering pump and bracket assembly near the battery tray. Do not disconnect the power steering hoses from the pump.

16. Remove the water pump attaching bolts. Note their location for reinstallation.

17. Remove the pump from the engine and discard the old gasket.

To install:

➡**Lightly oil all bolts and stud threads before installation except those retaining special sealant.**

18. Thoroughly clean the pump and engine mating surfaces.

19. Using an adhesive type sealer (Trim Adhesive D7AZ-19B508-B or equivalent), position a new gasket on the timing cover.

20. Position the water pump onto the engine, then install the retaining bolts. When all the bolts are started, tighten them to 84 inch lbs. (9 Nm).

21. Install the lower hose and connect the radiator hose.

22. If applicable, install the power steering pump support bracket to the engine. Tighten the fasteners to 30–40 ft. lbs. (40–54 Nm).

23. Install the water pump pulley, then hand tighten the 4 bolts.

24. Reinstall the engine accessory drive belt tensioner, then tighten the bolts to 27–33 ft. lbs. (35–45 Nm).

25. Install the heater water return hose at the water pump fitting, then tighten the hose clamp.

26. If equipped with an auxiliary heater, reinstall the auxiliary heater tube bracket at the power steering pump support bracket. Tighten the bolts to 6–8 ft. lbs. (8–12 Nm).

27. Install the alternator adjusting arm and brace. Tighten the retaining bolts to 30–40 ft. lbs. (40–54 Nm).

28. Attach the engine harness wiring to the alternator.

29. Install the oil fill tube bracket over the stud at the alternator, then tighten the retaining nuts to 32–37 ft. lbs. (42–50 Nm).

30. Install the accessory drive belts.

31. Tighten the 4 pulley bolts to 19 ft. lbs.

32. Install the fan/clutch assembly and the fan shroud.

33. Connect the negative battery cable.

34. Fill and bleed the cooling system.

35. Run the engine and check for leaks.

4.0L OHV Engine

1. Raise and safely support the vehicle so that access to the engine can be gained from both the top and the bottom of the engine compartment.

2. Drain the cooling system.
be reused unless it is contaminated or several years old.

3. Remove the lower radiator hose and heater return hose from the water pump.

4. Remove the fan and fan clutch assembly.

✳✳ WARNING

The following procedures for removing the fan clutch gives the factory recommended loosening and tightening directions for the fan hub nut. However, it has been our experience that certain aftermarket parts manufacturers have changed this to enable use of universal fit parts. We recommend trying the factory direction first, then, if the nut doesn't seem to be moving, reverse the direction. Placing too much load on the water pump snout will break it.

5. Loosen the alternator mounting bolts and remove the belt. On vans with air conditioning, remove the compressor and bracket without disconnecting the A/C lines. Support the compressor from the vehicle frame rail with strong cord or wire.

6. If equipped with power steering, remove the power steering pump. Set it aside as with the A/C compressor.

7. Remove the water pump pulley.

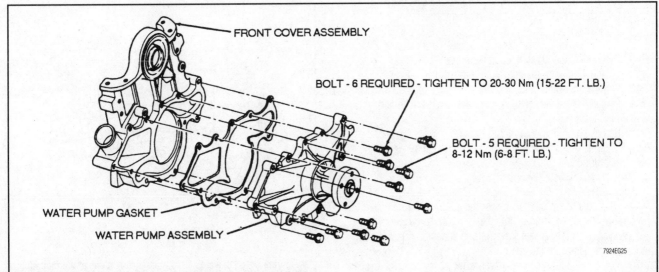

FRONT COVER ASSEMBLY

BOLT - 6 REQUIRED - TIGHTEN TO 20-30 Nm (15-22 FT. LB.)

BOLT - 5 REQUIRED - TIGHTEN TO 8-12 Nm (6-8 FT. LB.)

WATER PUMP GASKET

WATER PUMP ASSEMBLY

7924EG25

Exploded view of the water pump and front cover—Ford 3.0L engine

8. Remove the water pump attaching bolts, then remove the water pump.

To install:

9. Clean the mounting surfaces of the pump and front cover thoroughly. Remove all traces of gasket material.

10. Apply adhesive gasket sealer to both sides of a new gasket and place the gasket on the pump.

11. Position the pump on the cover and install the bolts finger-tight. When all bolts are in place, tighten them to 6–9 ft. lbs. (9–12 Nm).

12. Install the water pump pulley.

13. On vans with air conditioning, install the compressor and alternator with the bracket.

14. For vehicles with power steering, install the power steering pump.

15. Lower the vehicle.

16. Install and adjust the accessory drive belt.

17. Connect the coolant hoses to the water pump, then tighten the clamps.

18. Install the fan and clutch assembly.

➡ **The fan/clutch retaining nut is tightened counterclockwise.**

19. Fill and bleed the cooling system. Start the engine and check for leaks.

4.0L SOHC and 5.0L Engines

1. Disconnect the negative battery cable.
2. Drain the cooling system.

> ❄❄ **WARNING**
>
> **The following procedures for removing the fan clutch gives the factory recommended loosening and tightening directions for the fan hub nut. However, it has been our experience that certain aftermarket parts manufacturers have changed this to enable use of universal fit parts. We recommend trying the factory direction first, then, if the nut doesn't seem to be moving, reverse the direction. Placing too much load on the water pump snout will break it.**

3. Remove the fan and fan clutch assembly.
4. Remove the radiator if required for clearance.
5. Loosen the water pump pulley attaching bolts.
6. Remove the accessory drive belt and the idler pulley.
7. Slide the bypass hose clamp back, away from the pump.
8. Disconnect the heater hose at the pump.
9. On the 4.0L engine, remove the lower radiator hose.
10. Remove the water pump pulley.
11. On the 5.0L engine remove the water bypass hose and engine harness bracket.
12. Remove all of the water pump attaching bolts. Pay attention to the locations of any stud bolts. On the 5.0L engine, remove the lower radiator hose.
13. Remove the water pump.

To install:

14. Clean the mounting surfaces of the pump and front cover thoroughly. Remove all traces of gasket material.

15. Apply adhesive gasket sealer to both sides of a new gasket and place the gasket on the pump.

16. Position the pump on the cover, while connecting the bypass hose to the pump, and install the bolts finger–tight. On the 4.0L

SOHC engine, tighten the bolts to 72–108 inch lbs. (6–9 ft. lbs. or 8.5–12 Nm). On the 5.0L engine, tighten the bolts to 15–21 ft. lbs. (20–28 Nm).

17. Position the bypass hose clamp back to its original position.

18. Install the water pump pulley and its attaching bolts. Snug the bolts.

19. Connect the lower radiator and heater hoses to the water pump.

20. Install the belt idler pulley.

21. Lift the accessory drive belt tensioner and install the belt.

22. Securely tighten the water pump pulley attaching bolts.

23. If removed, install the radiator.

24. Install the engine fan/clutch assembly.

25. Refill and bleed the cooling system.

26. Connect the negative battery cable, start the engine and check for leaks.

FORD BRONCO, E-SERIES (VANS), EXPEDITION, F-SERIES (PICK-UPS) & LINCOLN NAVIGATOR

1. Disconnect the negative battery cable.
2. Remove the radiator, fan blade assembly and fan shroud.
3. Remove the accessory drive belt.
4. Remove the water pump pulley.
5. If equipped, disconnect the heater hose from the water pump.
6. Remove the water pump bolts and nuts. Note the locations of the bolts if different lengths.
7. Remove the water pump stud bolt, the water pump and the water pump housing gasket. Discard the water pump housing gasket.

To install:

8. Before water pump installation, be sure to aptly clean the water pump mounting surfaces of all dirt, grime and old gasket material.

➡ **All water pump housing bolts, nuts and studs are tightened to 15–22 ft. lbs. (20–30 Nm).**

9. Install the water pump onto the engine with a new gasket. Install the water pump stud bolt temporarily finger-tight.

10. Install the water pump mounting nuts and bolts temporarily finger-tight, then tighten all water pump housing fasteners to 15–22 ft. lbs. (20–30 Nm).

11. Install the heater water outlet tube, if equipped.

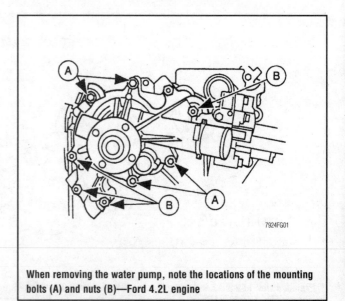

7924FG01

When removing the water pump, note the locations of the mounting bolts (A) and nuts (B)—Ford 4.2L engine

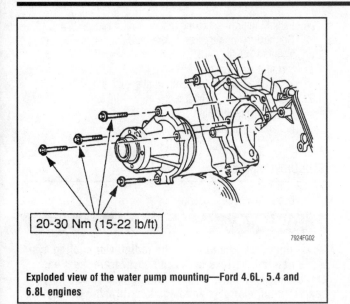

20-30 Nm (15-22 lb/ft)

7924FG02

Exploded view of the water pump mounting—Ford 4.6L, 5.4 and 6.8L engines

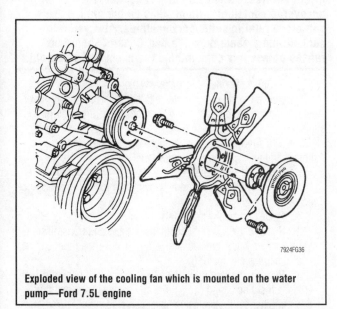

7924FG36

Exploded view of the cooling fan which is mounted on the water pump—Ford 7.5L engine

12. Install the water pump pulley and accessory drive belt.

13. Install the fan shroud, fan blade assembly and the radiator. Fill the cooling system, then connect the negative battery cable.

FORD WINDSTAR

3.0L Engine

1. Disconnect the negative battery cable.
2. Drain the engine coolant.
3. Loosen the four water pump pulley retaining bolts while the accessory drive belts are still tight.
4. Rotate the automatic tensioner down and to the left, then remove the accessory drive belt.
5. Remove the two nuts and bolt retaining the drive belt automatic tensioner to the engine, then remove the tensioner.
6. Disconnect and remove the lower radiator and heater hose from the water pump.
7. Remove the eleven water pump-to-engine retaining bolts, then lift the water pump and pulley up and out of the vehicle.

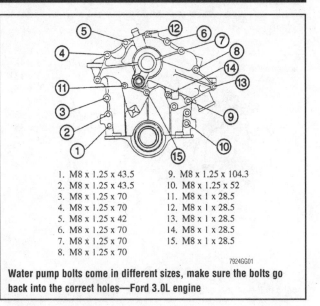

1. M8 x 1.25 x 43.5	9. M8 x 1.25 x 104.3
2. M8 x 1.25 x 43.5	10. M8 x 1.25 x 52
3. M8 x 1.25 x 70	11. M8 x 1 x 28.5
4. M8 x 1.25 x 70	12. M8 x 1 x 28.5
5. M8 x 1.25 x 42	13. M8 x 1 x 28.5
6. M8 x 1.25 x 70	14. M8 x 1 x 28.5
7. M8 x 1.25 x 70	15. M8 x 1 x 28.5
8. M8 x 1.25 x 70	

7924GG01

Water pump bolts come in different sizes, make sure the bolts go back into the correct holes—Ford 3.0L engine

➡Bolts 1, 2, 3 and 10 are for the front cover and are not removed for this procedure.

8. Remove the water pump pulley retaining bolts, then remove the pulley from the water pump.

To install:

⁂ WARNING

Be careful not to gouge the aluminum surfaces when scraping the old gasket material from the mating surfaces of the water pump and front cover.

9. Clean the gasket surfaces on the water pump and front cover. Lightly oil all bolt and stud threads, except those requiring special sealant.

10. Position a new water pump housing gasket on the water pump sealing surface using gasket sealant to hold the gasket in place.

11. With the water pump pulley and retaining bolts loosely installed on the water pump, align the water pump-to-engine front cover, then install the retaining bolts.

12. Tighten the bolts to the following specifications:
   a. Bolt numbers 4, 5, 6, 7, 8 and 9—22 ft. lbs. (20–30 Nm).
   b. Bolt numbers 11–15—71–106 inch lbs. (8–12 Nm).

13. Hand-tighten the water pump pulley retaining bolts.

14. Install the automatic belt tensioner assembly. Tighten the two retaining nuts and bolt to 35 ft. lbs. (47 Nm).

15. Install the alternator and power steering belts. Final tighten the water pump pulley retaining bolts to 15–22 ft. lbs. (22–30 Nm).

16. Position the hose clamps between the alignment marks on both ends of the hose, then slide the hose on the connection. Tighten the hose clamps to 20–30 inch lbs. (2.2–3.4 Nm).

17. Fill and bleed the cooling system.

18. Connect the negative battery cable.

19. Start the engine and check for leaks.

3.8L Engine

1. Disconnect the negative battery cable.
2. Drain the engine coolant.
3. Loosen the drive belt tensioner, then remove the drive belts.
4. Remove the lower radiator hose.

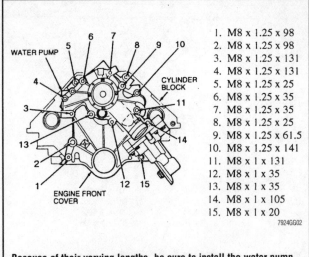

1. M8 x 1.25 x 98
2. M8 x 1.25 x 98
3. M8 x 1.25 x 131
4. M8 x 1.25 x 131
5. M8 x 1.25 x 25
6. M8 x 1.25 x 35
7. M8 x 1.25 x 35
8. M8 x 1.25 x 25
9. M8 x 1.25 x 61.5
10. M8 x 1.25 x 141
11. M8 x 1 x 131
12. M8 x 1 x 35
13. M8 x 1 x 35
14. M8 x 1 x 105
15. M8 x 1 x 20

7924GG02

Because of their varying lengths, be sure to install the water pump bolts in the correct bolt holes—Ford 3.8L engine

5. Remove the lower nut on both front engine supports.
6. Remove the alternator.
7. Position a drain pan under the power steering pump.
8. Disconnect power steering pressure line from the pump using a Fuel Line Disconnect tool (T90T-9550-S), or equivalent.
9. Remove the power steering reservoir filler cap.
10. Disconnect the water bypass hose and oil cooler hose from the heater water outlet tube.
11. Remove the retaining bolt and disconnect the heater water outlet tube from the water pump.
12. Remove the A/C bracket brace.
13. Raise the engine approximately 2 inches (51mm) to provide necessary clearance for water pump removal.
14. Remove the water pump pulley.
15. Remove the drive belt tensioner form the power steering pump brace.
16. Remove the power steering pump brace and place the pump and brace aside in the engine compartment.
17. Remove the water pump.
To install:
18. Clean all gasket mating surfaces thoroughly.

✳✳ WARNING

Be careful not to gouge the aluminum surfaces when scraping the old gasket material from the mating surfaces of the water pump and front cover.

19. Coat the threads of the No. 1 engine front cover stud with Teflon® pipe sealant, or equivalent.
20. Position a new water pump housing gasket on the water pump sealing surface using gasket sealant to hold the gasket in place.
21. Install the water pump and tighten the bolts to 15–22 ft. lbs. (20–30 Nm) and the nuts to 71–106 inch lbs. (8–12 Nm).
22. Install the power steering pump brace.
23. Install the drive belt tensioner.
24. Install the water pump pulley.
25. Lower the engine.
26. Install the A/C bracket brace.
27. Connect the heater water outlet tube.
28. Connect the water bypass hose and oil cooler hose.

29. Install the power steering reservoir filler cap.
30. Connect the power steering pressure line using Fuel Line Connect tool (T90T-9550-S), or equivalent.
31. Install the alternator.
32. Install the lower nut on both front engine supports.
33. Install the lower radiator hose.
34. Install the drive belts.
35. Fill and bleed the cooling system.
36. Connect the negative battery cable.

MERCURY VILLAGER

1. Drain the cooling system.

✳✳ CAUTION

Never open, service or drain the radiator or cooling system when hot; serious burns can occur from the steam and hot coolant. Also, when draining engine coolant, keep in mind that cats and dogs are attracted to ethylene glycol antifreeze and could drink any that is left in an uncovered container or in puddles on the ground. This will prove fatal in sufficient quantities. Always drain coolant into a sealable container. Coolant should be reused unless it is contaminated or is several years old.

2. Disconnect the negative battery cable.
3. Remove the alternator drive belt, the water pump and power steering pump drive belt and the A/C compressor drive belt (if equipped).
4. Use a strap wrench to hold the water pump pulley while removing the four water pump pulley bolts.
5. Remove the water pump pulley from the water pump.
6. Remove the crankshaft pulley using the following procedure.
   a. Raise and safely support the vehicle.
   b. Remove the five right side inner engine and transmission splash shield bolts and two screws and remove the inner engine and transmission shield.
   c. Remove the four right side outer engine and transmission splash shield bolts and two screws and remove the right side outer engine and transmission splash shields.

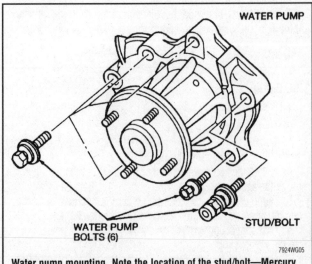

Water pump mounting. Note the location of the stud/bolt—Mercury Villager

7924WG05

d. Use a strap wrench to hold the crankshaft pulley while removing the crankshaft pulley bolt.

e. Use a crankshaft damper remover to draw the crankshaft pulley off the front of the crankshaft.

7. Remove the five lower engine front cover bolts and take of the front cover.

8. Remove the six water pump bolts. Make note of the locations of the bolts since one should be a stud/bolt and must be returned to its original location. Remove the water pump.

To install:

9. Clean all parts well. The bolt threads should be cleaned of any old sealer or corrosion. Be sure the mating surfaces between the water pump and the engine block are cleaned of any old sealant. Apply a continuous bead of gasket maker type sealer approximately ⅛ inch wide onto the water pump and position the water pump on the engine block.

10. Install the six water pump bolts. Refer to any notes made at removal so the bolts can be returned to their original locations. Do not over-tighten the water pump bolts. Tighten the water pump bolts evenly to 12–15 ft. lbs. (16–21 Nm).

11. Position the water pump pulley on the water pump and install the four pulley bolts. Use a strap wrench to hold the pulley as the bolts are tightened to 12–15 ft. lbs. (16–21 Nm).

12. Install the front engine cover and the five lower front cover bolts. Tighten to 27–44 inch lbs. (3–5 Nm).

13. Install the crankshaft pulley using the following procedure.

a. Install the crankshaft pulley and pulley bolt.

b. Hold the pulley with a strap wrench. Tighten the crankshaft pulley bolt to 90–98 ft. lbs. (123–132 Nm).

c. Install the inner and outer engine and transmission splash shields.

14. Install and adjust the drive belts.

15. Connect the negative battery cable.

16. Refill the cooling system.

17. Start the engine, bleed the cooling system and verify no leaks.

Geo and Suzuki

GEO TRACKER, SUZUKI SIDEKICK, SIDEKICK SPORT & X-90

1.6L Engines

1. Remove the timing belt cover, timing belt, tensioner, plate and spring.

2. Remove the water pump mounting bolts.

3. Remove the one (1.6L MFI engines) or two (1.6L TFI engines) small rubber seals from between the water pump and the oil pump, and the water pump and the cylinder head.

4. If necessary for clearance, remove the oil level dipstick tube retaining bolt from the engine block and the alternator adjusting brace.

✳✳ WARNING

Do NOT use a prybar between the water pump housing and the engine block to separate the two components; this can cause scratches and/or gouges, which can prevent proper sealing.

5. Pull the water pump off of the engine block. If the water pump is difficult to remove from the engine block, use a soft-faced mallet to tap the water pump housing until it loosens.

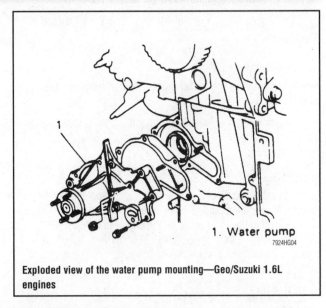

1. Water pump

Exploded view of the water pump mounting—Geo/Suzuki 1.6L engines

➡**Do not disassemble the water pump; if the water pump is damaged or defective, the entire unit is replaced.**

6. Thoroughly clean the water pump gasket mating surfaces of old gasket material and corrosion.

To install:

7. Along with a new gasket, install the water pump on the engine block. Tighten the water pump mounting bolts to 88–115 inch lbs. (10–13 Nm).

8. On TFI 1.6L engines, install two new rubber seals: one between the water pump and oil pump, and the other between the water pump and the cylinder head. The MFI 1.6L engines only use one rubber seal, located between the water and oil pumps.

9. If removed, install the alternator adjusting brace and the oil level dipstick retaining bolt.

10. Install the timing belt, tensioner, plate, spring and cover.

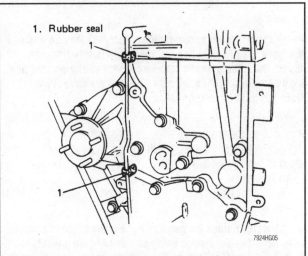

1. Rubber seal

During water pump installation, be sure to install two new rubber seals as shown—Geo/Suzuki 1.6L engines

1.8L Engine

1. Disconnect the negative battery cable.

2. Drain the engine cooling system.

3. Disconnect the upper radiator hose from the thermostat housing.

4. Remove the heater outlet pipe bolt.

5. Remove the alternator belt.

➡**When removing the water pump, do not misplace the dowel pin.**

6. Remove the four water pump mounting bolts, then remove the water pump from the engine. Discard the old water pump mounting bolts.

7. Remove the water pump O-ring and discard it.

To install:

8. Install a new O-ring on the water pump, and ensure that the dowel pins are still mounted in the water pump prior to installation.

9. Position the water pump on the engine and install NEW mounting bolts. Tighten the bolts to 221 inch lbs. (25 Nm). Failure to use four new bolts when installing the water pump may lead to coolant leakage.

10. Install the heater outlet pipe bolt.

11. Install the alternator drive belt.

12. Reattach the upper radiator hose to the thermostat housing.

13. Fill the cooling system.

14. Connect the negative battery cable.

General Motors Corporation

CHEVROLET BLAZER, S10, GMC ENVOY, JIMMY, SONOMA & OLDSMOBILE BRAVADA

1. Disconnect the negative battery cable, then drain the engine cooling system.

2. Relieve the belt tension, then remove the accessory drive belts or the serpentine drive belt, as applicable.

3. Remove the upper fan shroud, then remove the fan or fan and clutch assembly, as applicable.

4. Remove the water pump pulley.

5. Loosen the clamp and disconnect the coolant hose(s) from the water pump.

➡**For the hoses on some engines, removal may be easier if the hose is left attached until the pump is free from the block. Once the pump is removed from the engine, the pump may be pulled (giving a better grip and greater leverage) from the tight hose connection.**

6. Remove the retainers, then remove the water pump from the engine. Note the positions of all retainers as some engines will utilize different length fasteners in different locations and/or bolts and studs in different locations.

To install:

7. Using a gasket scraper, carefully clean the gasket mounting surfaces.

➡**The water pumps on some of the earlier engines covered may have been installed using sealer only, no gasket, at the factory. If a gasket is supplied with the replacement part, it should be used. Otherwise, a ⅛ in. (3mm) bead of RTV sealer should be used around the sealing surface of the pump.**

8. Apply GM 1052080 or equivalent sealant to the threads of the water pump retainers. Install the water pump to the engine using a new gasket, then thread the retainers in order to hold it in position.

9. Tighten the water pump retainers to specification:

  a. For 2.2L gasoline engines, tighten the water pump-to-engine retainers to 18 ft. lbs. (23 Nm).

  b. For the 4.3L engine, tighten the bolts and studs to 30 ft. lbs. (41 Nm).

10. Connect the coolant hose(s) and secure using the retaining clamp(s).

11. Install the water pump pulley, then install the fan or fan and clutch assembly.

12. If equipped with a serpentine drive belt, position the belt over the pulleys, then carefully allow the tensioner back into contact with the belt.

13. If equipped with V-belts, install the accessory drive belts and adjust the tension.

14. Install the upper fan shroud, then connect the negative battery cable.

15. Properly refill the engine cooling system, then run the engine and check for leaks.

CHEVROLET ASTRO & GMC SAFARI

1. Disconnect the negative battery cable, then drain the engine cooling system.

2. For 1996–99 vehicles, remove the Mass Air Flow (MAF) sensor clamp and the air cleaner housing.

3. Remove the upper fan shroud.

4. For 1996–99 vehicles, remove the drive belt and the fan and clutch assembly.

5. Remove the water pump pulley.

6. Loosen the clamps and disconnect any remaining hoses from the water pump, as applicable.

7. Remove the water pump retaining bolts. Remove the water pump assembly from the engine.

➡**On some engines, the pump retaining bolts will vary in size and thread. Be sure to note the positioning of all bolts during removal to assure proper installation.**

8. Clean gasket mounting surface.

To install:

9. Install water pump, and tighten to 33 ft. lbs. (45 Nm).

10. Install coolant hoses, and replace clamps.

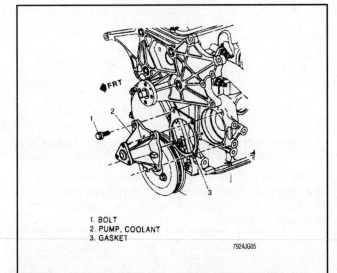

1. BOLT
2. PUMP, COOLANT
3. GASKET

7924JG05

Exploded view of the water pump mounting—GM 2.2L engine

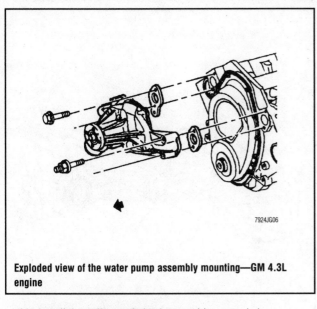

Exploded view of the water pump assembly mounting—GM 4.3L engine

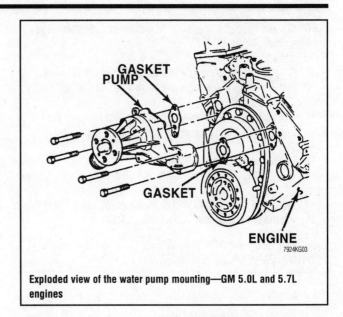

Exploded view of the water pump mounting—GM 5.0L and 5.7L engines

11. Install the pulley, and clutch assembly as needed.
12. Replace drive belt and fan shroud.
13. Install MAF sensor, if needed
14. Connect the negative battery cable.
15. Refill cooling system

C/K PICK-UPS, DENALI, ESCALADE, EXPRESS, G/P VANS, SAVANA, SIERRA, SUBURBAN, TAHOE & YUKON

4.3L, 5.0L, 5.7L and 7.4L Engines

1. Disconnect the negative battery cable.
2. Drain the radiator. Remove the fan shroud.
3. Remove the drive belt(s).
4. Remove the alternator and other accessories, if necessary.
5. Remove the fan, fan clutch and pulley.
6. Remove any accessory brackets that might interfere with water pump removal.
7. Disconnect the lower radiator hose from the water pump inlet and the heater hose from the nipple on the pump. On the 7.4L engine, remove the bypass hose.

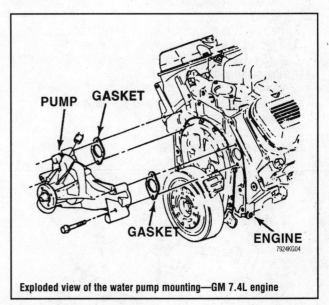

Exploded view of the water pump mounting—GM 7.4L engine

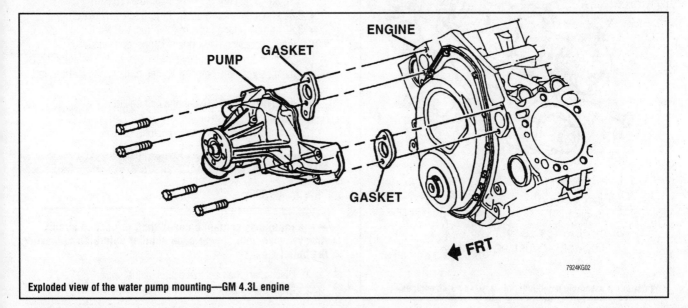

Exploded view of the water pump mounting—GM 4.3L engine

8. Remove the bolts, then pull the water pump assembly away from the timing cover.

To install:

9. Clean all old gasket material from the timing chain cover.

10. Install the pump assembly with a new gasket. Tighten the bolts to 30 ft. lbs. (41 Nm).

11. Connect the hose between the water pump inlet and the nipple on the pump. Connect the heater hose and the bypass hose (7.4L only).

12. Install the fan, fan clutch and pulley.

13. Install and adjust the alternator and other accessories, if necessary.

14. Install the drive belt(s). Install the upper radiator shroud

15. Fill the cooling system. Connect the battery.

CHEVROLET LUMINA APV, VENTURE, OLDSMOBILE SILHOUETTE, PONTIAC TRANS SPORT & MONTANA

All Engines

1. Disconnect the negative battery cable, then drain the engine cooling system.

2. On the 3.1L engine, disconnect the heater hose.

3. If needed, remove the serpentine drive belt shield.

4. For the 3.8L and 3.4L engines, loosen but do not remove the water pump pulley bolts.

5. Remove the serpentine drive belt. On the 3.1L engine, a ⅜ in. drive breaker bar may be used to pivot the belt tensioner.

6. Remove the water pump pulley.

7. Loosen the clamps and disconnect any remaining hoses from the water pump, as applicable.

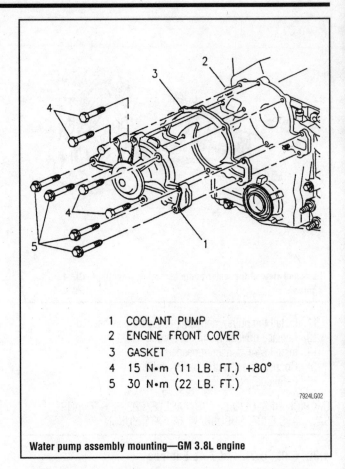

1	COOLANT PUMP
2	ENGINE FRONT COVER
3	GASKET
4	15 N•m (11 LB. FT.) +80°
5	30 N•m (22 LB. FT.)

7924LG02

Water pump assembly mounting—GM 3.8L engine

8. Remove the water pump retaining bolts. Remove the water pump assembly from the engine.

➡ On some engines, then pump retaining bolts will vary in size and thread. Be sure to note the positioning of all bolts during removal to assure proper installation.

To install:

9. Clean all sealing surfaces

10. Install a new gasket and the water pump.

11. Tighten all water pump retainers as follows:
- 3.1L and 3.4L engines—89 inch. lbs. (10 Nm)
- 3.8L engine—short bolts to 11 ft. lbs. (15 Nm) +80°
- 3.8L engine—long bolts to 22 ft. lbs. (30 Nm)

12. Install hoses and new hose clamps as needed.

13. Install water pump pulley and loosely install retaining bolts.

14. Install serpentine belt and tighten pulley bolts to 18 ft. lbs. (25 Nm).

15. If removed, install serpentine belt shield.

16. Refill cooling system.

17. Connect the negative battery cable.

Honda

CR-V & ODYSSEY

➡ The radio may contain a coded theft protection circuit. Always make note of your code number before disconnecting the battery.

1. Disconnect the negative battery cable.

2. Drain the coolant from the radiator.

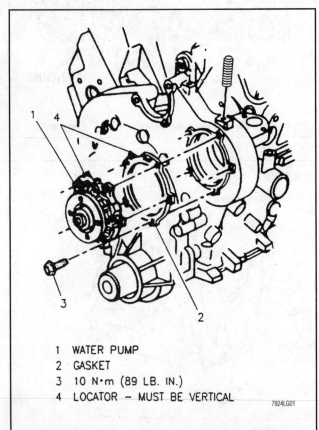

1	WATER PUMP
2	GASKET
3	10 N•m (89 LB. IN.)
4	LOCATOR – MUST BE VERTICAL

7924LG01

Water pump assembly mounting—GM 3.1L and 3.4L engines

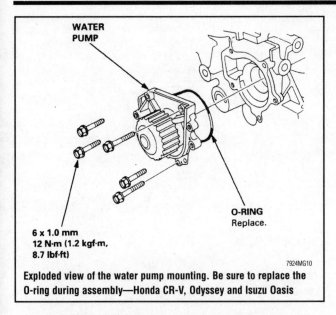

WATER PUMP

6 x 1.0 mm
12 N·m (1.2 kgf·m,
8.7 lbf·ft)

O-RING
Replace.

7924MG10

Exploded view of the water pump mounting. Be sure to replace the O-ring during assembly—Honda CR-V, Odyssey and Isuzu Oasis

3. Remove the timing belt. Refer to the Timing Belt unit repair section.

4. Remove the five bolts (6x1.0mm), that attach the pump to the cylinder block.

5. Remove the water pump.

To install:

6. Inspect and clean the O-ring mating surface on cylinder block.

7. Install a new O-ring on the water pump.

8. Install the water pump onto the cylinder block with the five 6x1.0mm bolts and tighten the bolts to 106 inch lbs. (12 Nm).

9. Install the timing belt.

10. Open the cooling system bleed bolt. It is located on the thermostat housing.

11. Refill the radiator with a coolant mixture containing 50–60% antifreeze. Use only antifreeze formulated to prevent the corrosion of aluminum parts. Fill the radiator until the coolant draining from the bleed bolt is free of air bubbles. Then, tighten the bleed bolt to 88 inch lbs. (10 Nm).

12. Install the radiator cap. Reconnect the negative battery cable.

13. Run the engine until it is at normal operating temperature. Turn the heater ON. Check for coolant leaks. Be sure the cooling fan turns ON.

14. Recheck the coolant level and add more if necessary.

15. Enter the radio security code.

PASSPORT

2.2L Engine

➡**Be sure to note the position of the mounting lug on the water pump. Failure to position the water pump correctly will cause difficulty in adjusting the timing belt and may cause overheating.**

1. Disconnect the negative battery cable.

2. Drain and recycle the engine coolant.

3. Remove the radiator hose on the inlet side of the water pump.

4. Remove the timing belt, refer to the timing belt unit repair section.

5. Remove the water pump mounting bolts, then the pump.

6. Clean the water pump mounting surface.

To install:

7. Coat the water pump sealing surface with silicone grease.

8. Install the water pump and O-ring and tighten the bolts to 18 ft. lbs. (25 Nm).

9. Install the timing belt.

10. Connect the radiator hose.

11. Fill and bleed the cooling system.

12. Connect the negative battery cable.

2.6L Engine

1. Disconnect the negative battery cable.

2. Drain the coolant from the radiator into a sealable container.

3. Disconnect the radiator hoses from the radiator.

4. Remove the air duct assembly.

5. Remove the lower fan guide clips and the bottom lock, then remove the lower fan shroud.

6. Remove the upper fan shroud bolts and remove the shroud.

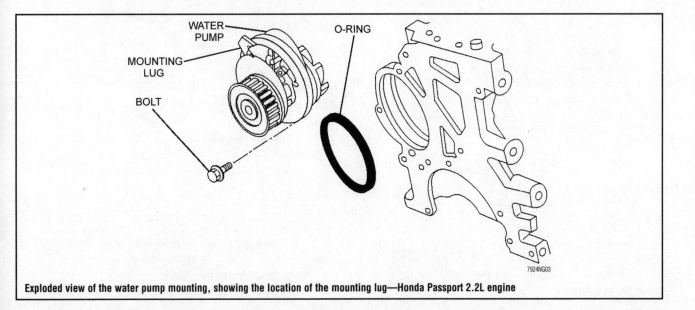

WATER PUMP

MOUNTING LUG

BOLT

O-RING

7924NG03

Exploded view of the water pump mounting, showing the location of the mounting lug—Honda Passport 2.2L engine

7. Remove the nuts attaching the fan to the water pump, then remove the fan.

8. If equipped with power steering, remove the drive belt.

9. If equipped with A/C, loosen the A/C idler pulley nuts, then remove the mounting bolts and idler pulley. Remove the A/C compressor belt.

10. Remove the alternator belt.

11. Remove the pulley from the water pump.

12. Rotate the crankshaft to align the crankshaft pulley timing marks.

13. Remove the starter and install flywheel holder (part No. J-38674) or equivalent.

14. Remove the crankshaft pulley bolt and pulley.

15. Remove the upper and lower timing belt covers.

16. Remove the four bolts and one nut from the water pump and remove the pump from the engine.

To install:

17. Clean the water pump mounting surface.

18. Install the water pump with a new gasket. Tighten the mounting bolts to 14 ft. lbs. (19 Nm), and the nut to 20 ft. lbs. (25 Nm).

19. Install the timing belt lower and upper covers. Tighten the timing belt cover bolts to 4 ft. lbs. (6 Nm).

20. Install the crankshaft pulley, tighten the bolt to 90 ft. lbs. (122 Nm).

21. Install the starter motor. Tighten the mounting bolts to 30 ft. lbs. (40 Nm).

22. Install the water pump pulley.

23. Install the alternator bracket and belt, do not tension the belt at this time.

24. If equipped with A/C, install the and idler pulley, then adjust the belt tension.

25. If equipped with power steering, install the bracket and belt, then adjust the drive belt.

26. Install the fan pulley to the water pump, and adjust the alternator belt tension. Tighten the fan attaching nuts to 20 ft. lbs. (27 Nm). Install the cooling fan.

27. Install the upper and lower fan shroud.

28. Install the air duct assembly.

29. Connect the radiator hoses.

30. Fill and bleed the cooling system.

31. Connect the negative battery cable.

3.2L Engines

1. Disconnect the negative battery cable.

2. Drain the engine coolant into a sealable container.

3. Remove the upper radiator hose.

4. Remove the timing belt and idler pulley. The timing belt must be replaced if it has been contaminated by oil or coolant.

5. Unbolt and remove the water pump. Clean any gasket material or sealant residue from the water pump mating sealing surfaces.

To install:

6. Install the water pump using a new gasket. Tighten the mounting bolts to 13 ft. lbs. (18 Nm) in a two-step crisscross sequence.

7. Install the idler pulley. Tighten the mounting bolt to 31 ft. lbs. (42 Nm).

8. Install and tension the timing belt.

9. Install the upper radiator hose.

10. Refill and bleed the cooling system.

11. Connect the negative battery cable. Start the engine and check for coolant leaks.

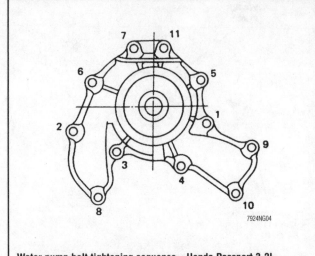

Water pump bolt tightening sequence—Honda Passport 3.2L engines

Infiniti

QX4

3.3L Engine

1. Disconnect the negative battery cable.

⁂ CAUTION

Never open, service or drain the radiator or cooling system when hot; serious burns can occur from the steam and hot coolant. Also, when draining engine coolant, keep in mind that cats and dogs are attracted to ethylene glycol antifreeze and could drink any that is left in an uncovered container or in puddles on the ground. This will prove fatal in sufficient quantities. Always drain coolant into a sealable container. Coolant should be reused unless it is contaminated or is several years old.

2. Drain the coolant from the radiator and the drain plugs on both sides of the engine block.

3. Remove the upper and lower radiator hoses.

4. Remove the fan shroud.

5. Remove the drive belts.

6. Remove the cooling fan and the water pump pulley.

7. Remove the crankshaft pulley.

8. Remove the upper and lower timing belt covers.

➡**Water pump mounting bolts are different sizes and must be reinstalled in their original locations.**

9. Remove the water pump. Don't let the engine coolant get on the timing belt.

To install:

10. Clean the gasket mating surfaces on the water pump and engine block.

11. Using a new gasket, install the water pump. Tighten the mounting bolts to 12–15 ft. lbs. (16–21 Nm).

12. Install the timing belt covers.

13. Install the crankshaft pulley.

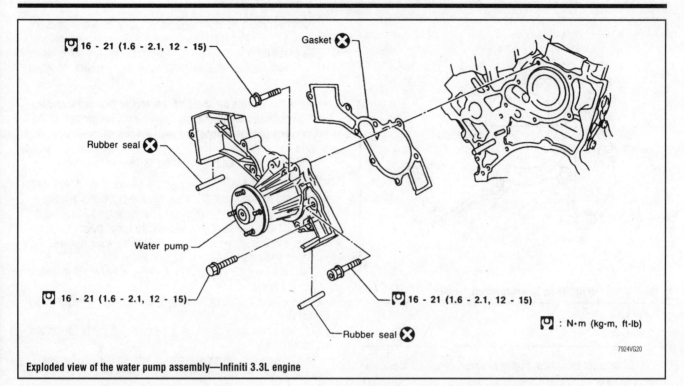

16 - 21 (1.6 - 2.1, 12 - 15)

Gasket ⊗

Rubber seal ⊗

Water pump

16 - 21 (1.6 - 2.1, 12 - 15)

16 - 21 (1.6 - 2.1, 12 - 15)

⊡ : N•m (kg-m, ft-lb)

Rubber seal ⊗

7924VG20

Exploded view of the water pump assembly—Infiniti 3.3L engine

14. Install the water pump pulley and the cooling fan.
15. Install the drive belts.
16. Install the fan shroud and the radiator hoses.
17. Connect the negative battery cable.
18. Refill the engine with coolant and bleed the system. Check for leaks.

Isuzu

TROOPER

1. Disconnect the negative battery cable.
2. Drain the engine coolant into a clean container.
3. Remove the upper radiator hose.
4. Remove the timing belt and the idler pulley.

➡**The timing belt must be replaced if it has been contaminated by oil or coolant.**

5. Unbolt and remove the water pump.
6. Remove the water pump gasket. Clean any gasket material or sealant residue from the water pump mating sealing surfaces.

To install:

7. Install the water pump using a new gasket. Tighten the mounting bolts to 13 ft. lbs. (18 Nm) in a two-step crisscross sequence.
8. Install the idler pulley. Tighten the mounting bolt to 31 ft. lbs. (42 Nm).
9. Install and tension the timing belt.
10. Install the upper radiator hose.
11. Refill and bleed the cooling system.
12. Connect the negative battery cable. Start the engine and check for coolant leaks.

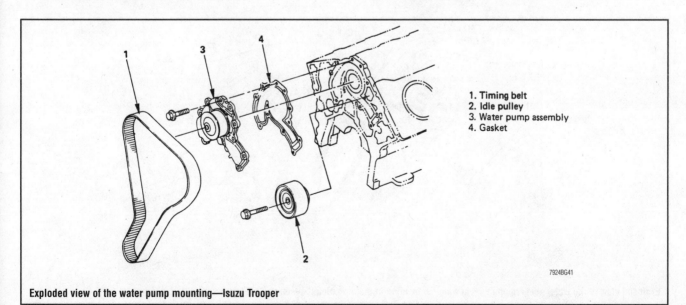

1
3
4

1. Timing belt
2. Idle pulley
3. Water pump assembly
4. Gasket

2

7924BG41

Exploded view of the water pump mounting—Isuzu Trooper

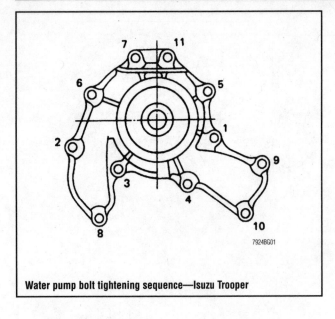

Water pump bolt tightening sequence—Isuzu Trooper

HOMBRE

1. Disconnect the negative battery cable, then drain the engine cooling system.

2. Relieve the belt tension, then remove the accessory drive belts or the serpentine drive belt, as applicable.

3. Remove the upper fan shroud, then remove the fan or fan and clutch assembly, as applicable.

4. Remove the water pump pulley.

5. Loosen the clamp and disconnect the coolant hose(s) from the water pump.

➡**For the hoses on some engines, removal may be easier if the hose is left attached until the pump is free from the block. Once the pump is removed from the engine, the pump may be pulled (giving a better grip and greater leverage) from the tight hose connection.**

6. Remove the retainers, then remove the water pump from the engine. Note the positions of all retainers as some engines will utilize different length fasteners in different locations and/or bolts and studs in different locations.

To install:

7. Using a gasket scraper, carefully clean the gasket mounting surfaces.

➡**The water pumps on some of the earlier engines covered may have been installed using sealer only, no gasket, at the factory. If a gasket is supplied with the replacement part, it should be used. Otherwise, a 1/8 in. (3mm) bead of RTV sealer should be used around the sealing surface of the pump.**

8. Apply GM 1052080 or equivalent sealant to the threads of the water pump retainers. Install the water pump to the engine using a new gasket, then thread the retainers in order to hold it in position.

9. Tighten the water pump retainers to specification:

   a. For 2.2L gasoline engines, tighten the water pump-to-engine retainers to 18 ft. lbs. (23 Nm).

   b. For the 4.3L engine, tighten the bolts and studs to 30 ft. lbs. (41 Nm).

10. Connect the coolant hose(s) and secure using the retaining clamp(s).

11. Install the water pump pulley, then install the fan or fan and clutch assembly.

12. If equipped with a serpentine drive belt, position the belt over the pulleys, then carefully allow the tensioner back into contact with the belt.

13. If equipped with V-belts, install the accessory drive belts and adjust the tension.

14. Install the upper fan shroud, then connect the negative battery cable.

15. Properly refill the engine cooling system, then run the engine and check for leaks.

RODEO & AMIGO

2.2L Engine

➡**Be sure to note the position of the mounting lug on the water pump. Failure to position the water pump correctly will cause difficulty in adjusting the timing belt and may cause overheating.**

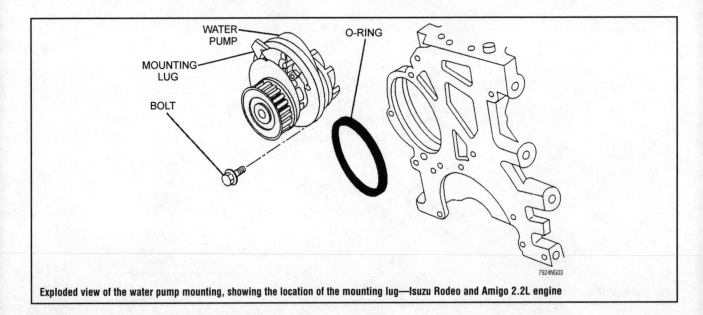

Exploded view of the water pump mounting, showing the location of the mounting lug—Isuzu Rodeo and Amigo 2.2L engine

1. Disconnect the negative battery cable.
2. Drain and recycle the engine coolant.
3. Remove the radiator hose on the inlet side of the water pump.
4. Remove the timing belt, refer to the timing belt unit repair section.
5. Remove the water pump mounting bolts, then the pump.
6. Clean the water pump mounting surface.

To install:

7. Coat the water pump sealing surface with silicone grease.
8. Install the water pump and O-ring and tighten the bolts to 18 ft. lbs. (25 Nm).
9. Install the timing belt.
10. Connect the radiator hose.
11. Fill and bleed the cooling system.
12. Connect the negative battery cable.

2.6L Engine

1. Disconnect the negative battery cable.
2. Drain the coolant from the radiator into a sealable container.
3. Disconnect the radiator hoses from the radiator.
4. Remove the air duct assembly.
5. Remove the lower fan guide clips and the bottom lock, then remove the lower fan shroud.
6. Remove the upper fan shroud bolts and remove the shroud.
7. Remove the nuts attaching the fan to the water pump, then remove the fan.
8. If equipped with power steering, remove the drive belt.
9. If equipped with A/C, loosen the A/C idler pulley nuts, then remove the mounting bolts and idler pulley. Remove the A/C compressor belt.
10. Remove the alternator belt.
11. Remove the pulley from the water pump.
12. Rotate the crankshaft to align the crankshaft pulley timing marks.
13. Remove the starter and install flywheel holder (part No. J-38674) or equivalent.
14. Remove the crankshaft pulley bolt and pulley.
15. Remove the upper and lower timing belt covers.
16. Remove the four bolts and one nut from the water pump and remove the pump from the engine.

To install:

17. Clean the water pump mounting surface.
18. Install the water pump with a new gasket. Tighten the mounting bolts to 14 ft. lbs. (19 Nm), and the nut to 20 ft. lbs. (25 Nm).
19. Install the timing belt lower and upper covers. Tighten the timing belt cover bolts to 4 ft. lbs. (6 Nm).
20. Install the crankshaft pulley, tighten the bolt to 90 ft. lbs. (122 Nm).
21. Install the starter motor. Tighten the mounting bolts to 30 ft. lbs. (40 Nm).
22. Install the water pump pulley.
23. Install the alternator bracket and belt, do not tension the belt at this time.
24. If equipped with A/C, install the and idler pulley, then adjust the belt tension.
25. If equipped with power steering, install the bracket and belt, then adjust the drive belt.
26. Install the fan pulley to the water pump, and adjust the alternator belt tension. Tighten the fan attaching nuts to 20 ft. lbs. (27 Nm). Install the cooling fan.

27. Install the upper and lower fan shroud.
28. Install the air duct assembly.
29. Connect the radiator hoses.
30. Fill and bleed the cooling system.
31. Connect the negative battery cable.

3.2L Engines

1. Disconnect the negative battery cable.
2. Drain the engine coolant into a sealable container.
3. Remove the upper radiator hose.
4. Remove the timing belt and idler pulley. The timing belt must be replaced if it has been contaminated by oil or coolant.
5. Unbolt and remove the water pump. Clean any gasket material or sealant residue from the water pump mating sealing surfaces.

To install:

6. Install the water pump using a new gasket. Tighten the mounting bolts to 13 ft. lbs. (18 Nm) in a two-step crisscross sequence.
7. Install the idler pulley. Tighten the mounting bolt to 31 ft. lbs. (42 Nm).
8. Install and tension the timing belt.
9. Install the upper radiator hose.
10. Refill and bleed the cooling system.
11. Connect the negative battery cable. Start the engine and check for coolant leaks.

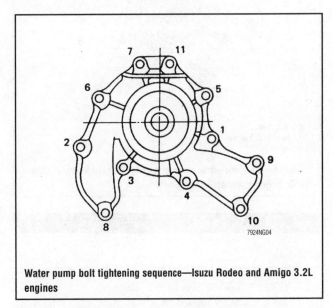

Water pump bolt tightening sequence—Isuzu Rodeo and Amigo 3.2L engines

OASIS

➡**The radio may contain a coded theft protection circuit. Always make note of your code number before disconnecting the battery.**

1. Disconnect the negative battery cable.
2. Drain the coolant from the radiator.
3. Remove the timing belt. Refer to the Timing Belt unit repair section.
4. Remove the five bolts (6x1.0mm), that attach the pump to the cylinder block.
5. Remove the water pump.

To install:

6. Inspect and clean the O-ring mating surface on cylinder block.

7. Install a new O-ring on the water pump.

8. Install the water pump onto the cylinder block with the five 6x1.0mm bolts and tighten the bolts to 106 inch lbs. (12 Nm).

9. Install the timing belt.

10. Open the cooling system bleed bolt. It is located on the thermostat housing.

11. Refill the radiator with a coolant mixture containing 50–60% antifreeze. Use only antifreeze formulated to prevent the corrosion of aluminum parts. Fill the radiator until the coolant draining from the bleed bolt is free of air bubbles. Then, tighten the bleed bolt to 88 inch lbs. (10 Nm).

12. Install the radiator cap. Reconnect the negative battery cable.

13. Run the engine until it is at normal operating temperature. Turn the heater ON. Check for coolant leaks. Be sure the cooling fan turns ON.

14. Recheck the coolant level and add more if necessary.

15. Enter the radio security code.

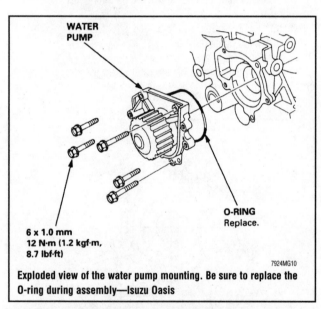

WATER PUMP

O-RING
Replace.

6 x 1.0 mm
12 N·m (1.2 kgf·m,
8.7 lbf·ft)

7924MG10

Exploded view of the water pump mounting. Be sure to replace the O-ring during assembly—Isuzu Oasis

Kia

SPORTAGE

1. Disconnect the negative battery cable.

2. Raise and safely support the vehicle.

3. Remove the lower splash plate.

4. Remove the radiator drain cock and drain the engine coolant into a suitable container.

5. Lower the vehicle.

6. Disconnect the coolant reservoir tank hose.

7. Remove the fresh air duct.

8. Remove the five shroud bolts.

9. Remove the radiator bracket bolts and lift the radiator upward.

10. Remove the cooling fan with the shroud.

11. Loosen the alternator mounting and adjusting bolts.

12. Remove the alternator belt.

13. Remove the fan pulley.

14. Remove the fan bracket assembly.

15. Remove the upper and lower timing belt covers.

16. Turn the crankshaft until No. 1 cylinder is at TDC.

17. Loosen the tensioner lockbolt and pry the tensioner away from the belt.

18. Remove the timing belt.

19. Loosen the tensioner bolt to allow the tensioner to rest.

20. Remove the five attaching bolts and remove the water pump.

21. Remove the tensioners from the water pump.

To install:

22. Clean the surface of any old gasket material.

23. Install the tensioners on the water pump.

24. Install the water pump and gasket. Tighten the bolts to 14–19 ft. lbs. (19–25 Nm).

25. Install the timing belt, as described in the Timing Belt unit repair section located in the front of this manual.

26. Loosen the tensioner lockbolt and allow the tensioner to rest against the belt.

27. Tighten the tensioner lockbolt 32 ft. lbs. (43 Nm).

28. Install the upper and lower timing belt covers.

29. Install the fan bracket assembly.

30. Install the fan pulley.

31. Install the alternator drive belt.

32. Install the cooling fan with the shroud.

33. Position the radiator and tighten the bracket bolts to 89 inch lbs. (10 Nm).

34. Tighten the five shroud bolts to 89 inch lbs. (10 Nm).

35. Tighten the alternator adjusting and mounting bolts.

36. Position the fresh air duct over the radiator and tighten the retaining bolt 89 inch lbs. (10 Nm).

37. Raise and safely support the vehicle.

38. Tighten the drain cock.

39. Connect the radiator hoses and tighten the clamps.

40. Install the lower splash plate.

41. Lower the vehicle.

42. Fill the radiator with a 50/50 mixture of water and coolant.

43. Connect the negative battery cable.

44. Start the vehicle and bring the engine to operating temperature. Add coolant as required.

45. Secure the radiator cap and check the cooling system for leaks.

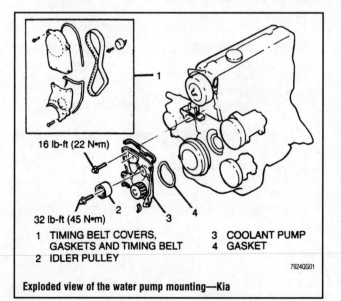

16 lb-ft (22 N•m)

32 lb-ft (45 N•m)

| 1 | TIMING BELT COVERS, GASKETS AND TIMING BELT | 3 | COOLANT PUMP |
| 2 | IDLER PULLEY | 4 | GASKET |

7924QG01

Exploded view of the water pump mounting—Kia

Land Rover

DEFENDER 90, DISCOVERY & RANGE ROVER

3.9L and 4.0L Engines

1. Disconnect the negative battery cable and drain the cooling system. Remove the lower hose.
2. Loosen the nut that attaches the fan clutch to the water pump. The nut has left-hand thread and must be turned clockwise to remove. Remove the fan and clutch assembly.
3. Remove the accessory drive belt(s).
4. Remove the alternator adjustment bracket.
5. Remove the water pump attaching bolts, noting their positions for reinstallation.
6. Remove the water pump. Clean all gasket mating surfaces.

To install:

7. Position the water pump and new gasket in place.
8. Apply sealer to the long bolts. Tighten all the bolts to 21 ft. lbs. (28 Nm).
9. Attach the lower radiator to the water pump.
10. Install the alternator bracket.
11. Install and adjust the drive belts.
12. Install the fan, fan clutch and shroud. Tighten the fan clutch nut to 30–100 ft. lbs. (40–135 Nm).
13. Connect the negative battery cable. Fill and bleed the cooling system.

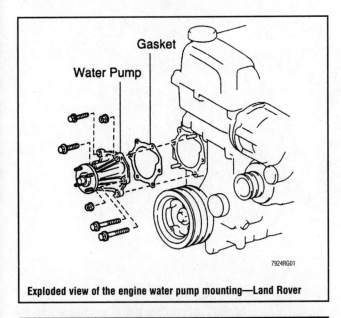

Exploded view of the engine water pump mounting—Land Rover

(labels: Gasket, Water Pump, 7924RG01)

Lexus

RX300

1. Disconnect the negative battery cable.
2. Drain the engine coolant.
3. Remove the wiper and blade assembly.
4. Remove the top cowl seal and panel.
5. Label and disconnect the window washer hoses from the ventilator louvers.
6. Remove the left and right ventilator louvers.
7. Remove the heater air duct.
8. On RX300 models, remove the front upper suspension brace.
9. Remove the timing belt.
10. Mark the left and right camshaft pulleys with a touch of paint. Using a spanner wrench, remove the bolts to the right and left camshaft pulleys. Separate the pulleys from the engine. Be sure not to mix up the pulleys.
11. Remove the No. 2 idler pulley by removing the bolt.
12. Disconnect the three clamps and engine wire from the rear timing belt cover.
13. Remove the six bolts holding the No. 3 timing belt cover to the engine block.
14. Remove the bolts and nuts to the extract the water pump.
15. Raise the engine slightly and remove the water pump and the gasket from the engine.

To install:

16. Check that the water pump turns smoothly. Also check the air hole for coolant leakage.
17. Using a new gasket, apply liquid sealer to the gasket, water pump and engine block.
18. Install the gasket and pump to the engine and install the four bolts and two nuts. Tighten the nuts and bolts to 53 inch lbs. (6 Nm).
19. Install the rear timing belt cover and tighten the six bolts to 74 inch lbs. (9 Nm).
20. Connect the engine wire with the three clamps to the rear timing belt cover.
21. Install the No. 2 idler pulley with the bolt. Tighten the bolt to 32 ft. lbs. (43 Nm). After tightening the bolt, be sure the idler pulley moves smoothly.
22. With the flange side **outward**, install the right-hand camshaft pulley to the engine. Be sure to align the knock pin hole on the camshaft pulley with the knock pin on the camshaft. Using the same tools as removal, tighten the camshaft bolt to 65 ft. lbs. (88 Nm).
23. With the flange side **inward**, install the left-hand camshaft pulley to the engine. Be sure to align the knock pin hole on the camshaft pulley with the knock pin on the camshaft. Using the same tools as removal, tighten the camshaft bolt to 94 ft. lbs. (125 Nm).
24. Install the timing belt to the engine.
25. On RX300 models, install the front upper suspension brace and tighten the mounting nuts to 59 ft. lbs. (80 Nm).
26. Fill the engine coolant.
27. Install the heater air duct.
28. Install the left and right ventilator louvers.
29. Connect the window washer hoses from the ventilator louvers.
30. Install the top cowl seal and panel.
31. Install the wiper and blade assembly.
32. Connect the negative battery cable to the battery and start the engine.
33. Top off the engine coolant and check for leaks.

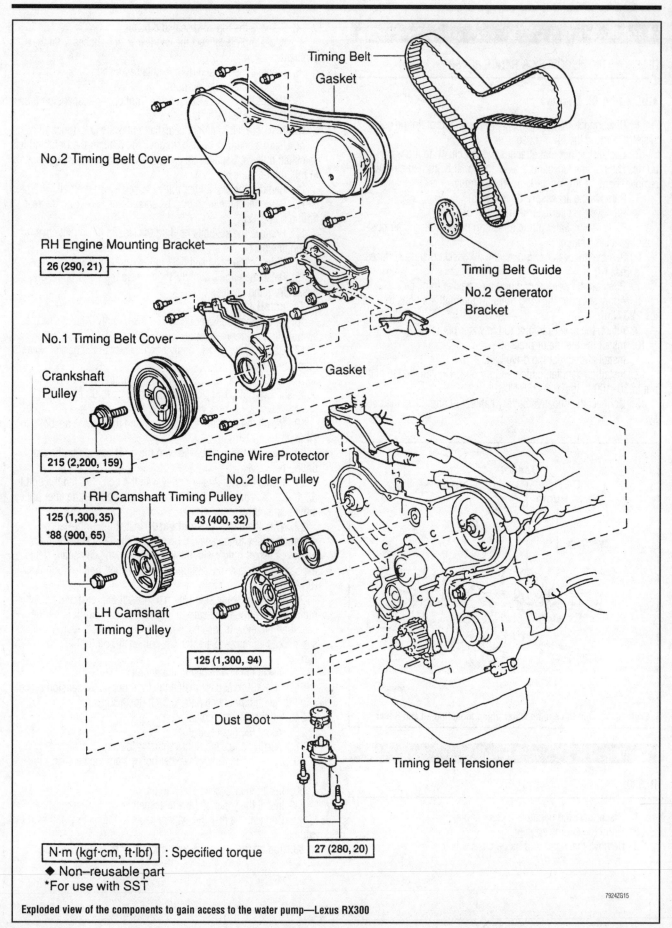

Timing Belt

Gasket

No.2 Timing Belt Cover

RH Engine Mounting Bracket

26 (290, 21)

Timing Belt Guide

No.2 Generator Bracket

No.1 Timing Belt Cover

Gasket

Crankshaft Pulley

215 (2,200, 159)

Engine Wire Protector

No.2 Idler Pulley

RH Camshaft Timing Pulley

125 (1,300, 35)
*88 (900, 65)

43 (400, 32)

LH Camshaft Timing Pulley

125 (1,300, 94)

Dust Boot

Timing Belt Tensioner

27 (280, 20)

N·m (kgf·cm, ft·lbf) : Specified torque

◆ Non–reusable part
*For use with SST

7924ZG15

Exploded view of the components to gain access to the water pump—Lexus RX300

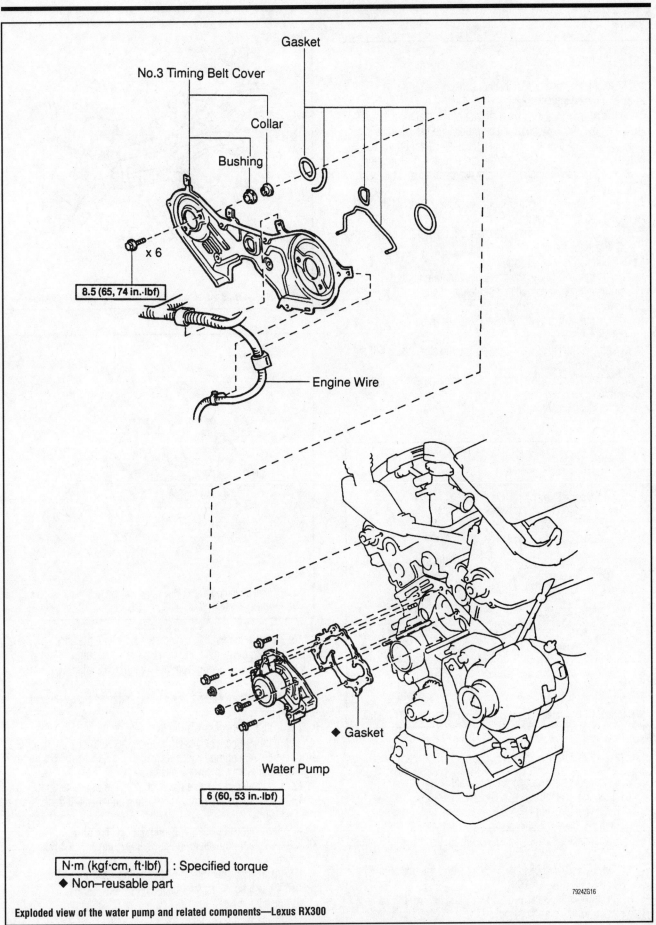

Gasket

No.3 Timing Belt Cover

Collar

Bushing

x 6

8.5 (65, 74 in.·lbf)

Engine Wire

◆ Gasket

Water Pump

6 (60, 53 in.·lbf)

N·m (kgf·cm, ft·lbf) : Specified torque

◆ Non–reusable part

7924ZG16

Exploded view of the water pump and related components—Lexus RX300

LX450 & LX470

4.5L (1FZ-FE) Engine

1. Disconnect the negative battery cable.
2. Drain the engine coolant.
3. Disconnect the No. 3 water bypass and radiator inlet hoses.
4. Remove the drive belts, fan assembly and the fan shroud.
5. Disconnect the oil cooler hose from the clamp on the fan shroud. Remove the bolts holding the fan shroud to the radiator.
6. Remove the four bolts, two nuts, water pump and the gasket.

To install:

7. Install the water pump using a new gasket. Tighten the fasteners to 15 ft. lbs. (21 Nm).
8. Install the water pump pulley, fan shroud and the drive belts.
   a. Place the fan with the fluid coupling, water pump pulley and the fan shroud in position.
   b. Temporarily install the fan pulley mounting nuts.
   c. Install the fan shroud and tighten the bolts to 4.9 ft. lbs. (5.4 Nm).
   d. Connect the oil cooler hose to the clamp on the fan shroud.
9. Connect the No. 3 water bypass and radiator hoses. Fill the cooling system.
10. Connect the negative battery cable, start the engine and check for leaks.
11. Recheck the coolant level.

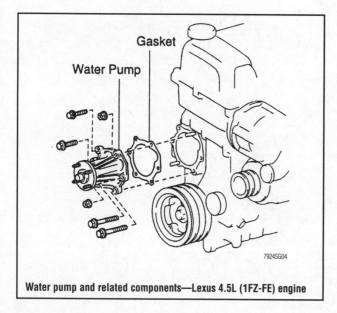

Water pump and related components—Lexus 4.5L (1FZ-FE) engine

4.7L (2UZ-FE) Engine

1. Disconnect the negative battery cable.
2. Drain the coolant.
3. Remove the timing belt.
4. Remove the No. 2 idler pulley.
5. Disconnect the water bypass hose from the water inlet housing.
6. Remove the two bolts attaching the water inlet housing to the water pump.
7. Disconnect the water bypass pipe, then remove the water inlet housing from the water pump.

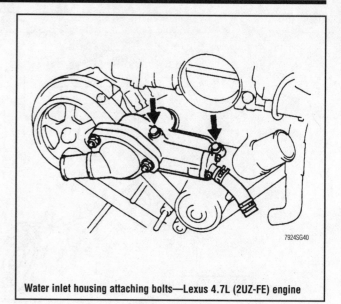

Water inlet housing attaching bolts—Lexus 4.7L (2UZ-FE) engine

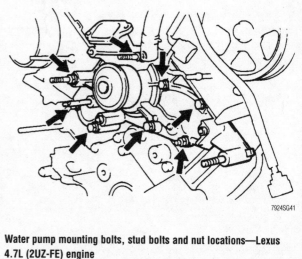

Water pump mounting bolts, stud bolts and nut locations—Lexus 4.7L (2UZ-FE) engine

8. Remove the five bolts, two stud bolts and nut attaching the water pump to the engine and remove the water pump.
9. Remove the O-ring from the water bypass pipe.

To install:

10. Apply soapy water a new O-ring and install it on the water bypass pipe.
11. Install the bypass pipe in the water pump and install the water pump using a new gasket. Tighten the bolts to 15 ft. lbs. (21 Nm) and the remaining fasteners to 13 ft. lbs. (18 Nm). Be sure to tighten them evenly in several passes.
12. Clean all old silicone material off of the water inlet housing.
13. Apply soapy water to a new O-ring and install it on the water inlet housing.
14. Apply a bead of Seal Packing (RTV silicone) No. 08826–00100 or equivalent in the sealing groove on the water inlet housing.
15. Install the water inlet housing on the water pump. Alternately tighten the two bolts to 13 ft. lbs. (18 Nm).
16. Install the No. 2 idler pulley.
17. Install the timing belt.

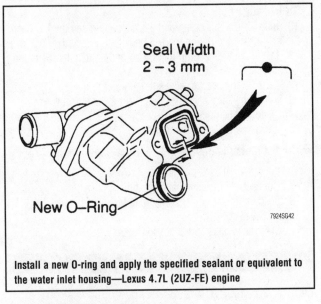

Seal Width
2 – 3 mm

New O–Ring

7924SG42

Install a new O-ring and apply the specified sealant or equivalent to the water inlet housing—Lexus 4.7L (2UZ-FE) engine

18. Refill the engine with coolant.
19. Start the engine and check for leaks.
20. Allow the engine to cool down, then recheck the coolant level.

Mazda

MPV

1. Position the engine at TDC on the compression stroke.
2. Disconnect the negative battery cable.
3. Remove the air cleaner assembly.
4. Drain the cooling system.
5. Remove the spark plug wires.
6. Remove the fresh air duct assembly.
7. Remove the cooling fan and radiator cowling.
8. Remove the drive belts.

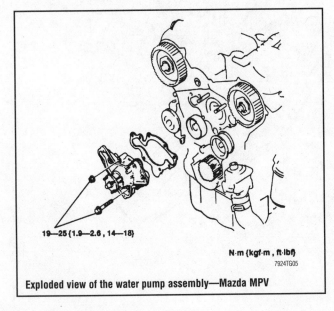

19–25 {1.9–2.6 , 14–18}

N·m {kgf·m , ft·lbf}
7924TG05

Exploded view of the water pump assembly—Mazda MPV

9. Remove the air conditioning compressor idler pulley. Remove the compressor, and position it to the side without disconnecting the hose lines.
10. Remove the crankshaft pulley and baffle plate.
11. Remove the coolant bypass hose.
12. Remove the upper radiator hose.
13. Remove the timing belt cover retaining bolts. Remove the timing belt covers and gasket.
14. Turn the crankshaft to align the mating marks of the pulleys.
15. Remove the upper idler pulley.
16. Remove the timing belt. If reusing the belt be sure to mark the direction of rotation.
17. Remove the timing belt auto tensioner.
18. Unbolt and remove the water pump. Discard the gasket.
19. Thoroughly clean the mating surfaces of the pump and engine.

To install:

20. Position the pump and a new gasket, coated with sealer, on the engine. Tighten the bolts to 14–18 ft. lbs. (19–25 Nm).
21. The automatic tensioner must be pre-loaded. To load the tensioner, place a flat washer on the bottom of the tensioner body to prevent damage to the body and position the unit on an arbor press. Press the rod into the tensioner body. Do not use more than 2000 lbs. (8900N) of pressure. Once the rod is fully inserted into the body, insert a suitable L-shaped pin or a small Allen wrench through the body and the rod to hold the rod in place. Remove the unit from the press and install onto the block and tighten the mounting bolt to 14–19 ft. lbs. (19–26 Nm) Leave the pin in place, it will be removed later.
22. Be sure all the timing marks are aligned properly. With the upper idler pulley removed, hang the timing belt on each pulley in the proper order. Install the upper idler pulley and tighten the mounting bolt to 27–38 ft. lbs. (37–51 Nm). Rotate the crankshaft twice in the normal direction of rotation to align all the timing marks.
23. Be sure all the marks are aligned correctly. If not, repeat the previous step.
24. Remove the pin from the auto tensioner. Again turn the crankshaft twice in the normal direction of rotation and be sure all the timing marks are aligned properly.
25. Check the timing belt deflection by applying 22 lbs. of force (98N). If the deflection is not 0.20–0.28 inch (5–7mm), repeat the adjustment procedure.

➡**Excessive belt deflection is caused by auto tensioner failure or an excessively stretched timing belt.**

26. Install the timing belt covers and new gasket.
27. Install the timing belt cover retaining bolts.
28. Install the upper radiator hose.
29. Install the coolant bypass hose.
30. Install the crankshaft pulley and baffle plate.
31. Install the compressor.
32. Install the air conditioning compressor idler pulley.
33. Install and adjust the drive belts.
34. Install the cooling fan and radiator cowling.
35. Install the fresh air duct assembly.
36. Install the spark plug wires.
37. Fill the cooling system.
38. Install the air cleaner assembly.
39. Connect the negative battery cable.

Mercedes-Benz

ML320

✳✳ CAUTION

Never open, service or drain the radiator or cooling system when hot; serious burns can occur from the steam and hot coolant. Also, when draining engine coolant, keep in mind that cats and dogs are attracted to ethylene glycol antifreeze and could drink any that is left in an uncovered container or in puddles on the ground. This will prove fatal in sufficient quantities. Always drain coolant into a sealable container. Coolant should be reused unless it is contaminated or is several years old.

1. Disconnect the negative battery cable.
2. Remove the engine cooling fan and clutch, then the fan shroud.

➡**The fan clutch is equipped with right-hand thread.**

3. Drain and recycle the engine coolant.
4. Remove the engine cover.
5. Lock the automatic belt tensioner by rotating the tensioner counterclockwise until a 5mm drift or pin fits through the tensioner, then remove the serpentine belt.
6. Disconnect the coolant hoses from the water pump.
7. Remove the belt pulley.
8. Remove the water pump mounting bolts, then the water pump.
9. Clean and dry the gasket mating surface for the water pump.

To install:

10. Install the water pump and gasket, and tighten the M6 bolts to 88 inch lbs. (10 Nm) and the M8 bolts to 177 inch lbs. (20 Nm).
11. Install the water pump belt pulley and tighten the mounting bolts to 88 inch lbs. (10 Nm).
12. Connect the coolant hoses to the water pump.
13. Install the serpentine belt and remove the locking pin.
14. Install the engine cover.
15. Install the fan shroud and fan.

➡**The fan clutch is equipped with right-hand thread.**

16. Fill the engine with coolant.
17. Connect the negative battery cable.
18. Read fault memory, encode the radio and normalize the power windows.
19. Start the vehicle and check for leaks.

Mitsubishi

MIGHTY MAX, MONTERO & MONTERO SPORT

1. If necessary, properly release the fuel pressure.
2. Disconnect the negative battery cable.

✳✳ CAUTION

Wait at least 90 seconds after the negative battery cable is disconnected to prevent possible deployment of the air bag.

3. Drain the cooling system.

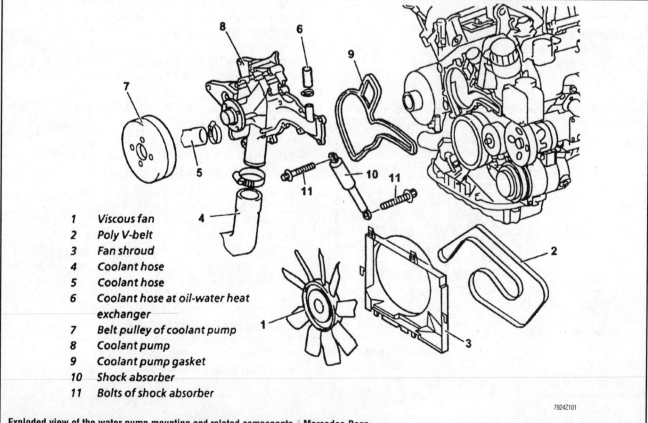

1	Viscous fan
2	Poly V-belt
3	Fan shroud
4	Coolant hose
5	Coolant hose
6	Coolant hose at oil-water heat exchanger
7	Belt pulley of coolant pump
8	Coolant pump
9	Coolant pump gasket
10	Shock absorber
11	Bolts of shock absorber

7924Z101

Exploded view of the water pump mounting and related components—Mercedes-Benz

4. Remove the upper radiator shroud.

5. Remove all accessory belts. Remove the air conditioning compressor tensioner pulley, if equipped.

6. Remove the cooling fan and clutch assembly and remove the water pump pulley.

7. Disconnect the radiator hose from the water pump.

8. Remove the crankshaft pulley(s).

9. Remove the timing belt covers. If the same timing belt will be reused, mark the direction of the timing belt's rotation, for installation in the same direction. Be sure the engine is positioned so the No. 1 cylinder is at the TDC of its compression stroke and the sprockets timing marks are aligned with the engine's timing mark indicators. Remove the timing belt.

10. The water pump bolts are different lengths, note their positions before removing. Remove the water pump mounting bolts and remove the pump from the block and the water pipe connection. Remove the O-ring from the water pipe connection.

To install:

11. Clean and dry the mating surfaces of the block and water pump. Install a new O-ring to the water pipe connection. Coat the new O-ring with water to aid in installation.

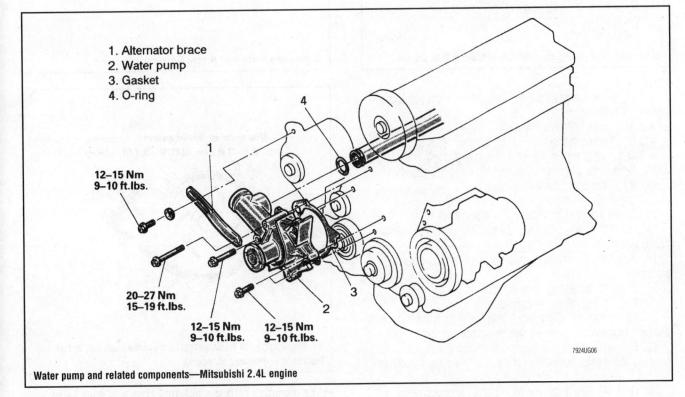

1. Alternator brace
2. Water pump
3. Gasket
4. O-ring

12–15 Nm
9–10 ft.lbs.

20–27 Nm
15–19 ft.lbs.

12–15 Nm
9–10 ft.lbs.

12–15 Nm
9–10 ft.lbs.

7924UG06

Water pump and related components—Mitsubishi 2.4L engine

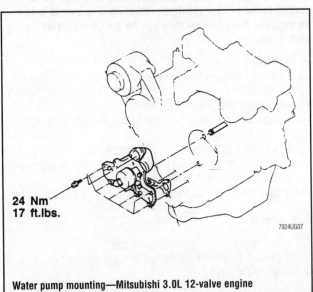

24 Nm
17 ft.lbs.

7924UG07

Water pump mounting—Mitsubishi 3.0L 12-valve engine

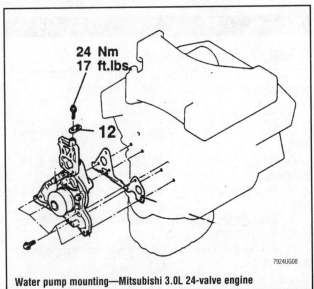

24 Nm
17 ft.lbs.

12

7924UG08

Water pump mounting—Mitsubishi 3.0L 24-valve engine

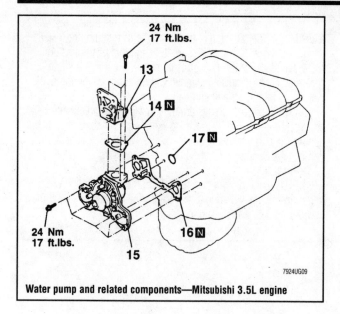

Water pump and related components—Mitsubishi 3.5L engine

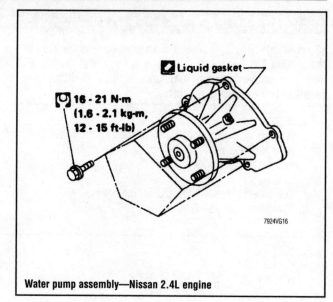

Water pump assembly—Nissan 2.4L engine

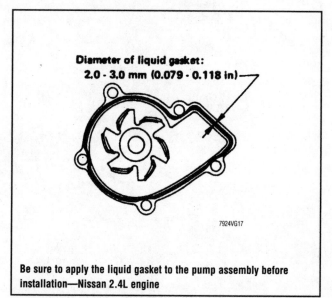

Be sure to apply the liquid gasket to the pump assembly before installation—Nissan 2.4L engine

12. Install the water pump with a new gasket to the block and tighten the bolts to:
- 2.4L engine: 14 ft. lbs. (19 Nm)
- 3.0L 12 valve and 3.5L engines: 17 ft. lbs. (23 Nm)
- 3.0L 24 valve engine: 12–14 ft. lbs. (17–20 Nm)

13. Tighten the alternator bracket bolt to 17 ft. lbs. (23 Nm).

14. Install the timing belt(s) and covers.

15. Install the crankshaft pulley(s).

16. Connect the radiator hose to the water pump.

17. Install the water pump pulley. Install the cooling fan and clutch assembly.

18. Install the air conditioning compressor tensioner pulley, if equipped.

19. Install the accessory belts, adjust if necessary.

20. Install the upper radiator shroud.

21. Fill the radiator with coolant. This cooling system has a self-bleeding thermostat, so system bleeding is not required.

22. Connect the negative battery cable, run the vehicle until the thermostat opens and fill the overflow tank. Check for leaks.

23. Once the vehicle has cooled, recheck the coolant level.

Nissan

PATHFINDER, PICK-UP & FRONTIER

2.4L Engine

1. Disconnect the negative battery cable.

2. Drain the cooling system and engine block, using the block drain.

3. Remove the upper radiator hose to provide working room and remove the drive belt(s) from the pulleys.

4. Remove the retaining screws, and lift the fan shroud from the engine.

5. While holding the pulley, remove the nuts retaining the fan and pulley to the water pump.

6. Remove the mounting bolts and pull the water pump from the engine.

➡The mounting bolts are different sizes and must be reinstalled in the correct location, therefore it is a good idea to arrange the bolts so that they can be easily identified during installation.

To install:

7. Be sure all gasket surfaces are clean and properly apply silicone sealer to the pump. Install the pump to the engine and tighten the bolts to 12–15 ft. lbs. (16–21 Nm).

8. Install the fan clutch, fan, and pulley and tighten the nuts or bolts to 5–6 ft. lbs. (7–8 Nm).

9. Install the fan shroud and drive belt(s).

10. Connect the upper hose, then fill and bleed the cooling system.

11. Connect the negative battery cable.

12. Start the engine to check for leaks.

3.0L Engine

1. Disconnect the negative battery cable.

2. Drain the coolant from the radiator and the drain plugs on both sides of the engine block.

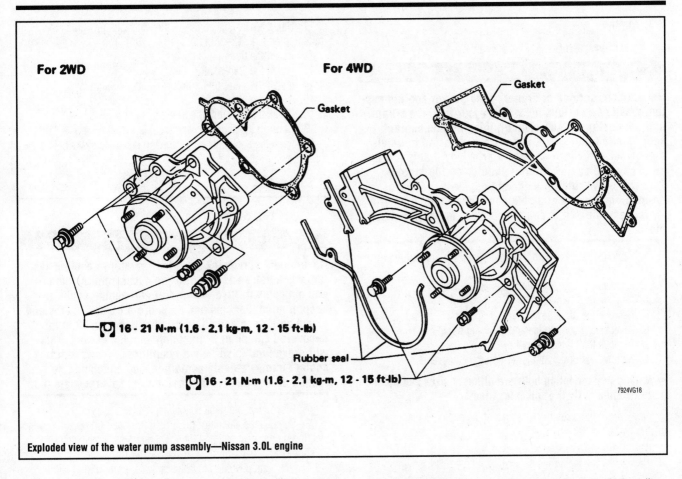

For 2WD

For 4WD

Gasket

Gasket

🔧 16 - 21 N·m (1.6 - 2.1 kg-m, 12 - 15 ft-lb)

Rubber seal

🔧 16 - 21 N·m (1.6 - 2.1 kg-m, 12 - 15 ft-lb)

7924VG18

Exploded view of the water pump assembly—Nissan 3.0L engine

RIGHT SIDE:

DRAIN PLUG

LEFT SIDE:

DRAIN PLUG

7924VG19

Remove the plugs to drain the coolant from the engine block—Nissan 6-cylinder engines

3. Remove the radiator hoses, on automatic transmission, disconnect and plug the fluid cooling lines.

4. Remove the lower section of the fan shroud and remove the screws to lift the shroud from the engine. Remove the bracket bolts and lift the radiator out of the vehicle.

5. Remove all the accessory drive belts.

6. Hold the pulley and remove the nuts to remove the fan and pulley from the water pump.

7. Remove the timing belt covers.

8. Remove the bolts to remove the water pump from the engine.

➡**Water pump mounting bolts are different sizes and must be reinstalled in their original locations.**

To install:

9. Be sure all gasket surfaces are clean and use a new gasket or silicone sealer when installing the pump to the engine. Tighten the bolts to 15 ft. lbs. (21 Nm).

10. Install the timing belt covers. On 4WD models, be sure the sealing surfaces are clean and carefully install the rubber seal when installing the cover. The timing belt must be properly protected from dirt and oil.

11. Install the pulley, fan clutch, and the fan.

12. Install the accessory drive belts and adjust the tension.

13. Install the radiator and fan shroud; connect the cooling system hoses.

14. If equipped with an automatic transmission, connect the A/T oil cooler lines.

15. Connect the negative battery cable.

16. Fill and bleed the cooling system and check for leaks.

3.3L Engine

1. Disconnect the negative battery cable.

> **✳✳ CAUTION**
>
> **Never open, service or drain the radiator or cooling system when hot; serious burns can occur from the steam and hot coolant. Also, when draining engine coolant, keep in mind that cats and dogs are attracted to ethylene glycol antifreeze and could drink any that is left in an uncovered container or in puddles on the ground. This will prove fatal in sufficient quantities. Always drain coolant into a sealable container. Coolant should be reused unless it is contaminated or is several years old.**

2. Drain the coolant from the radiator and the drain plugs on both sides of the engine block.
3. Remove the upper and lower radiator hoses.
4. Remove the fan shroud.
5. Remove the drive belts.
6. Remove the cooling fan and the water pump pulley.
7. Remove the crankshaft pulley.
8. Remove the upper and lower timing belt covers.

➡ **Water pump mounting bolts are different sizes and must be reinstalled in their original locations.**

9. Remove the water pump. Don't let the engine coolant get on the timing belt.

To install:

10. Clean the gasket mating surfaces on the water pump and engine block.

11. Using a new gasket, install the water pump. Tighten the mounting bolts to 12–15 ft. lbs. (16–21 Nm).
12. Install the timing belt covers.
13. Install the crankshaft pulley.
14. Install the water pump pulley and the cooling fan.
15. Install the drive belts.
16. Install the fan shroud and the radiator hoses.
17. Connect the negative battery cable.
18. Refill the engine with coolant and bleed the system. Check for leaks.

QUEST

1. Drain the cooling system.

> **✳✳ CAUTION**
>
> **Never open, service or drain the radiator or cooling system when hot; serious burns can occur from the steam and hot coolant. Also, when draining engine coolant, keep in mind that cats and dogs are attracted to ethylene glycol antifreeze and could drink any that is left in an uncovered container or in puddles on the ground. This will prove fatal in sufficient quantities. Always drain coolant into a sealable container. Coolant should be reused unless it is contaminated or is several years old.**

2. Disconnect the negative battery cable.
3. Remove the alternator drive belt, the water pump and power steering pump drive belt and the A/C compressor drive belt (if equipped).
4. Use a strap wrench to hold the water pump pulley while removing the four water pump pulley bolts.

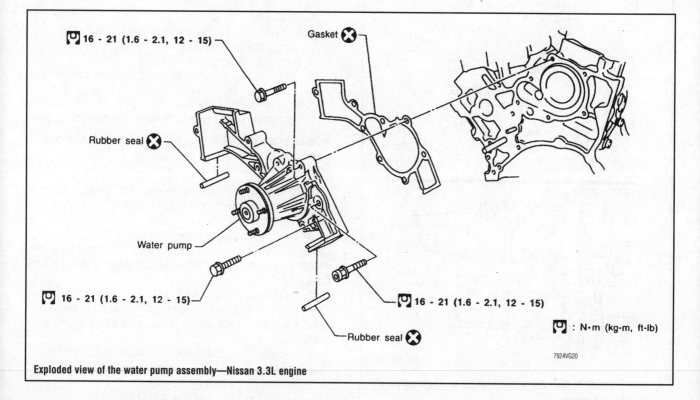

Exploded view of the water pump assembly—Nissan 3.3L engine

16 - 21 (1.6 - 2.1, 12 - 15)
Gasket ✖
Rubber seal ✖
Water pump
16 - 21 (1.6 - 2.1, 12 - 15)
16 - 21 (1.6 - 2.1, 12 - 15)
Rubber seal ✖
: N·m (kg-m, ft-lb)

7924VG20

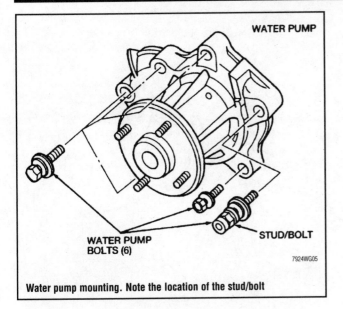

WATER PUMP

WATER PUMP
BOLTS (6)

STUD/BOLT

7924WG05

Water pump mounting. Note the location of the stud/bolt

5. Remove the water pump pulley from the water pump.
6. Remove the crankshaft pulley using the following procedure.
   a. Raise and safely support the vehicle.
   b. Remove the five right side inner engine and transmission splash shield bolts and two screws and remove the inner engine and transmission shield.
   c. Remove the four right side outer engine and transmission splash shield bolts and two screws and remove the right side outer engine and transmission splash shields.
   d. Use a strap wrench to hold the crankshaft pulley while removing the crankshaft pulley bolt.
   e. Use a crankshaft damper remover to draw the crankshaft pulley off the front of the crankshaft.
7. Remove the five lower engine front cover bolts and take of the front cover.
8. Remove the six water pump bolts. Make note of the locations of the bolts since one should be a stud/bolt and must be returned to its original location. Remove the water pump.

To install:

9. Clean all parts well. The bolt threads should be cleaned of any old sealer or corrosion. Be sure the mating surfaces between the water pump and the engine block are cleaned of any old sealant. Apply a continuous bead of gasket maker type sealer approximately ⅛ inch wide onto the water pump and position the water pump on the engine block.
10. Install the six water pump bolts. Refer to any notes made at removal so the bolts can be returned to their original locations. Do not over-tighten the water pump bolts. Tighten the water pump bolts evenly to 12–15 ft. lbs. (16–21 Nm).
11. Position the water pump pulley on the water pump and install the four pulley bolts. Use a strap wrench to hold the pulley as the bolts are tightened to 12–15 ft. lbs. (16–21 Nm).
12. Install the front engine cover and the five lower front cover bolts. Tighten to 27–44 inch lbs. (3–5 Nm).

13. Install the crankshaft pulley using the following procedure.
   a. Install the crankshaft pulley and pulley bolt.
   b. Hold the pulley with a strap wrench. Tighten the crankshaft pulley bolt to 90–98 ft. lbs. (123–132 Nm).
   c. Install the inner and outer engine and transmission splash shields.
14. Install and adjust the drive belts.
15. Connect the negative battery cable.
16. Refill the cooling system.
17. Start the engine, bleed the cooling system and verify no leaks.

Subaru

FORESTER

2.5L Engine

1. Disconnect the negative battery cable.
2. Raise and safely support the vehicle and remove the engine undercover.
3. Drain the coolant into a suitable container.
4. Disconnect the radiator outlet hose.
5. Remove the radiator fan motor assembly.
6. Remove the accessory drive belts.
7. Remove the timing belt, tensioner and camshaft angle sensor. Refer to the procedure in this section for removal and installation steps.
8. Remove the left side camshaft pulley(s) and left side rear timing belt cover. Remove the tensioner bracket.
9. Disconnect the radiator hose and heater hose from the water pump.
10. Remove the water pump retainer bolts.
11. Remove the water pump.

To install:

12. Clean the gasket mating surfaces thoroughly. Always use new gaskets during installation.
13. Install the water pump and tighten the bolts, in sequence, to 7–10 ft. lbs. (10–14 Nm). After tightening the bolts once, retighten to the same specification again.
14. Inspect the radiator hoses for deterioration and replace as necessary. Connect the radiator hose and heater hose to the water pump.
15. Install the left side rear timing belt cover, left side camshaft pulley(s) and tensioner bracket.
16. Install the camshaft angle sensor, tensioner and timing belt.
17. Install the accessory drive belts.
18. Install the radiator fan motor assembly.
19. Install the radiator outlet hose.
20. Install the engine undercover.
21. Fill the system with coolant.
22. Connect the negative battery cable.
23. Start the engine and allow it to reach operating temperature.
24. Check for leaks.

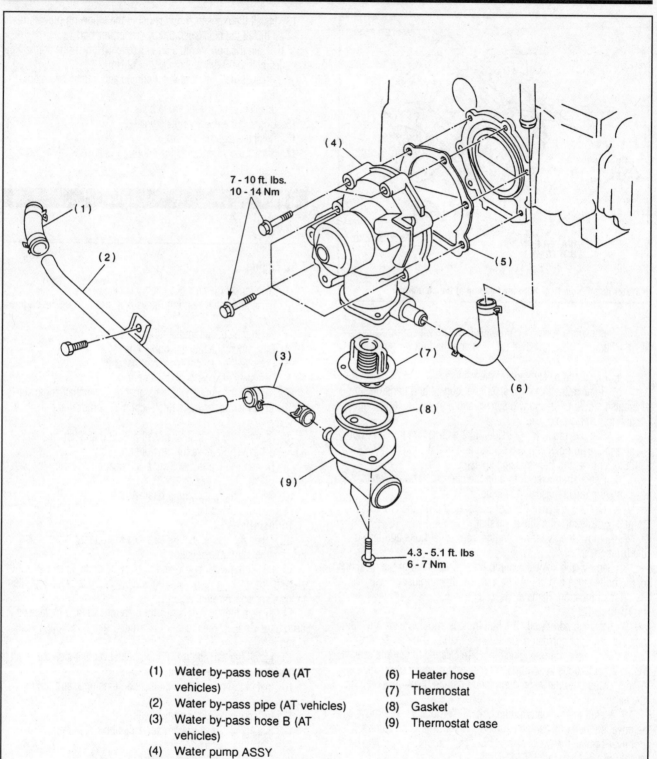

(1)

(2)

7 - 10 ft. lbs.
10 - 14 Nm

(4)

(5)

(3)

(7)

(6)

(8)

(9)

4.3 - 5.1 ft. lbs
6 - 7 Nm

(1) Water by-pass hose A (AT vehicles)
(2) Water by-pass pipe (AT vehicles)
(3) Water by-pass hose B (AT vehicles)
(4) Water pump ASSY
(5) Gasket

(6) Heater hose
(7) Thermostat
(8) Gasket
(9) Thermostat case

7924XG01

Exploded view of the water pump mounting and related components—Subaru 2.5L engine

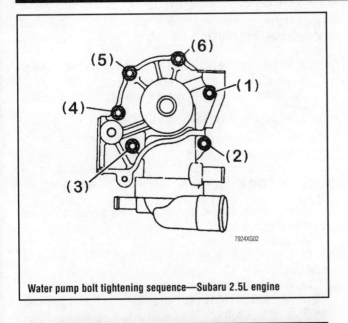

Water pump bolt tightening sequence—Subaru 2.5L engine

Toyota

T-100, TACOMA & 4RUNNER

2.4L (22R-E) Engine

1. Disconnect the negative battery cable.
2. Drain the cooling system.
3. If equipped with an air conditioning compressor or power steering pump drive belts, it may be necessary to loosen the adjusting bolt, remove the drive belt(s) and move the component(s) out of the way.

4. Remove the fluid coupling with the fan and water pump pulley.
5. Remove the water pump.
6. Clean the gasket mounting surfaces.

To install:

7. Install the replacement water pump using a new gasket.
8. Install the water pump pulley and fluid coupling with the fan.
9. Install the removed engine drive belts.
10. Fill the cooling system.
11. Connect the negative battery cable. Start the engine and check for leaks.
12. Bleed the cooling system.

3.0L (3VZ-E) and 3.4L (5VZ-FE) Engines

4RUNNER

1. Disconnect the negative battery cable.
2. Drain the cooling system.
3. Remove the timing belt.
4. Remove the thermostat.
5. Disconnect the No. 2 oil cooler hose from the water pump.
6. Remove the water pump by removing the bolts.
7. Thoroughly clean the mating surfaces.

To install:

8. Apply sealant (PN 08826–00100 or equivalent) to the water pump. Parts must be assembled within five minutes of application. Otherwise the material must be removed and reapplied.
9. Install the water pump and tighten the bolts to 14 ft. lbs. (20 Nm).
10. Connect the No. 2 oil cooler hose.
11. Install the thermostat.
12. Install the timing belt.
13. Connect the negative battery cable.

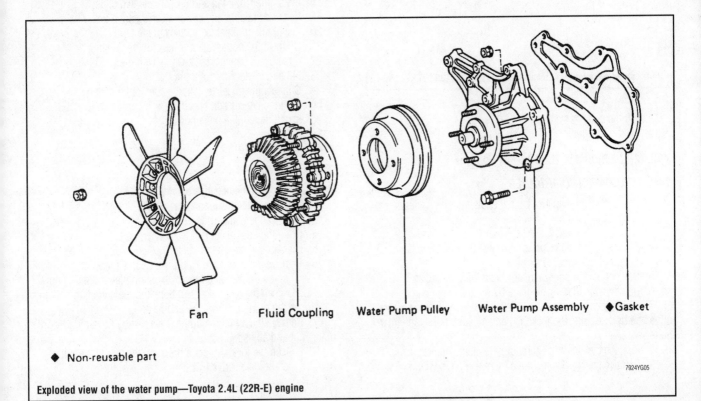

Fan   Fluid Coupling   Water Pump Pulley   Water Pump Assembly   ◆Gasket

◆ Non-reusable part

Exploded view of the water pump—Toyota 2.4L (22R-E) engine

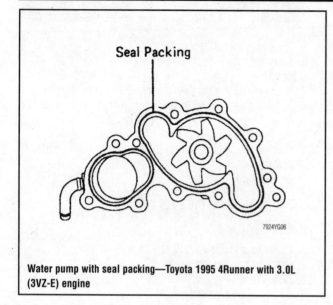

Water pump with seal packing—Toyota 1995 4Runner with 3.0L (3VZ-E) engine

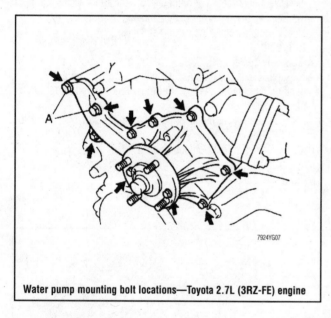

Water pump mounting bolt locations—Toyota 2.7L (3RZ-FE) engine

14. Fill the cooling system.
15. Start the engine and check for leaks.

2.4L (2RZ-FE) and 2.7L (3RZ-FE) Engines

T-100, TACOMA AND 4RUNNER

1. Disconnect the negative battery cable from the battery.
2. Remove the engine undercover.
3. Drain the cooling system.
4. For the California vehicles with 3RZ-FE engine, remove the two bolts and disconnect the air pipe.
5. Disconnect the upper radiator hose from the radiator.
6. Remove the oil dipstick guide by removing the bolt.
7. If equipped with power steering, remove the power steering drive belt by loosening the lockbolt and adjusting bolt to the idler pulley.
8. Remove the No. 2 fan shroud by removing the two clips.
9. Remove the No. 1 fan shroud by removing the four bolts.

10. If equipped with air conditioning, loosen the idler pulley nut and adjusting bolt. Remove the air conditioning drive belt from the engine.
11. Remove the alternator drive belt, fan (with fan clutch), water pump pulley, and the fan shroud as follows:
    a. Stretch the belt and loosen the water pump pulley mounting nuts.
    b. Loosen the lock, pivot, and the adjusting bolts for the alternator and remove the alternator drive belt from the engine.
    c. Remove the four water pump pulley mounting nuts.
    d. Remove the fan (with fan clutch) and the water pump pulley.
12. Remove the 10 bolts and remove the water pump and gasket from the engine.

To install:
13. Clean all surfaces and apply a thin layer of liquid sealant to a new gasket.
14. Place the gasket and water pump into position. Tighten the 14mm head bolts **A** to 18 ft. lbs. (25 Nm) and the 12mm head bolts to 78 inch lbs. (9 Nm).
15. Install the water pump pulley, fan shroud, fan (with fan clutch), and the alternator drive belt as follows:
    a. Place the fan (with the fan clutch), water pump pulley, and the fan shroud in position.
    b. Install the water pump pulley mounting nuts but do not tighten the nuts at this time.
    c. Install the alternator drive belt to the engine.
    d. Stretch the alternator belt tight and tighten the fan nuts to 16 ft. lbs. (21 Nm).
    e. Adjust the drive belt for the alternator.
16. If equipped with air conditioning, install and adjust the drive belt.
17. Install the No. 1 fan shroud by installing the four bolts.
18. Install the No. 2 fan shroud with the two clips.
19. Install and adjust the power steering drive belt.
20. Install the oil dipstick guide with the bolt.
21. Connect the upper radiator hose to the radiator.
22. If removed, connect the air pipe with the two bolts.
23. Fill and bleed the cooling system.
24. Connect the negative battery cable to the battery.
25. Start the engine and check for leaks.
26. Install the engine undercover.

3.4L (5VZ-FE) Engine

T-100 AND TACOMA

1. Disconnect the negative battery cable.
2. Raise and safely support the vehicle.
3. Remove the engine undercover.
4. Drain the engine coolant.
5. Disconnect the upper radiator hose from the engine.
6. Remove the power steering drive belt as follows:
    a. Stretch the belt and loosen the fan pulley mounting nuts.
    b. Loosen the lockbolt, pivot bolt, and the adjusting bolt and remove the drive belt from the engine.
7. Remove the air conditioning drive belt by loosening the idler pulley nut and adjusting bolt.
8. Loosen the lockbolt, pivot bolt, and the adjusting bolt. Remove the alternator drive belt.

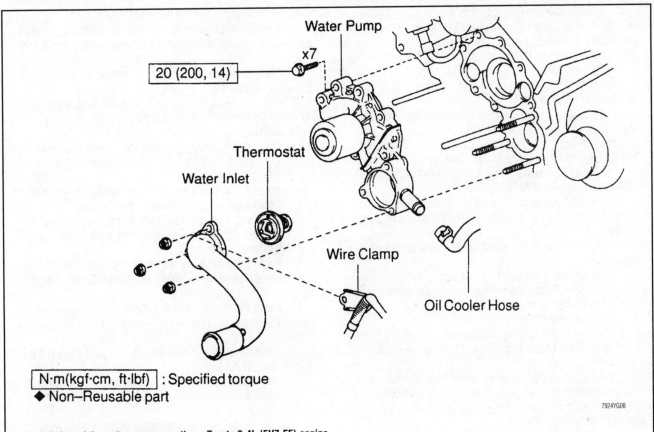

20 (200, 14)

Water Pump

x7

Thermostat

Water Inlet

Wire Clamp

Oil Cooler Hose

N·m(kgf·cm, ft·lbf) : Specified torque
◆ Non–Reusable part

7924YG08

Exploded view of the water pump mounting—Toyota 3.4L (5VZ-FE) engine

9. Remove the No. 2 fan shroud by removing the two clips.

10. Remove the fan with the fluid coupling and fan pulleys.

11. Disconnect the power steering pump from the engine and set aside. Do not disconnect the lines from the pump.

12. If equipped with air conditioning, disconnect the compressor from the engine and set aside. Do not disconnect the lines from the compressor.

13. If equipped with air conditioning, disconnect the air conditioning bracket.

14. Remove the No. 2 timing belt cover as follows:

a. Detach the camshaft position sensor connector from the No. 2 timing belt cover.

b. Disconnect the three spark plug wire clamps from the No. 2 timing belt cover.

c. Remove the six bolts and remove the timing belt cover.

15. Remove the fan bracket as follows:

a. Remove the power steering adjusting strut by removing the nut.

b. Remove the fan bracket by removing the bolt and nut.

16. Set the No. 1 cylinder at TDC of the compression stroke.

a. Turn the crankshaft pulley and align its groove with the timing mark **O** of the No. 1 timing belt cover.

b. Check that the timing marks of the camshaft timing pulleys and the No. 3 timing belt cover are aligned. If not, turn the crankshaft pulley one revolution (360°).

➡ **If re-using the timing belt, be sure that you can still read the installation marks. If not, place new installation marks**

on the timing belt to match the timing marks of the camshaft timing pulleys.

17. Remove the timing belt tensioner by alternately loosening the two bolts.

18. Remove the camshaft timing pulleys.

a. Using Variable Wrench Set No. 09960–10010 or equivalent, remove the pulley bolt, the timing pulley, and the knock pin. Remove the two timing pulleys with the timing belt.

19. Remove the thermostat.

20. Disconnect the No. 2 oil cooler hose from the water pump.

21. Remove the water pump by removing the seven bolts.

22. Thoroughly clean the mating surfaces.

To install:

23. Apply sealant (PN 08826–00100 or equivalent) to the water pump. Parts must be assembled within five minutes of application. Otherwise the material must be removed and reapplied.

24. Install the water pump. Tighten the bolts to 14 ft. lbs. (20 Nm).

25. Connect the No. 2 oil cooler hose.

26. Install the thermostat.

27. Install the left camshaft timing pulley. Tighten the pulley bolt to 81 ft. lbs. (110 Nm).

28. Set the No. 1 cylinder to TDC of the compression stroke.

29. Connect the timing belt to the left camshaft timing pulley. Check that the installation mark on the timing belt is aligned with the end of the No. 1 timing belt cover.

a. Using Variable Pin Wrench Set 09960–01000 or equivalent, slightly turn the left camshaft timing pulley clockwise. Align the installation mark on the timing belt with the timing mark of the camshaft timing pulley, and hang the timing belt on the left camshaft timing pulley.

b. Align the timing marks of the left camshaft pulley and the No. 3 timing belt cover.

c. Check that the timing belt has tension between the crankshaft timing pulley and the left camshaft timing pulley.

30. Install the right camshaft timing pulley and the timing belt.

31. Set the timing belt tensioner as follows:

a. Using a press, slowly press in the pushrod using 220–2,205 lbs. (981–9,807 N) of force.

b. Align the holes of the pushrod and housing, pass a 1.5mm hexagon wrench through the holes to keep the setting position of the pushrod.

c. Release the press and install the dust boot to the tensioner.

32. Install the timing belt tensioner and alternately tighten the bolts to 20 ft. lbs. (28 Nm). Using pliers, remove the 1.5mm hexagon wrench from the belt tensioner.

33. Check the valve timing.

a. Slowly turn the crankshaft pulley two revolutions from the TDC to TDC. Always turn the crankshaft pulley clockwise.

b. Check that each pulley aligns with the timing marks. If the timing marks do not align, remove the timing belt and reinstall it.

34. Install the fan bracket with the bolt and nut. Install the remaining components.

35. Fill with engine coolant.

36. Connect the negative battery cable.

37. Start the engine and check for leaks.

RAV4

1. Disconnect the negative battery cable.

❊❊ CAUTION

Wait 90 seconds from the time the key is turned to LOCK and the negative battery cable is disconnected to begin work. This allows the SRS capacitor to discharge and prevent deployment of the air bag(s).

2. Remove the right-hand engine undercover.

3. Drain the engine coolant from the radiator and engine.

4. Remove the timing belt.

5. Disconnect the lower radiator hose from the water inlet.

6. Remove the timing belt tension spring and the No. 2 idler pulley.

7. Disconnect the crankshaft position sensor connector clamp.

8. Remove the alternator drive belt adjusting bar.

9. Remove the two nuts holding the water pump to the water bypass pipe.

10. Remove the three bolts in the sequence shown.

11. Disconnect the water pump cover from the water bypass pipe and remove the water pump and water pump cover assembly.

12. Remove the gasket and two O-rings from the water pump and water bypass pipe.

13. Remove the water pump from the water pump cover by removing the three bolts, water pump and gasket.

To install:

14. Using a new gasket, install the water pump to the water pump cover. Install the three bolts and tighten the bolts to 78 inch lbs. (8.8 Nm).

15. Install a new O-ring and gasket to the water pump cover.

16. Install a new O-ring to the water bypass pipe.

17. Apply soapy water to the O-ring on the water bypass pipe.

18. Connect the water pump cover to the water bypass pipe. Do not install the nuts at this time.

19. Install the water pump with the three bolts. Tighten the bolts in sequence shown. Tighten the bolts to 78 inch lbs. (8.8 Nm).

20. Install the two nuts holding the water pump cover to the water pump pipe. Tighten the two bolts to 82 inch lbs. (9.3 Nm).

21. Install the alternator drive belt adjusting bar. Tighten the bolt to 20 ft. lbs. (27 Nm).

22. Attach the crankshaft position sensor connector clamp.

23. Install the No. 2 idler pulley and timing belt tension spring.

24. Connect the lower radiator hose.

25. Install the timing belt.

26. Fill the engine and radiator with engine coolant.

27. Connect the negative battery cable.

28. Start the engine and check for leaks.

29. Install the right-hand engine undercover.

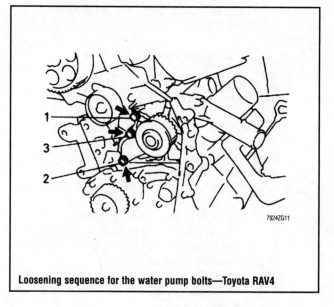

Loosening sequence for the water pump bolts—Toyota RAV4

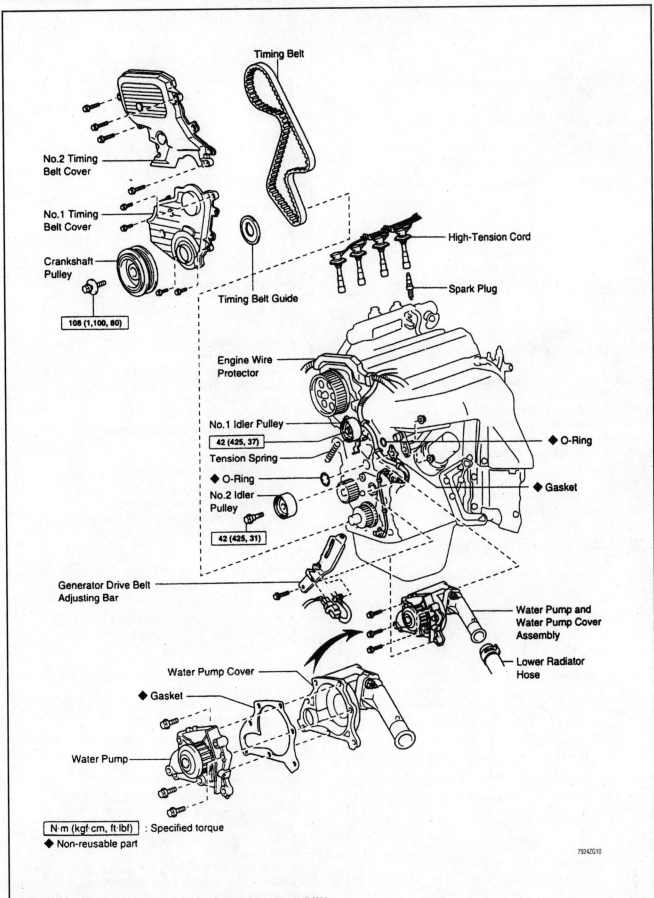

No.2 Timing Belt Cover

No.1 Timing Belt Cover

Crankshaft Pulley

`108 (1,100, 80)`

Timing Belt

Timing Belt Guide

High-Tension Cord

Spark Plug

Engine Wire Protector

No.1 Idler Pulley

`42 (425, 37)`

Tension Spring

◆ **O-Ring**

No.2 Idler Pulley

`42 (425, 31)`

◆ **O-Ring**

◆ **Gasket**

Generator Drive Belt Adjusting Bar

Water Pump and Water Pump Cover Assembly

Lower Radiator Hose

Water Pump Cover

◆ **Gasket**

Water Pump

`N·m (kgf·cm, ft·lbf)` : Specified torque
◆ Non-reusable part

7924ZG10

Exploded view of the water pump and related components—Toyota RAV4

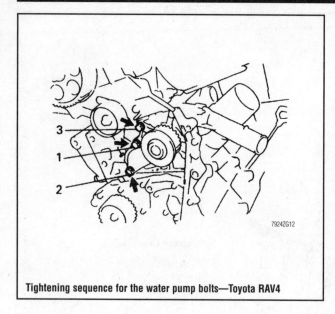

Tightening sequence for the water pump bolts—Toyota RAV4

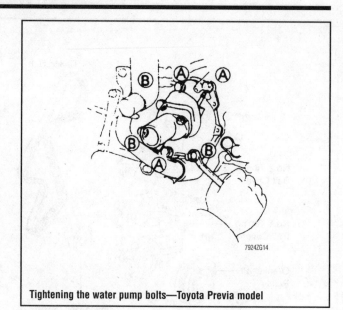

Tightening the water pump bolts—Toyota Previa model

PREVIA

1. Disconnect the negative battery cable from the battery.
2. Raise the vehicle and support safely.
3. Remove the engine undercovers.
4. Drain the engine coolant.
5. Drain the engine oil.
6. Disconnect the heater hose and radiator outlet hoses.
7. Remove the oil filter bracket.
8. Disconnect the water hose from the water pump.
9. Remove the water pump retaining bolts and pump from the timing cover.
10. Remove the O-ring from the water pump.
11. Remove the water pump from the housing by removing the two bolts.

To install:

12. Install the water pump with a new gasket and tighten the bolts to 14 ft. lbs. (20 Nm).
13. Install the water pump to the timing cover and install the bolts. Tighten the bolts for the water pump as follows:
- Bolt A: 14 ft. lbs. (20 Nm)
- Bolt B: 21 ft. lbs. (28 Nm)
14. Connect the water hose to the water pump.
15. Install the oil filter bracket to the engine using a new O-ring.
16. Connect the heater hose and radiator outlet hose.
17. Fill the engine with oil.
18. Fill the engine and radiator with coolant.
19. Connect the negative battery cable to the battery.
20. Start the engine and check for leaks.

SIENNA

1. Disconnect the negative battery cable.
2. Drain the engine coolant.
3. Remove the wiper and blade assembly.
4. Remove the top cowl seal and panel.
5. Label and disconnect the window washer hoses from the ventilator louvers.
6. Remove the left and right ventilator louvers.
7. Remove the heater air duct.
8. On RX300 models, remove the front upper suspension brace.
9. Remove the timing belt.
10. Mark the left and right camshaft pulleys with a touch of paint. Using a spanner wrench, remove the bolts to the right and left camshaft pulleys. Separate the pulleys from the engine. Be sure not to mix up the pulleys.
11. Remove the No. 2 idler pulley by removing the bolt.
12. Disconnect the three clamps and engine wire from the rear timing belt cover.
13. Remove the six bolts holding the No. 3 timing belt cover to the engine block.
14. Remove the bolts and nuts to the extract the water pump.
15. Raise the engine slightly and remove the water pump and the gasket from the engine.

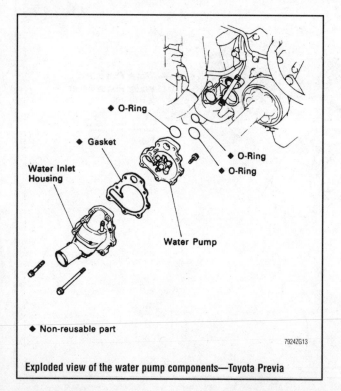

Exploded view of the water pump components—Toyota Previa

- O-Ring
- Gasket
- O-Ring
- O-Ring
- Water Inlet Housing
- Water Pump
- Non-reusable part

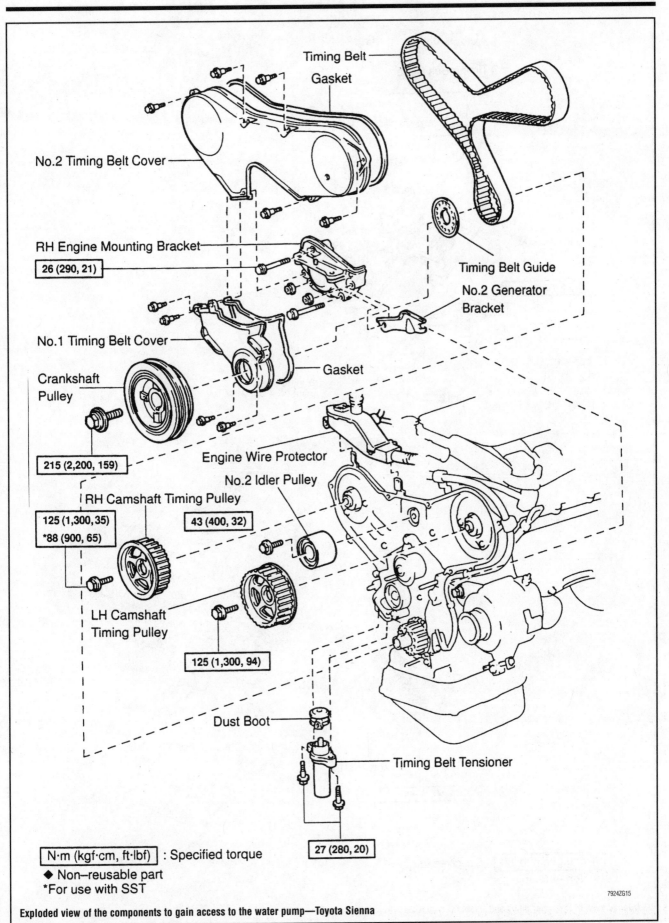

Timing Belt

Gasket

No.2 Timing Belt Cover

Timing Belt Guide

No.2 Generator Bracket

RH Engine Mounting Bracket

26 (290, 21)

No.1 Timing Belt Cover

Gasket

Crankshaft Pulley

215 (2,200, 159)

Engine Wire Protector

No.2 Idler Pulley

RH Camshaft Timing Pulley

43 (400, 32)

125 (1,300, 35)
*88 (900, 65)

LH Camshaft Timing Pulley

125 (1,300, 94)

Dust Boot

Timing Belt Tensioner

27 (280, 20)

N·m (kgf·cm, ft·lbf) : Specified torque

◆ Non–reusable part
*For use with SST

Exploded view of the components to gain access to the water pump—Toyota Sienna

7924ZG15

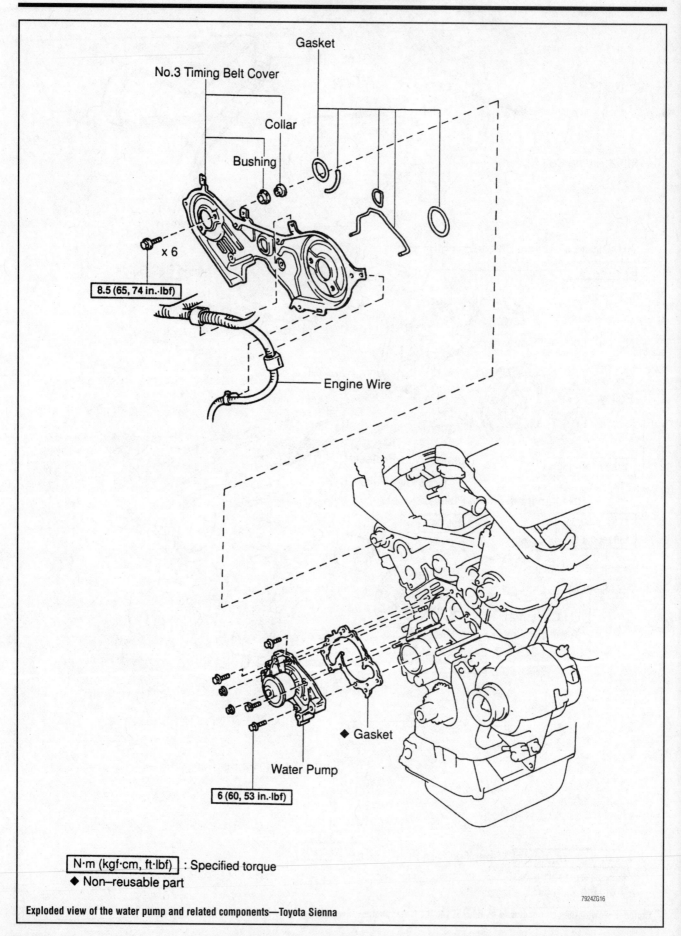

Gasket

No.3 Timing Belt Cover

Collar

Bushing

8.5 (65, 74 in.·lbf)

x 6

Engine Wire

◆ Gasket

Water Pump

6 (60, 53 in.·lbf)

N·m (kgf·cm, ft·lbf) : Specified torque
◆ Non–reusable part

Exploded view of the water pump and related components—Toyota Sienna

7924ZG16

To install:

16. Check that the water pump turns smoothly. Also check the air hole for coolant leakage.

17. Using a new gasket, apply liquid sealer to the gasket, water pump and engine block.

18. Install the gasket and pump to the engine and install the four bolts and two nuts. Tighten the nuts and bolts to 53 inch lbs. (6 Nm).

19. Install the rear timing belt cover and tighten the six bolts to 74 inch lbs. (9 Nm).

20. Connect the engine wire with the three clamps to the rear timing belt cover.

21. Install the No. 2 idler pulley with the bolt. Tighten the bolt to 32 ft. lbs. (43 Nm). After tightening the bolt, be sure the idler pulley moves smoothly.

22. With the flange side **outward**, install the right-hand camshaft pulley to the engine. Be sure to align the knock pin hole on the camshaft pulley with the knock pin on the camshaft. Using the same tools as removal, tighten the camshaft bolt to 65 ft. lbs. (88 Nm).

23. With the flange side **inward**; install the left-hand camshaft pulley to the engine. Be sure to align the knock pin hole on the camshaft pulley with the knock pin on the camshaft. Using the same tools as removal, tighten the camshaft bolt to 94 ft. lbs. (125 Nm).

24. Install the timing belt to the engine.

25. On RX300 models, install the front upper suspension brace and tighten the mounting nuts to 59 ft. lbs. (80 Nm).

26. Fill the engine coolant.

27. Install the heater air duct.

28. Install the left and right ventilator louvers.

29. Connect the window washer hoses from the ventilator louvers.

30. Install the top cowl seal and panel.

31. Install the wiper and blade assembly.

32. Connect the negative battery cable to the battery and start the engine.

33. Top off the engine coolant and check for leaks.

LAND CRUISER

4.5L (1FZ-FE) Engine

1. Disconnect the negative battery cable.
2. Drain the engine coolant.
3. Disconnect the No. 3 water bypass and radiator inlet hoses.
4. Remove the drive belts, fan assembly and the fan shroud.
5. Disconnect the oil cooler hose from the clamp on the fan shroud. Remove the bolts holding the fan shroud to the radiator.
6. Remove the four bolts, two nuts, water pump and the gasket.

To install:

7. Install the water pump using a new gasket. Tighten the fasteners to 15 ft. lbs. (21 Nm).
8. Install the water pump pulley, fan shroud and the drive belts.

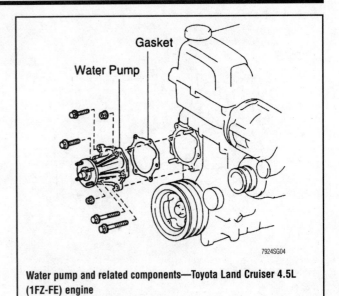

Water pump and related components—Toyota Land Cruiser 4.5L (1FZ-FE) engine

a. Place the fan with the fluid coupling, water pump pulley and the fan shroud in position.

b. Temporarily install the fan pulley mounting nuts.

c. Install the fan shroud and tighten the bolts to 4.9 ft. lbs. (5.4 Nm).

d. Connect the oil cooler hose to the clamp on the fan shroud.

9. Connect the No. 3 water bypass and radiator hoses. Fill the cooling system.

10. Connect the negative battery cable, start the engine and check for leaks.

11. Recheck the coolant level.

4.7L (2UZ-FE) Engine

1. Disconnect the negative battery cable.
2. Drain the coolant.
3. Remove the timing belt.
4. Remove the No. 2 idler pulley.
5. Disconnect the water bypass hose from the water inlet housing.
6. Remove the two bolts attaching the water inlet housing to the water pump.
7. Disconnect the water bypass pipe, then remove the water inlet housing from the water pump.
8. Remove the five bolts, two stud bolts and nut attaching the water pump to the engine and remove the water pump.
9. Remove the O-ring from the water bypass pipe.

To install:

10. Apply soapy water a new O-ring and install it on the water bypass pipe.

11. Install the bypass pipe in the water pump and install the water pump using a new gasket. Tighten the bolts to 15 ft. lbs. (21 Nm) and the remaining fasteners to 13 ft. lbs. (18 Nm). Be sure to tighten them evenly in several passes.

12. Clean all old silicone material off of the water inlet housing.

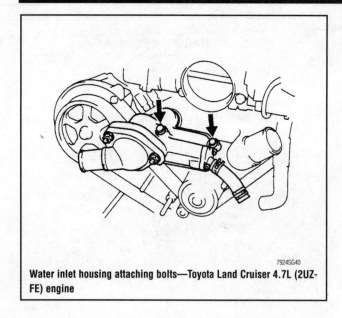

Water inlet housing attaching bolts—Toyota Land Cruiser 4.7L (2UZ-FE) engine

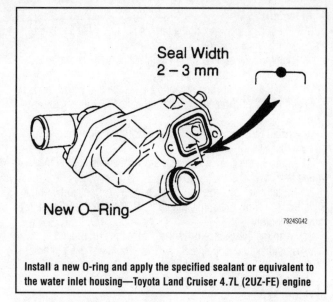

Seal Width
2 – 3 mm

New O–Ring

Install a new O-ring and apply the specified sealant or equivalent to the water inlet housing—Toyota Land Cruiser 4.7L (2UZ-FE) engine

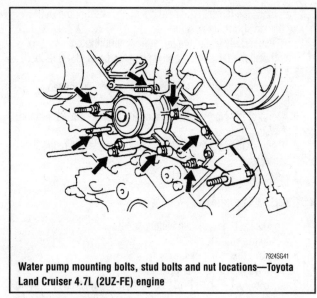

Water pump mounting bolts, stud bolts and nut locations—Toyota Land Cruiser 4.7L (2UZ-FE) engine

13. Apply soapy water to a new O-ring and install it on the water inlet housing.

14. Apply a bead of Seal Packing (RTV silicone) No. 08826–00100 or equivalent in the sealing groove on the water inlet housing.

15. Install the water inlet housing on the water pump. Alternately tighten the two bolts to 13 ft. lbs. (18 Nm).

16. Install the No. 2 idler pulley.

17. Install the timing belt.

18. Refill the engine with coolant.

19. Start the engine and check for leaks.

20. Allow the engine to cool down, then recheck the coolant level.

CONVERSION FACTORS

LENGTH–DISTANCE

Inches (in.)	x 25.4	= Millimeters (mm)	x .0394	= Inches
Feet (ft.)	x .305	= Meters (m)	x 3.281	= Feet
Miles	x 1.609	= Kilometers (km)	x .0621	= Miles

VOLUME

Cubic Inches (in3)	x 16.387	= Cubic Centimeters	x .061	= in3
IMP Pints (IMP pt.)	x .568	= Liters (L)	x 1.76	= IMP pt.
IMP Quarts (IMP qt.)	x 1.137	= Liters (L)	x .88	= IMP qt.
IMP Gallons (IMP gal.)	x 4.546	= Liters (L)	x .22	= IMP gal.
IMP Quarts (IMP qt.)	x 1.201	= US Quarts (US qt.)	x .833	= IMP qt.
IMP Gallons (IMP gal.)	x 1.201	= US Gallons (US gal.)	x .833	= IMP gal.
Fl. Ounces	x 29.573	= Milliliters	x .034	= Ounces
US Pints (US pt.)	x .473	= Liters (L)	x 2.113	= Pints
US Quarts (US qt.)	x .946	= Liters (L)	x 1.057	= Quarts
US Gallons (US gal.)	x 3.785	= Liters (L)	x .264	= Gallons

MASS–WEIGHT

Ounces (oz.)	x 28.35	= Grams (g)	x .035	= Ounces
Pounds (lb.)	x .454	= Kilograms (kg)	x 2.205	= Pounds

PRESSURE

Pounds Per Sq. In. (psi)	x 6.895	= Kilopascals (kPa)	x .145	= psi
Inches of Mercury (Hg)	x .4912	= psi	x 2.036	= Hg
Inches of Mercury (Hg)	x 3.377	= Kilopascals (kPa)	x .2961	= Hg
Inches of Water (H_2O)	x .07355	= Inches of Mercury	x 13.783	= H_2O
Inches of Water (H_2O)	x .03613	= psi	x 27.684	= H_2O
Inches of Water (H_2O)	x .248	= Kilopascals (kPa)	x 4.026	= H_2O

TORQUE

Pounds–Force Inches (in–lb)	x .113	= Newton Meters (N·m)	x 8.85	= in–lb
Pounds–Force Feet (ft–lb)	x 1.356	= Newton Meters (N·m)	x .738	= ft–lb

VELOCITY

Miles Per Hour (MPH)	x 1.609	= Kilometers Per Hour (KPH)	x .621	= MPH

POWER

Horsepower (Hp)	x .745	= Kilowatts	x 1.34	= Horsepower

FUEL CONSUMPTION*

Miles Per Gallon IMP (MPG)	x .354	= Kilometers Per Liter (Km/L)
Kilometers Per Liter (Km/L)	x 2.352	= IMP MPG
Miles Per Gallon US (MPG)	x .425	= Kilometers Per Liter (Km/L)
Kilometers Per Liter (Km/L)	x 2.352	= US MPG

*It is common to covert from miles per gallon (mpg) to liters/100 kilometers (1/100 km), where mpg (IMP) x 1/100 km = 282 and mpg (US) x 1/100 km = 235.

TEMPERATURE

Degree Fahrenheit (°F)	= (°C x 1.8) + 32
Degree Celsius (°C)	= (°F – 32) x .56

TCCS1044

Standard Torque Specifications and Fastener Markings

In the absence of specific torques, the following chart can be used as a guide to the maximum safe torque of a particular size/grade of fastener.
- There is no torque difference for fine or coarse threads.
- Torque values are based on clean, dry threads. Reduce the value by 10% if threads are oiled prior to assembly.
- The torque required for aluminum components or fasteners is considerably less.

U.S. Bolts

SAE Grade Number	1 or 2			5			6 or 7		
Number of lines always 2 less than the grade number.									
Bolt Size (Inches)—(Thread)	Ft./Lbs.	Kgm	Nm	Ft./Lbs.	Kgm	Nm	Ft./Lbs.	Kgm	Nm
¼—20	5	0.7	6.8	8	1.1	10.8	10	1.4	13.5
—28	6	0.8	8.1	10	1.4	13.6			
5/16—18	11	1.5	14.9	17	2.3	23.0	19	2.6	25.8
—24	13	1.8	17.6	19	2.6	25.7			
3/8—16	18	2.5	24.4	31	4.3	42.0	34	4.7	46.0
—24	20	2.75	27.1	35	4.8	47.5			
7/16—14	28	3.8	37.0	49	6.8	66.4	55	7.6	74.5
—20	30	4.2	40.7	55	7.6	74.5			
½—13	39	5.4	52.8	75	10.4	101.7	85	11.75	115.2
—20	41	5.7	55.6	85	11.7	115.2			
9/16—12	51	7.0	69.2	110	15.2	149.1	120	16.6	162.7
—18	55	7.6	74.5	120	16.6	162.7			
5/8—11	83	11.5	112.5	150	20.7	203.3	167	23.0	226.5
—18	95	13.1	128.8	170	23.5	230.5			
¾—10	105	14.5	142.3	270	37.3	366.0	280	38.7	379.6
—16	115	15.9	155.9	295	40.8	400.0			
7/8—9	160	22.1	216.9	395	54.6	535.5	440	60.9	596.5
—14	175	24.2	237.2	435	60.1	589.7			
1—8	236	32.5	318.6	590	81.6	799.9	660	91.3	894.8
—14	250	34.6	338.9	660	91.3	849.8			

Metric Bolts

Relative Strength Marking	4.6, 4.8			8.8		
Bolt Markings						
Bolt Size Thread Size x Pitch (mm)	Ft./Lbs.	Kgm	Nm	Ft./Lbs.	Kgm	Nm
6 x 1.0	2–3	.2–.4	3–4	3–6	.4–.8	5–8
8 x 1.25	6–8	.8–1	8–12	9–14	1.2–1.9	13–19
10 x 1.25	12–17	1.5–2.3	16–23	20–29	2.7–4.0	27–39
12 x 1.25	21–32	2.9–4.4	29–43	35–53	4.8–7.3	47–72
14 x 1.5	35–52	4.8–7.1	48–70	57–85	7.8–11.7	77–110
16 x 1.5	51–77	7.0–10.6	67–100	90–120	12.4–16.5	130–160
18 x 1.5	74–110	10.2–15.1	100–150	130–170	17.9–23.4	180–230
20 x 1.5	110–140	15.1–19.3	150–190	190–240	26.2–46.9	160–320
22 x 1.5	150–190	22.0–26.2	200–260	250–320	34.5–44.1	340–430
24 x 1.5	190–240	26.2–46.9	260–320	310–410	42.7–56.5	420–550

TCCS1098

ENGLISH TO METRIC CONVERSION: TORQUE

To convert foot-pounds (ft. lbs.) to Newton-meters (Nm), multiply the number of ft. lbs. by 1.36
To convert Newton-meters (Nm) to foot-pounds (ft. lbs.), multiply the number of Nm by 0.7376

ft. lbs.	Nm	ft. lbs.	Nm	ft. lbs.	Nm	ft. lbs.	Nm
0.1	0.1	34	46.2	76	103.4	118	160.5
0.2	0.3	35	47.6	77	104.7	119	161.8
0.3	0.4	36	49.0	78	106.1	120	163.2
0.4	0.5	37	50.3	79	107.4	121	164.6
0.5	0.7	38	51.7	80	108.8	122	165.9
0.6	0.8	39	53.0	81	110.2	123	167.3
0.7	1.0	40	54.4	82	111.5	124	168.6
0.8	1.1	41	55.8	83	112.9	125	170.0
0.9	1.2	42	57.1	84	114.2	126	171.4
1	1.4	43	58.5	85	115.6	127	172.7
2	2.7	44	59.8	86	117.0	128	174.1
3	4.1	45	61.2	87	118.3	129	175.4
4	5.4	46	62.6	88	119.7	130	176.8
5	6.8	47	63.9	89	121.0	131	178.2
6	8.2	48	65.3	90	122.4	132	179.5
7	9.5	49	66.6	91	123.8	133	180.9
8	10.9	50	68.0	92	125.1	134	182.2
9	12.2	51	69.4	93	126.5	135	183.6
10	13.6	52	70.7	94	127.8	136	185.0
11	15.0	53	72.1	95	129.2	137	186.3
12	16.3	54	73.4	96	130.6	138	187.7
13	17.7	55	74.8	97	131.9	139	189.0
14	19.0	56	76.2	98	133.3	140	190.4
15	20.4	57	77.5	99	134.6	141	191.8
16	21.8	58	78.9	100	136.0	142	193.1
17	23.1	59	80.2	101	137.4	143	194.5
18	24.5	60	81.6	102	138.7	144	195.8
19	25.8	61	83.0	103	140.1	145	197.2
20	27.2	62	84.3	104	141.4	146	198.6
21	28.6	63	85.7	105	142.8	147	199.9
22	29.9	64	87.0	106	144.2	148	201.3
23	31.3	65	88.4	107	145.5	149	202.6
24	32.6	66	89.8	108	146.9	150	204.0
25	34.0	67	91.1	109	148.2	151	205.4
26	35.4	68	92.5	110	149.6	152	206.7
27	36.7	69	93.8	111	151.0	153	208.1
28	38.1	70	95.2	112	152.3	154	209.4
29	39.4	71	96.6	113	153.7	155	210.8
30	40.8	72	97.9	114	155.0	156	212.2
31	42.2	73	99.3	115	156.4	157	213.5
32	43.5	74	100.6	116	157.8	158	214.9
33	44.9	75	102.0	117	159.1	159	216.2

9300BA01

METRIC TO ENGLISH CONVERSION: TORQUE

To convert foot-pounds (ft. lbs.) to Newton-meters (Nm), multiply the number of ft. lbs. by 1.36
To convert Newton-meters (Nm) to foot-pounds (ft. lbs.), multiply the number of Nm by 0.7376

Nm	ft. lbs.	Nm	ft. lbs.	Nm	ft. lbs.	Nm	ft. lbs.	Nm	ft. lbs.
0.1	0.1	34	25.0	76	55.9	118	86.8	160	117.6
0.2	0.1	35	25.7	77	56.6	119	87.5	161	118.4
0.3	0.2	36	26.5	78	57.4	120	88.2	162	119.1
0.4	0.3	37	27.2	79	58.1	121	89.0	163	119.9
0.5	0.4	38	27.9	80	58.8	122	89.7	164	120.6
0.6	0.4	39	28.7	81	59.6	123	90.4	165	121.3
0.7	0.5	40	29.4	82	60.3	124	91.2	166	122.1
0.8	0.6	41	30.1	83	61.0	125	91.9	167	122.8
0.9	0.7	42	30.9	84	61.8	126	92.6	168	123.5
1	0.7	43	31.6	85	62.5	127	93.4	169	124.3
2	1.5	44	32.4	86	63.2	128	94.1	170	125.0
3	2.2	45	33.1	87	64.0	129	94.9	171	125.7
4	2.9	46	33.8	88	64.7	130	95.6	172	126.5
5	3.7	47	34.6	89	65.4	131	96.3	173	127.2
6	4.4	48	35.3	90	66.2	132	97.1	174	127.9
7	5.1	49	36.0	91	66.9	133	97.8	175	128.7
8	5.9	50	36.8	92	67.6	134	98.5	176	129.4
9	6.6	51	37.5	93	68.4	135	99.3	177	130.1
10	7.4	52	38.2	94	69.1	136	100.0	178	130.9
11	8.1	53	39.0	95	69.9	137	100.7	179	131.6
12	8.8	54	39.7	96	70.6	138	101.5	180	132.4
13	9.6	55	40.4	97	71.3	139	102.2	181	133.1
14	10.3	56	41.2	98	72.1	140	102.9	182	133.8
15	11.0	57	41.9	99	72.8	141	103.7	183	134.6
16	11.8	58	42.6	100	73.5	142	104.4	184	135.3
17	12.5	59	43.4	101	74.3	143	105.1	185	136.0
18	13.2	60	44.1	102	75.0	144	105.9	186	136.8
19	14.0	61	44.9	103	75.7	145	106.6	187	137.5
20	14.7	62	45.6	104	76.5	146	107.4	188	138.2
21	15.4	63	46.3	105	77.2	147	108.1	189	139.0
22	16.2	64	47.1	106	77.9	148	108.8	190	139.7
23	16.9	65	47.8	107	78.7	149	109.6	191	140.4
24	17.6	66	48.5	108	79.4	150	110.3	192	141.2
25	18.4	67	49.3	109	80.1	151	111.0	193	141.9
26	19.1	68	50.0	110	80.9	152	111.8	194	142.6
27	19.9	69	50.7	111	81.6	153	112.5	195	143.4
28	20.6	70	51.5	112	82.4	154	113.2	196	144.1
29	21.3	71	52.2	113	83.1	155	114.0	197	144.9
30	22.1	72	52.9	114	83.8	156	114.7	198	145.6
31	22.8	73	53.7	115	84.6	157	115.4	199	146.3
32	23.5	74	54.4	116	85.3	158	116.2	200	147.1
33	24.3	75	55.1	117	86.0	159	116.9	201	147.8

9300BA02

ENGLISH/METRIC CONVERSION: LENGTH

To convert inches (in.) to millimeters (mm), multiply the number of inches by 25.4
To convert millimeters (mm) to inches (in.), multiply the number of millimeters by 0.04

Inches		Millimeters	Inches		Millimeters	Inches		Millimeters
Fraction	Decimal	Decimal	Fraction	Decimal	Decimal	Fraction	Decimal	Decimal
1/64	0.016	0.397	11/32	0.344	8.731	11/16	0.688	17.463
1/32	0.031	0.794	23/64	0.359	9.128	45/64	0.703	17.859
3/64	0.047	1.191	3/8	0.375	9.525	23/32	0.719	18.256
1/16	0.063	1.588	25/64	0.391	9.922	47/64	0.734	18.653
5/64	0.078	1.984	13/32	0.406	10.319	3/4	0.750	19.050
3/32	0.094	2.381	27/64	0.422	10.716	49/64	0.766	19.447
7/64	0.109	2.778	7/16	0.438	11.113	25/32	0.781	19.844
1/8	0.125	3.175	29/64	0.453	11.509	51/64	0.797	20.241
9/64	0.141	3.572	15/32	0.469	11.906	13/16	0.813	20.638
5/32	0.156	3.969	31/64	0.484	12.303	53/64	0.828	21.034
11/64	0.172	4.366	1/2	0.500	12.700	27/32	0.844	21.431
3/16	0.188	4.763	33/64	0.516	13.097	55/64	0.859	21.828
13/64	0.203	5.159	17/32	0.531	13.494	7/8	0.875	22.225
7/32	0.219	5.556	35/64	0.547	13.891	57/64	0.891	22.622
15/64	0.234	5.953	9/16	0.563	14.288	29/32	0.906	23.019
1/4	0.250	6.350	37/64	0.578	14.684	59/64	0.922	23.416
17/64	0.266	6.747	19/32	0.594	15.081	15/16	0.938	23.813
9/32	0.281	7.144	39/64	0.609	15.478	61/64	0.953	24.209
19/64	0.297	7.541	5/8	0.625	15.875	31/32	0.969	24.606
5/16	0.313	7.938	41/64	0.641	16.272	63/64	0.984	25.003
21/64	0.328	8.334	21/32	0.656	16.669	1/1	1.000	25.400
			43/64	0.672	17.066			

Inches	Millimeters	Inches	Millimeters	Inches	Millimeters	Inches	Millimeters
0.0001	0.00254	0.005	0.1270	0.09	2.286	4	101.6
0.0002	0.00508	0.006	0.1524	0.1	2.54	5	127.0
0.0003	0.00762	0.007	0.1778	0.2	5.08	6	152.4
0.0004	0.01016	0.008	0.2032	0.3	7.62	7	177.8
0.0005	0.01270	0.009	0.2286	0.4	10.16	8	203.2
0.0006	0.01524	0.01	0.254	0.5	12.70	9	228.6
0.0007	0.01778	0.02	0.508	0.6	15.24	10	254.0
0.0008	0.02032	0.03	0.762	0.7	17.78	11	279.4
0.0009	0.02286	0.04	1.016	0.8	20.32	12	304.8
0.001	0.0254	0.05	1.270	0.9	22.86	13	330.2
0.002	0.0508	0.06	1.524	1	25.4	14	355.6
0.003	0.0762	0.07	1.778	2	50.8	15	381.0
0.004	0.1016	0.08	2.032	3	76.2	16	406.4

9300BA03

CHILTON'S ...

For Professionals, From The Professionals